Ranks of Groups

Ranks of Groups

The Tools, Characteristics, and Restrictions

Martyn R. Dixon
Leonid A. Kurdachenko
Igor Ya Subbotin

WILEY

This edition first published 2017

© 2017 John Wiley & Sons, Inc.

The right of Martyn R Dixon, Leonid A Kurdachenko and Igor Ya Subbotin to be identified as the author(s) of this work has been asserted in accordance with law.

Registered Offices
John Wiley & Sons, Inc., 111 River Street, Hoboken, NJ 07030, USA

Editorial Office
111 River Street, Hoboken, NJ 07030, USA

For details of our global editorial offices, customer services, and more information about Wiley products visit us at www.wiley.com.

Wiley also publishes its books in a variety of electronic formats and by print-on-demand. Some content that appears in standard print versions of this book may not be available in other formats.

Library of Congress Cataloguing-in-Publication Data is available.

Hardback: 9781119080275

Cover image: © nadla/Gettyimages
Cover design: Wiley

10 9 8 7 6 5 4 3 2 1

Contents

Preface

Mathematics is a fluid subject always building on results that, while relatively modern, quickly become classical. It is then important for researchers to present the results of a particular body of work in a coherent, organized way, giving the general techniques and objectives used and hopefully providing a resource for new generations of mathematicians. This should be done only for well-developed theories that are central to a subject which have proved to be influential and amenable to new ideas. Typically the ideas collected in a book are enhanced by appropriate brief historical notes, but need not be in chronological order. Everyone has their own favourite survey book which has played a major role in the formation of their basic knowledge and which has heavily influenced their future career. Such books are not only essential sources for graduate courses, but are also useful references for experts in different areas.

For successful study in mathematics, efficient tools are needed for discovering and cracking sophisticated problems and a sandbox saturated with concrete, important examples must be created. This is especially important when we face such very abstract areas as modern algebra and group theory, the latter now pervading various branches of mathematics, on a par with number theory in the quantity and quality of solved and yet unsolved problems. Group theory however is a relative youngster of some two hundred years compared to the role that number theory has enjoyed through twenty five centuries of its history. The rapid manner in which group theory has become established can be explained by the crucial role that groups play not only in mathematics but in other areas such as physics, chemistry, astrophysics, cosmology, particle theory, crystallography and other disciplines. The main reason for this success is that groups are the best tools for studying and describing symmetry.

In many cases, the presence of numerical invariants, more or less closely related to groups, makes it possible to achieve significant progress in the study of the structure of these groups. Classical examples here concern groups with different kinds of topological structures where it is possible to employ powerful tools from analysis. In cases when the group does not have an additional topological structure, the numerical invariants defined are generally integers, giving numerical characteristics of certain properties of the group. Many such invariants have their origin in the classical notion of the dimension of a space, which can be described as the least of the orders of the subsets generating the space. This characterization was the basis for the notion of a minimal generating set of a finitely generated group G and the associated number $\mathbf{d}(G)$, the number of elements in a minimal generating set of G. The situation in arbitrary finitely generated groups is much more complicated than in vector spaces however, presenting the opportunity for much innovative research. Growth functions of finitely generated groups, for example, were introduced in connection with the question of growth of the universal covering spaces of Riemannian spaces, and in order to study the fundamental groups of the compact Riemannian manifolds with restrictions on curvature. Growth functions were in demand in the theory of invariant means on groups, spectral operator theory, statistical physics, differential geometry and the theory of random walks. M. Gromov proved that a finitely generated group has polynomial growth if and only if it is an almost nilpotent group. At the same time, it turned out that there are simply described groups, which have polynomial growth; for instance, the semi-direct product of the additive group $\mathbb{Q}_p$ of p-ary fractions by the infinite cyclic group. R. I. Grigorchuk has developed a series of impressive examples of infinite periodic finitely generated groups with intermediate growth function, the structure of which is much more complex than the group structure of $\mathbb{Q}_p \rtimes \mathbb{Z}$.

There are a number of integer invariants associated with groups in which the word "rank" appears. Most of these have their origin in the dimension of a vector space. One of the first extensions of the concept of dimension was the concept of the R-rank of a module A over a (commutative) ring R. Since every abelian group is a module over the ring $\mathbb{Z}$ of integers, the notion of $\mathbb{Z}$-rank (in the theory of abelian groups, the more familiar term 0-rank is used) has worked quite effectively in describing abelian groups. While vector spaces of finite dimension have quite a simple structure, the same cannot be said for torsion-free abelian groups of finite 0-rank, and the problem of describing such torsion-free abelian groups has led to the creation of a well-established theory rich in interesting results.

It was natural to extend the concept of 0-rank to certain classes of non-abelian groups close to being abelian. One such class is the class of hypercentral groups. S. N. Chernikov observed that a torsion-free hypercentral group has an ascending central series whose factors are torsion-free locally cyclic groups, and the length of every such series is an invariant of the group. He started studying groups which have a finite subnormal series whose factors are torsion-free locally cyclic. Later, D. J. S. Robinson called such groups *polyrational*. The length of such a series is an invariant of a polyrational groups called the *rational rank*. The study of polyrational groups has been actively continued by such prominent students of Chernikov as V. M. Glushkov, V. S. Charin, N. F. Sesekin, and D. I. Zaitcev. During that time, A. I. Maltsev began studying the class of soluble A_1-groups, the class of groups having a finite subnormal series whose factors are torsion-free locally cyclic or periodic abelian. The number of torsion-free locally cyclic factors in every such series is an invariant of an A_1-group. Here we see some parallels with polycyclic groups, investigated by K. A. Hirsch, who observed that in every subnormal series of a polycyclic group with cyclic factors, the number of infinite cyclic factors is an invariant of the group. The combination of all these approaches and the interesting results generated led to the formation of the modern concept of the 0-rank (or torsion-free rank), which proved to be very useful in the study of generalized soluble groups. Some of these results are presented in this book.

One important feature of a finite dimensional vector space V of dimension d is that every subspace of V has dimension at most d and thus in a finitely generated vector space, every subspace is finitely generated, a property that is not enjoyed by finitely generated groups (and other algebraic structures). Consequently, the exploration of groups in which an analogue of this important property holds is of high interest. Such groups were first studied by A. I. Maltsev, and they were named groups of finite special rank. Groups of finite special rank became popular in the theory of both infinite groups and finite groups, and studies related to the special rank have been very extensive. Many deep, fundamental, interesting results were discovered for such groups which have had an enormous influence on the development of group theory. The success of this theory emboldened group theorists to create various generalizations of the notion of special rank. The groups of finite special rank r have the basic property that every elementary abelian p-section is finite of order at most p^r, for each prime p. This property and that of the p-rank of an abelian group were the starting points for the emergence of the concept of the section p-rank of a group. As noted above, the notion of 0-rank has been effective in the theory of abelian groups, but it really works only for non-periodic groups. For periodic groups, the concept of p-rank is more useful and is also based on the notion of the dimension of a vector space. If A is a periodic abelian group, then A is a direct product of its Sylow p-subgroups (or p-components). In other words, the study of periodic abelian groups can be naturally reduced to the study of primary abelian groups. In turn, the properties of a primary abelian group depend on its important subgroup, the lower layer, the set of all elements of prime order together with the identity element. The lower layer of a p-group is an elementary abelian p-subgroups which can be thought of as a vector space over the prime field $\mathbb{F}_p$. The dimension of this vector space is called the p-rank of the abelian p-group. If an abelian p-group has finite p-rank r, then every elementary abelian p-section has order at most p^r.

The emergence of the concept of groups of finite section p-rank is rather interesting. In 1968, D. J. S. Robinson introduced the class of groups G having *finite abelian section rank*. In such groups, every abelian section of G has finite 0-rank and finite p-rank for all primes p. In 1986, D. J. S. Robinson modified this definition in the following way. A group G is said to have *finite abelian sectional rank* if it has no infinite elementary abelian p-sections for all primes p. The interesting point here is that the word "rank" generally implies the existence of some numerical characteristics and in the above definitions such characteristics are not visible. This feature appears in the article of S. Franciosi, F. de Giovanni and L. A. Kurdachenko [**87**], which leads to the formation of the concept of section p-rank in the following natural form: Let p be a prime. A group G has finite section p-rank r if every elementary abelian p-section of G is finite of order at most p^r and there is an elementary abelian p-section A/B of G such that $|A/B| = p^r$.

A group G has finite section rank if the section p-rank of G is finite for each prime p. As can be seen, this concept is very closely related to the special rank, but, at the same time, it is a rather significant generalization. As we shall see later, the section p-rank and special rank coincide for locally finite p-groups and in general groups of finite section rank have some good properties inherent in groups of finite special rank. For example, every locally finite group, whose Sylow p-subgroups are finite for all primes p, has finite section rank, but can otherwise be quite complicated. On the other hand, a locally finite group having finite special rank is almost hyperabelian. Groups of finite special rank are essentially groups that have section p-rank which are finite and bounded for all primes p.

This book will deal with these and other ranks. Among the vast diversity of results, we hope that we have chosen and illustrated the roles of those that are the most striking and central to the subject. These results were proved at different times using many methods. In this book, we have tried to carry out their presentation from a unified point of view without narrowing the framework. This has meant that we often have to discuss certain essential results from other branches of algebra and involve them in our considerations. The choice of topics ultimately follows the interests of the authors, and there are many results that we would have liked to include, but due to limitation of space they were omitted. A sample of such results is included in the last chapter, but without proof. The theory of groups of finite rank is a vast subject, parts of which we have not really touched on. The reader who studies this book can use it as a stepping stone to focus on such areas, as well as associated areas of the theory of groups.

This book is intended for students who have studied a standard university course in abstract algebra and learned the basics of group theory. We hope that it will be useful to both graduate students and experts in group theory, not only as a course for graduate seminars in mathematics, but also as a reference book. We include many key results on the topic which have been carefully selected from hundreds of original research articles and presented in a unified, concise way.

We would like to extend our sincere appreciation to the University of Alabama (Tuscaloosa, USA) and National University (California, USA) for their support in this work. Portions of this book were given as a course of lectures at the University of Alabama in 2009 and the authors are grateful to Rachel Bishop-Ross, Michael Green and Yalcin Karatas for reading very early versions of part of this manuscript. The authors would also like to thank their families for all their love and much needed support while this work was in progress. An endeavour such as this is made lighter by the joy that they bring. Finally, it is a pleasure to thank the staff of our publishers for their co-operation and dedication.

Martyn R. Dixon
Leonid A. Kurdachenko
Igor Ya Subbotin

CHAPTER 1

Essential Toolbox

In this book, we are concerned with the structure and properties of groups satisfying different rank conditions in various classes of generalized soluble groups. There are a variety of monographs in which the properties of different classes of generalized soluble groups are considered. Nevertheless, we have found it useful to collect together, in this first chapter, some concepts that will be useful in later chapters together with some standard results from the theory of generalized soluble groups. Some of these results are supplied with proof whilst others, which require specific techniques that may not be germane to the discussion, we just exhibit without proof. For such results, complete references are supplied for the book or article where the reader can find the proof.

1.1. Ascending and Descending Series in Groups

In this first section, we obtain some of the standard results concerning ascending and descending series in groups. We shall omit some of the proofs here since many of these results have been well documented in various other books.

Let H be a subgroup of a group G and suppose that γ is an ordinal number. An *ascending series* from H to G is a set of subgroups

$$(1.1) \qquad H = V_0 \leq V_1 \leq \ldots V_\alpha \leq V_{\alpha+1} \leq \ldots V_\gamma = G$$

such that V_α is normal in $V_{\alpha+1}$ and $V_\lambda = \bigcup_{\beta < \alpha} V_\beta$ for all limit ordinals $\lambda < \gamma$. In this case, we call H an *ascendant subgroup* of G unless the series involved is of finite length in which case, of course, H is said to be *subnormal* in G. The subgroups V_α, for $\alpha \leq \gamma$, are called the *terms* of this series and the factor groups $V_{\alpha+1}/V_\alpha$ are called the *factors* of the series. An ascending series is called *normal* if each term of the series (1.1) is a normal subgroup of G.

Let $\mathfrak{X}$ be a class of groups (which can also be defined by means of a group theoretical property). A group G is called a *hyper-$\mathfrak{X}$ group* if it has an ascending normal series, which is infinite in general, starting

1

Ranks of Groups: The Tools, Characteristics, and Restrictions
By Martyn R. Dixon, Leonid A. Kurdachenko and Igor Ya Subbotin
Copyright © 2017 John Wiley & Sons, Inc.

with 1, terminating in G itself, whose factors belong to $\mathfrak{X}$. We denote this class of groups by $\acute{\text{P}}_n\mathfrak{X}$. Thus a group G is called *hyperabelian* if G has an ascending normal series whose factors are abelian. This class of groups is one of the many generalizations of the class of soluble groups.

We shall also meet the class of *hyperfinite groups*. Such groups have an ascending series of normal subgroups whose factors are all finite.

There are also generalizations of the notion of nilpotency. We let $\zeta(G)$ denote the centre of the group G. The ascending series

(1.2) $$1 = Z_0 \leq Z_1 \leq \ldots Z_\alpha \leq Z_{\alpha+1} \leq \ldots Z_\gamma = G$$

is called *central* if the corresponding factors $Z_{\alpha+1}/Z_\alpha$ are central, that is $Z_{\alpha+1}/Z_\alpha \leq \zeta(G/Z_\alpha)$ for all $\alpha < \gamma$.

On the other hand, the *upper central series* of a group G is the ascending central series

$$1 = Z_0 \leq Z_1 \leq \ldots Z_\alpha \leq Z_{\alpha+1} \leq \ldots Z_\gamma$$

of characteristic subgroups such that $Z_{\alpha+1}/Z_\alpha = \zeta(G/Z_\alpha)$, for all ordinals $\alpha < \gamma$, $Z_\lambda = \bigcup_{\beta<\lambda} Z_\beta$, for all limit ordinals λ and $\zeta(G/Z_\gamma) = 1$. For the upper central series, we shall often write $\zeta_\alpha(G)$ for the subgroup Z_α and we call $\zeta_\alpha(G)$ the αth *hypercentre* of G. The last term $\zeta_\gamma(G)$ of the upper central series is called the *upper hypercentre* of G and will be denoted by $\zeta_\infty(G)$. The ordinal γ is called the *central length* of G and is denoted by $\mathbf{zl}(G)$. A group G is called *hypercentral* if and only if it coincides with some term of its upper central series and so $\zeta_\infty(G) = G$ in this case. A hypercentral group having finite central length leads to the usual notion of nilpotence. We note that a group G is hypercentral if and only if it has an ascending central series and G is nilpotent if and only if it has a finite central series.

The following lemma shows one way to characterize hyperabelian and hypercentral groups. Its proof is quite straightforward and we leave it as an exercise.

1.1.1. LEMMA. *Let G be a group.*

(i) *G is hyperabelian if and only if every nontrivial homomorphic image of G contains a nontrivial normal abelian subgroup.*

(ii) *G is hypercentral if and only if every nontrivial homomorphic image of G has nontrivial centre.*

EXERCISE 1.1. *Prove Lemma 1.1.1.*

The following corollary is also easy to prove.

1.1.2. COROLLARY. *Let G be a hyperabelian (respectively hypercentral) group.*

(i) *If H is a subgroup of G, then H is hyperabelian (respectively hypercentral);*

(ii) *If H is a normal subgroup of G, then G/H is hyperabelian (respectively hypercentral);*

(iii) *If H, K are subgroups of G such that H is a normal subgroup of K, then K/H is hyperabelian (respectively hypercentral).*

If H is a subgroup of a group G and if γ is an ordinal, then a *descending series* from G to H is a series of subgroups

$$G = V_0 \geq V_1 \geq V_2 \geq \ldots V_\alpha \geq V_{\alpha+1} \geq \ldots V_\gamma = H$$

such that $V_{\alpha+1}$ is normal in V_α and $V_\lambda = \bigcap_{\beta<\lambda} V_\beta$, for all limit ordinals λ. The subgroup H is then said to be a *descendant* subgroup of G; when β is finite then, of course, H is called subnormal in G. Again, the subgroups V_α, for $\alpha \leq \gamma$, are called the *terms* of this series and the factor groups $V_\alpha/V_{\alpha+1}$ are called the *factors* of the series. A descending series is called normal if each term of the series is normal in G.

If $\mathfrak{X}$ is a class of groups then a group G is called a *hypo-$\mathfrak{X}$ group* if it has a descending normal series, which is infinite in general, terminating in 1, whose factors belong to $\mathfrak{X}$. We denote this class of groups by $\grave{P}_n\mathfrak{X}$. Thus a group G is called *hypoabelian* if G has a descending normal series, whose factors are abelian.

We continue to set up notation and terminology by defining, for subsets X, Y of the group G, the subgroup

$$[X, Y] = \langle [x, y] = x^{-1}y^{-1}xy \mid x \in X, y \in Y \rangle.$$

If $Y = \{y\}$, then we simply write $[X, y]$ instead of $[X, \{y\}]$. For the subsets $X_1, \ldots, X_n$ we inductively define

$$[X_1, X_2, \ldots, X_n] = [[X_1, \ldots, X_{n-1}], X_n]$$

and if $X = X_1 = X_2 = \cdots = X_n$, then we write $[Y, X_1, X_2, \ldots, X_n]$ as $[Y, {}_n X]$. We note that $H \leq \zeta_n(G)$ if and only if $[H, {}_n G] = 1$. With these definitions we now define the *lower central series* of a group G to be the descending central series

$$G = \gamma_1(G) \geq \gamma_2(G) \geq \ldots \gamma_\alpha(G) \geq \gamma_{\alpha+1}(G) \geq \ldots \gamma_\delta(G)$$

of characteristic subgroups defined by $\gamma_1(G) = G$, $\gamma_2(G) = [G, G]$ and recursively, $\gamma_{\alpha+1}(G) = [\gamma_\alpha(G), G]$, for all ordinals α and $\gamma_\lambda(G) = \bigcap_{\beta<\lambda} \gamma_\beta(G)$, for all limit ordinals λ. If this series terminates in 1, then the group is called *hypocentral*.

As we will see shortly, when a group is defined by one or other infinite ascending series, this can have a significant effect on the structure and properties of the group, while the presence of an infinite descending series is usually less influential. For example, by a theorem of Magnus (see, for example, [**172**, Chapter IX, §36]), every non-abelian free group is hypocentral of length ω, the first infinite ordinal. Furthermore, every free group has a descending chain of normal subgroups whose factors are finite p-groups, for each prime p, a result first obtained by K. Iwasawa [**119**].

1.1.3. LEMMA. *Let G be a group and let A be an abelian normal subgroup of G. Then*

(i) $[A, x] = \{[a, x] | a \in A\}$;

(ii) $[A, \langle x \rangle] = [A, x]$, *for each $x \in G$;*

(iii) *If $M \subseteq G$ then $[A, \langle M \rangle]$ is the product of the subgroups $[A, x]$, for $x \in M$;*

(iv) *If $M \subseteq G$ and if $G = \langle C_G(A), M \rangle$ then $[A, G]$ is the product of the subgroups $[A, x]$ where $x \in M$.*

PROOF. (i) Let $S = \{[a, x] | a \in A\}$, so that $[A, x] = \langle S \rangle$, by definition and it now suffices to prove that S is a subgroup. Using the well-known properties of commutators and the fact that A is an abelian normal subgroup if $a, b \in A$, we have

$$[ab, x] = [a, x]^b[b, x] = [a, x][b, x], \text{ and}$$
$$[a^{-1}, x] = (a[a, x]a^{-1})^{-1} = [a, x]^{-1},$$

which proves (i).

(ii) It is clear that $[A, x] \leq [A, \langle x \rangle]$. Furthermore, for each natural number $n \geq 2$, we have

$$[a, x^n] = [a, x^{n-1}x] = [a, x][a, x^{n-1}]^x = [a, x][a^x, x^{n-1}] \in [A, x],$$

since A is normal in G, using a natural induction argument. Also

$$[a, x^{-1}] = (x[a, x]x^{-1})^{-1} = [xax^{-1}, x]^{-1} \in [A, x],$$

since $[A, x]$ is a subgroup and it is now easy to see that (ii) holds.

(iii) It is clearly sufficient to prove that the assertion holds in the case when $M = \{x_1, x_2, \ldots, x_n\}$ is a finite set. We note that a general element of $\langle M \rangle$ can be written in the form $u = x_{i_1}^{n_1} x_{i_2}^{n_2} \ldots x_{i_k}^{n_k}$, where $n_i \in \mathbb{Z}$, for $i = 1, \ldots, k$. If $k = 1$ then $[a, u] \in [A, x_{i_1}]$, for all $a \in A$, which now forms the basis for an induction on k. Indeed we have

$$(1.3) \qquad [a, u] = [a, x_{i_2}^{n_2} \ldots x_{i_k}^{n_k}][a, x_{i_1}^{n_1}]^{x_{i_2}^{n_2} \ldots x_{i_k}^{n_k}}.$$

The first term on the right-hand side of (1.3) lies in the product $[A, x_1][A, x_2] \ldots [A, x_n]$ by the induction hypothesis. The second term can be written as $[a, x_{i_1}^{n_1}][a, x_{i_1}^{n_1}, x_{i_2}^{n_2} \ldots x_{i_k}^{n_k}]$. We have $[a, x_{i_1}^{n_1}] \in [A, x_{i_1}]$ by (ii) and, since A is normal in G, we also have $[a, x_{i_1}^{n_1}, x_{i_2}^{n_2} \ldots x_{i_k}^{n_k}] \in$

$[A, x_{i_2}^{n_2} \ldots x_{i_k}^{n_k}]$. Again the induction hypothesis can be applied and the result then follows.

(iv) Since A is normal in G, we have $C_G(A) \triangleleft G$ and hence $G = C_G(A)\langle M \rangle$. If $g \in G$ then $g = cx$, where $c \in C_G(A), x \in \langle M \rangle$, and we have for each $a \in A$,

$$[a, g] = [a, cx] = [a, x][a, c]^x = [a, x].$$

Hence $[A, G] = [A, \langle M \rangle]$ and the result now follows by (iii). $\qquad \square$

It is an easy exercise to prove the following result due, we believe, to D. H. McLain.

1.1.4. COROLLARY. *Let G be a group and let M be a subset of G such that $G = G'\langle M \rangle$. Then $\gamma_n(G) = \langle \gamma_{n+1}(G), [x_1, \ldots, x_n] | x_i \in M$, for all $i \rangle$, for every positive integer n.*

EXERCISE 1.2. *Prove Corollary 1.1.4.*

1.1.5. COROLLARY. *Let G be a group. If G/G' is finitely generated, then $\gamma_n(G)/\gamma_{n+1}(G)$ is finitely generated for all natural numbers n.*

We shall need a number of properties concerning the terms of the upper and lower central series. We collect some of these well-known results in the following lemma.

1.1.6. PROPOSITION. *Let G be a group. Then*

(i) *$[x, y^{-1}, z]^y [y, z^{-1}, x]^z [z, x^{-1}, y]^x = 1$, for all $x, y, z \in G$;*

(ii) *Let X, Y, Z be subgroups of G and let L be a normal subgroup of G. If L contains two of the subgroups $[X, Y, Z], [Y, Z, X]$ and $[Z, X, Y]$, then L also contains the third of these subgroups;*

(iii)) *If m, n are natural numbers and if $n \geq m$, then $[\gamma_m(G), \zeta_n(G)] \leq \zeta_{n-m}(G)$;*

(iv) *If m, n are natural numbers, then $\zeta_n(G/\zeta_m(G)) = \zeta_{m+n}(G)/\zeta_m(G)$.*

We use this to prove the following fact.

1.1.7. LEMMA. *Let G be a group and let K be a normal subgroup of G. Then $[\gamma_n(G), K] \leq [K, {}_n G]$ for all natural numbers n. Furthermore, if H is a subgroup of G such that $G = HK$ then $\gamma_{n+1}(G) = \gamma_{n+1}(H)[K, {}_n G]$, for all natural numbers n.*

PROOF. We prove the first statement using induction on n. If $n = 1$, then $\gamma_1(G) = G$ and $[\gamma_1(G), K] = [G, K] = [K, G]$. Suppose that $n > 1$ and that we have proved that $[\gamma_{n-1}(G), K] \leq [K, {}_{n-1} G]$, for all normal subgroups K of G. Note that $[\gamma_n(G), K] = [[\gamma_{n-1}(G), G], K]$. We consider the subgroups $[[G, K], \gamma_{n-1}(G)] = [\gamma_{n-1}(G), [G, K]]$ and

$[[K, \gamma_{n-1}(G)], G] = [[\gamma_{n-1}(G), K], G]$. Let $L = [G, K] = [K, G]$ and note that L is normal in G. Hence, by the induction hypothesis,

$$[\gamma_{n-1}(G), L] \leq [L, {}_{n-1}G] = [[K, G], {}_{n-1}G] = [K, {}_{n}G].$$

Using the induction hypothesis we also have

$$[[\gamma_{n-1}(G), K], G] \leq [[K, {}_{n-1}G], G] = [K, {}_{n}G].$$

Hence $[K, {}_{n}G]$ contains both $[[G, K], \gamma_{n-1}(G)]$ and $[[K, \gamma_{n-1}(G)], G]$. Proposition 1.1.6(ii) now implies that $[K, {}_{n}G]$ contains $[\gamma_n(G), K]$, which completes the inductive step.

The second statement will also be proved using induction on n. When $n = 0$ we have $\gamma_1(G) = G = HK = \gamma_1(H)[K, {}_{0}G]$. For the inductive step, we let $n \geq 1$ and set $K_j = [K, {}_{j}G]$ so that $K_{j+1} = [K_j, G]$, for $j \in \mathbb{N}$. We note that $K_n \leq \gamma_{n+1}(G)$ and $\gamma_{n+1}(H) \leq \gamma_{n+1}(G)$ so $\gamma_{n+1}(H)K_n \leq \gamma_{n+1}(G)$.

Next we note that $[K, {}_{n}G]$ is normal in G. Let $x \in \gamma_{n+1}(H), y \in K$. Then $[x, y] \in [\gamma_n(G), K] \leq [K, {}_{n}G]$, by the first part of the proof. It follows that $x^y = x[x, y] \in \gamma_{n+1}(H)[K, {}_{n}G]$. Also, if $z \in H$ then $x^z \in \gamma_{n+1}(H)$. Since $G = HK$, it follows that $\gamma_{n+1}(H)[K, {}_{n}G] \triangleleft G$. Finally, using the induction hypothesis, we note that

$$\gamma_{n+1}(G) = [\gamma_n(G), G] = [\gamma_n(H)[K, {}_{n-1}G], G] = [\gamma_n(H)[K, {}_{n-1}G], HK].$$

Now we have

$$[\gamma_n(H), H] = \gamma_{n+1}(H) \leq \gamma_{n+1}(H)[K, {}_{n}G];$$
$$[\gamma_n(H), K] \leq [\gamma_n(G), K] \leq [K, {}_{n}G] \leq \gamma_{n+1}(H)[K, {}_{n}G] \text{ and}$$
$$[[K, {}_{n-1}G], HK] = [[K, {}_{n-1}G], G] = [K, {}_{n}G] \leq \gamma_{n+1}(H)[K, {}_{n}G].$$

Let $a \in \gamma_n(H), b \in [K, {}_{n-1}G], y \in K, z \in H$. Then

$$[ab, zy] = [ab, y][ab, z]^y = [a, y]^b[b, y][a, z]^{by}[b, z]^y.$$

Since $\gamma_{n+1}(H)[K, {}_{n}G]$ is normal in G, these equalities and the preceding inclusions imply that $\gamma_{n+1}(G) \leq \gamma_{n+1}(H)[K, {}_{n}G]$. Thus $\gamma_{n+1}(G) = \gamma_{n+1}(H)[K, {}_{n}G]$, which completes the induction step and the proof. $\square$

The following very useful consequence is very easy to deduce.

1.1.8. COROLLARY. *Let G be a group and let H be a subgroup of G such that $G = H\zeta_n(G)$. Then $\gamma_{n+1}(G) = \gamma_{n+1}(H)$.*

1.2. Generalized Soluble Groups

In this section, we shall collect together some well-known facts concerning generalized soluble groups. We begin with the class of hypercentral groups. We let $N_G(H)$ denote the normalizer of the subgroup H of G.

1.2.1. LEMMA. *Let G be a hypercentral group.*

(i) *If H is a proper subgroup of G then $H \neq N_G(H)$;*
(ii) *Every subgroup of G is ascendant.*

In particular, every subgroup of a nilpotent group is subnormal.

EXERCISE 1.3. *Prove Lemma 1.2.1.*

The next two properties of hypercentral groups are very important. Using Zorn's Lemma it is easy to see that every group has maximal abelian normal subgroups.

1.2.2. LEMMA. *Let G be a hypercentral group and suppose that A is a nontrivial normal subgroup of G. Then $A \cap \zeta(G) \neq 1$. If A is a maximal abelian normal subgroup of G, then $C_G(A) = A$.*

PROOF. Let

$$1 = Z_0 \leq Z_1 \leq \ldots Z_\alpha \leq Z_{\alpha+1} \leq \ldots Z_\gamma = G$$

be the upper central series of G.

Since $A \neq 1$ there is a least ordinal α such that $A \cap Z_\alpha \neq 1$. If α is a limit ordinal, then using the definition of α, we have

$$A \cap Z_\alpha = A \cap \left(\bigcup_{\beta < \alpha} Z_\beta \right) = \bigcup_{\beta < \alpha} (A \cap Z_\beta) = 1,$$

which is a contradiction. Hence $\alpha - 1$ exists, so $A \cap Z_{\alpha-1} = 1$ by definition of α. However, it then follows that

$$[A \cap Z_\alpha, G] \leq A \cap Z_{\alpha-1} = 1,$$

so that $A \cap Z_\alpha \leq \zeta(G)$. Hence $A \cap \zeta(G) \neq 1$, which proves the first part of the lemma.

Suppose now that A is a maximal normal abelian subgroup of G. It is clear that $A \leq C_G(A)$. We let $C = C_G(A)$ and suppose that $A \neq C$. Then C/A is a nontrivial normal subgroup of the hypercentral group G/A, so there exists $xA \in C/A \cap \zeta(G/A)$. Hence $\langle x, A \rangle$ is an abelian normal subgroup of G, so that $x \in A$, by choice of A, a contradiction which proves the lemma. $\square$

A different type of characterization of hypercentral groups is illustrated in the following useful result due to S. N. Chernikov [**39**].

1.2.3. LEMMA. *A group G is hypercentral if and only if for each element $a \in G$ and every countable subset $\{x_n | n \in \mathbb{N}\}$ of elements of G there exists an integer k such that*

$$[\ldots [[a, x_1], x_2], \ldots, x_k] = 1.$$

PROOF. Let G be a hypercentral group and let

$$1 = Z_0 \leq Z_1 \leq \ldots Z_\alpha \leq Z_{\alpha+1} \leq \ldots Z_\gamma = G$$

be the upper central series of G. Put $a_n = [\ldots [[a, x_1], x_2], \ldots, x_n]$, for $n \in \mathbb{N}$, and suppose, for a contradiction, that $a_n \neq 1$ for all $n \in \mathbb{N}$. Since $a_n \neq 1$ for each n, it follows that there is a least ordinal α such that Z_α contains a_t for some t. Clearly α cannot be a limit ordinal so $\alpha - 1$ exists. Then, by definition of α, $a_t \notin Z_{\alpha-1}$ for all $t \in \mathbb{N}$. However, this implies the desired contradiction since $a_{t+1} = [a_t, x_{t+1}] \in Z_{\alpha-1}$, contrary to the choice of α.

To prove sufficiency of the condition, we first note that the condition is inherited by each factor group of G. Hence, by Lemma 1.1.1, it suffices to prove that $\zeta(G) \neq 1$. We suppose, for a contradiction, that this is false. Let $a = a_0 \in G$ and suppose that we have constructed elements $x_1, \ldots, x_n \in G$ such that $[a_j, x_{j+1}] = a_{j+1} \neq 1$, for $0 \leq j \leq n - 1$. Then, by assumption, $a_n \notin \zeta(G)$ so there exists $x_{n+1} \in G$ such that $[a_n, x_{n+1}] = a_{n+1} \neq 1$. In this way we construct a countable subset $\{x_n | n \in \mathbb{N}\}$ such that $[\ldots [[a, x_1], x_2], \ldots, x_n] \neq 1$ for all $n \in \mathbb{N}$. This contradicts our hypothesis and it follows that $\zeta(G) \neq 1$. The proof is complete.   □

1.2.4. PROPOSITION (Chernikov [**39**]). *A group G is hypercentral if and only if each countable subgroup of G is hypercentral.*

PROOF. If G is hypercentral, then each subgroup of G is hypercentral and hence every countable subgroup is certainly hypercentral.

Conversely, suppose that every countable subgroup of G is hypercentral, but G is not hypercentral. Factoring by the hypercentre we may suppose that $\zeta(G) = 1$. By Lemma 1.2.3 there exists $a \in G$ and a countable subset $\{x_n | n \in \mathbb{N}\}$ such that $[\ldots [[a, x_1], x_2], \ldots, x_n] \neq 1$, for each $n \in \mathbb{N}$. The subgroup $\langle a, x_n | n \in \mathbb{N} \rangle$ is countable and hence is hypercentral. Lemma 1.2.3 shows that there exists $k \in \mathbb{N}$ such that $[\ldots [[a, x_1], x_2], \ldots, x_k] = 1$ and we obtain a contradiction. Consequently G is hypercentral.   □

We note that a class $\mathfrak{X}$ of groups is a *countably recognizable* class if the group G is an $\mathfrak{X}$-group whenever every countable subset of elements of G is contained in some $\mathfrak{X}$-subgroup of G. Thus Proposition 1.2.4 asserts that the class of hypercentral groups is a countably recognizable class and there is a similar result, due to Baer [12], for the class of hyperabelian groups.

In a similar vein, if $\mathfrak{X}$ is a class of groups then a group G is called *locally* $\mathfrak{X}$ if every finite subset of G is contained in some $\mathfrak{X}$-subgroup of G. In particular, a group is called locally nilpotent, locally soluble or locally finite in the cases when $\mathfrak{X}$ denotes, respectively, the class $\mathfrak{N}$ of all nilpotent groups, the class $\mathfrak{S}$ of all soluble groups and the class $\mathfrak{F}$ of all finite groups. We denote the class of locally $\mathfrak{X}$-groups by $\mathrm{L}\mathfrak{X}$.

Of course, every locally nilpotent group is locally soluble. It is a well-known result of A. I. Maltsev [187] that every hypercentral group is locally nilpotent, a result which we now prove.

1.2.5. PROPOSITION. *Every finitely generated hypercentral group is nilpotent.*

PROOF. Let G be a finitely generated hypercentral group. Let

$$1 = Z_0 \leq Z_1 \leq Z_2 \leq \ldots Z_\alpha \leq Z_{\alpha+1} \leq \ldots Z_\eta = G$$

be the upper central series of G and suppose that $G = \langle M \rangle$, where $M = \{g_1, g_2, \ldots, g_n\}$. Let $\lambda(j)$ denote the least ordinal such that $g_j \in Z_{\lambda(j)}$, for $1 \leq j \leq n$. It is easy to see that none of the $\lambda(j)$ are limit ordinals. Thus η is the maximal ordinal in the set $\{\lambda(1), \lambda(2), \ldots, \lambda(n)\}$. Suppose, for a contradiction, that η is infinite. Then there is a limit ordinal $\beta \geq \omega$ such that $\eta = \beta + m$, for some positive integer m. Since G/Z_β is then nilpotent of class at most m we have $\gamma_{m+1}(G) \leq Z_\beta$. It follows that $[x_1, \ldots, x_{m+1}] \in Z_\beta$ for all elements $x_1, x_2, \ldots, x_{m+1} \in M$. Since M is finite there is non-limit ordinal $\nu < \beta$ such that Z_ν contains all commutators $[x_1, x_2, \ldots, x_{m+1}]$, for all $x_i \in M$. Corollary 1.1.4 now shows that $\gamma_{m+1}(G) \leq Z_\nu$, and since $G/\gamma_{m+1}(G)$ is nilpotent of class at most m it follows that $G = Z_{\nu+m}$. However, $\nu + m < \beta$ and we obtain a contradiction. Hence η is finite so that G is nilpotent.   □

1.2.6. COROLLARY. *Every hypercentral group is locally nilpotent.*

However, the classes of hyperabelian and locally soluble groups are distinct as can be seen from [193]. In particular, a finitely generated hyperabelian group need not be soluble and a locally soluble group need not be hyperabelian.

We shall need some properties of torsion-free locally nilpotent groups.

1.2.7. LEMMA. *Let G be a torsion-free locally nilpotent group. If $x, y \in G$ are elements such that $x^n = y^n$, for some $n \in \mathbb{N}$ then $x = y$.*

EXERCISE 1.4. *Prove Lemma 1.2.7.*

From this we deduce another important result of Maltsev [187].

1.2.8. COROLLARY. *Let G be a torsion-free locally nilpotent group and let $g, x \in G$. If there exist natural numbers k, t such that $g^t x^k = x^k g^t$ then $gx = xg$.*

PROOF. Let $y = x^k$, so that $g^t y = y g^t$. It follows that $g^t = y^{-1} g^t y = (y^{-1} gy)^t$ and by Lemma 1.2.7 we deduce that $g = y^{-1} gy$. Thus $x^k g = g x^k$ and the same argument allows us to deduce that $xg = gx$, as required. $\square$

Let G be a group. We recall that a subgroup H of G is called *pure*, or *isolated*, in G if either $\langle g \rangle \cap H = \langle g \rangle$ or $\langle g \rangle \cap H = 1$, for each $g \in G$. This leads us to a further result of Maltsev [**187**].

1.2.9. COROLLARY. *Let G be a torsion-free locally nilpotent group.*

(i) *If M is a subset of G then $C_G(M)$ is a pure subgroup of G;*
(ii) *If A is a normal subgroup of G then A is pure in G if and only if G/A is torsion-free;*
(iii) *The intersection of every family of pure subgroups is pure;*
(iv) *Every term of the upper central series of G is pure.*

In particular, if G is hypercentral, then the factors of the upper central series of G are torsion-free.

For our next results we need the following notation. If X, Y are subsets of a group G, then $\langle X^Y \rangle = \langle x^y \mid x \in X, y \in Y \rangle$ is the *normal closure* of X by Y.

The following result appears in Plotkin [**216**, Corollary to Theorem 1].

1.2.10. PROPOSITION. *Let G be a group and let π be a set of primes.*

(i) *The product of the periodic normal π-subgroups of G is a periodic normal π-subgroup of G;*
(ii) *If H is a periodic ascendant π-subgroup of G, then H^G is a periodic normal π-subgroup of G.*

PROOF. (i) Let $\mathcal{M}$ be a family of normal periodic π-subgroups of G and let $M = \langle T \mid T \in \mathcal{M} \rangle$. Clearly $M \triangleleft G$. If $x \in M$ then there are subgroups $T_1, T_2, \ldots, T_n \in \mathcal{M}$ such that $x \in T_1 T_2 \ldots T_n$. We use induction on n to prove that $T_1 \ldots T_n$ is a periodic π-subgroup. Suppose first that $n = 2$. Then $T_1 \cap T_2$ is a periodic π-subgroup and $T_1 T_2 / (T_1 \cap T_2) = T_1 / (T_1 \cap T_2) \times T_2 / (T_1 \cap T_2)$ is also a periodic π-group. It follows that $T_1 T_2$ is a periodic π-subgroup. Suppose now that $n > 2$ and we have already proved that $U = T_1 \ldots T_{n-1}$ is a periodic π-subgroup. Then $T_1 \ldots T_n = U T_n$, and by the case $n = 2$, the result now follows.

(ii) Let

$$H = V_0 \leq V_1 \leq V_2 \leq \ldots V_\alpha \leq V_{\alpha+1} \leq \ldots V_\gamma = G$$

be an ascending series from H to G. We prove, using transfinite induction, that H^{V_α} is a normal π-subgroup of V_α for each ordinal $\alpha \leq \gamma$. This is very easy to see when α is a limit ordinal, so we suppose that the result is true when $\alpha - 1$ exists and then prove it for α. Let $H_{\alpha-1} = H^{V_{\alpha-1}}$ and let $x \in V_\alpha$. Then $(H_{\alpha-1})^x \lhd V_{\alpha-1}$, since $V_{\alpha-1} \lhd V_\alpha$, and hence $H_\alpha = H^{V_\alpha}$ is a π-subgroup of V_α, using (i). Since normality is clear, we have that H^{V_α} is a periodic normal π-subgroup of V_α and the result follows. □

If π is a set of primes and G is a group, then we let $\mathbf{O}_\pi(G)$ denote the largest normal π-subgroup of G. If $\pi = \Pi(G)$, the set of prime divisors of the orders of elements of G, then $\mathbf{O}_\pi(G)$ is the largest normal periodic subgroup of G, which we denote by $\mathbf{Tor}(G)$. This subgroup is called the *periodic part* of G. In some cases, when $\mathbf{Tor}(G)$ consists of all the elements of G of finite order, then it is called the *torsion subgroup* of G.

Locally nilpotent groups enjoy some of the properties of abelian groups as the following result shows.

1.2.11. PROPOSITION. *Let G be a locally nilpotent group and let π be a set of primes. Then*

(i) *The set of elements of G whose order is a π-number corresponds to $\mathbf{O}_\pi(G)$;*

(ii) *The set of elements of finite order in G is a characteristic subgroup of G and coincides with $\mathbf{Tor}(G)$. Also $G/\mathbf{Tor}(G)$ is torsion-free;*

(ii)i *For each prime p, the set $\mathbf{Tor}_p(G)$, consisting of all elements of p-power order, is a characteristic subgroup of $\mathbf{Tor}(G)$ and $\mathbf{Tor}(G) = \underset{p \in \Pi(G)}{Dr} \mathbf{Tor}_p(G)$.*

PROOF. (i) Suppose first that G is nilpotent. Let P be a maximal π-subgroup of G and note that, by Lemma 1.2.1, P is subnormal in G. Then Proposition 1.2.10 implies that P^G is a π-group and hence $P = P^G$. Consequently, P is normal in G. Let x be a π-element. Then, as above, $\langle x \rangle^G$ is a π-subgroup of G and Proposition 1.2.10 shows that $\langle x \rangle^G P$ is a π-subgroup of G, containing P. By definition of P we see that $x \in P$. Hence P contains all the π-elements of G, and coincides with $\mathbf{O}_\pi(G)$.

Suppose now that G is locally nilpotent and let T denote the set of π-elements of G. If $y, z \in T$ then $\langle y, z \rangle$ is nilpotent and a π-subgroup, by our work above. It follows that T is a π-subgroup, coinciding with $\mathbf{O}_\pi(G)$, which proves (i).

(ii) follows upon setting $\pi = \Pi(G)$ in (i) and observing that if $x \in G$ is such that $x\mathbf{Tor}\,(G)$ has finite order n then $x^n \in \mathbf{Tor}\,(G)$, so that $x \in \mathbf{Tor}\,(G)$.

(iii) follows upon setting $\pi = \{p\}$ in (i). Furthermore we note that if $x \in \mathbf{Tor}_p(G)$ and $y \in \mathbf{Tor}_q(G)$, for distinct primes p, q, then $[x, y] \in \mathbf{Tor}_p(G) \cap \mathbf{Tor}_q(G) = 1$ and the result now follows.   □

It is a well-known theorem due to K. A. Hirsch [116] and B. I. Plotkin [215] that the subgroup generated by the normal locally nilpotent subgroups of a group is also locally nilpotent. Thus every group G has a unique maximal normal locally nilpotent subgroup called the *locally nilpotent radical* or the *Hirsch–Plotkin radical* of G and denoted in this book by $\mathbf{Ln}(G)$. Of course $\mathbf{Ln}(G)$ is a characteristic subgroup of G.

We now recall some facts which we often use and which are derived from [216, Corollary to Theorem 1]. If H is an ascendant locally nilpotent subgroup of a group G, then it turns out that H^G is also locally nilpotent. It follows that if G contains a non-trivial ascendant locally nilpotent subgroup, then G contains a non-trivial normal locally nilpotent subgroup.

A group G is called *radical* if it is hyper (locally nilpotent) and our remarks above show that the class of radical groups is precisely the class of groups with an ascending series each factor of which is locally nilpotent. Every locally nilpotent group and every hyperabelian group is radical and it is often convenient to subsume the classes of hyperabelian groups and locally nilpotent groups into the single larger class of radical groups.

We define the *radical series*

$$1 = R_0 \le R_1 \le \ldots R_\alpha \le R_{\alpha+1} \le \ldots R_\gamma$$

of a group G by

$$R_1 = \mathbf{Ln}(G)$$
$$R_{\alpha+1}/R_\alpha = \mathbf{Ln}(G/R_\alpha) \text{ for all ordinals } \alpha < \gamma$$
$$R_\lambda = \bigcup_{\beta < \lambda} R_\beta \text{ for limit ordinals } \lambda < \gamma$$
$$\mathbf{Ln}(G/R_\gamma) = 1.$$

It may happen that $\mathbf{Ln}(G)$ is trivial. The group G is radical if and only if $G = R_\gamma$ for some term R_γ of this series. Since the Hirsch–Plotkin radical is always a characteristic subgroup, a group G is radical if and only if G has an ascending series of *characteristic* subgroups with locally nilpotent factors. Here are some other useful properties of radical groups.

1.2.12. LEMMA. *Let G be a radical group.*
(i) *If H is a subgroup of G, then H is a radical group;*
(ii) *If H is a normal subgroup of G, then G/H is a radical group;*
(iii) *If H, K are subgroups of G, such that H is normal in K, then K/H is a radical group;*
(iv) $C_G(\mathbf{Ln}(G)) \le \mathbf{Ln}(G)$.

PROOF. The assertions (i)–(iii) are left as an exercise. To prove (iv) we let $L = \mathbf{Ln}(G)$ and $C = C_G(L)$. Suppose, for a contradiction, that $C \nleq L$. Then CL/L is a nontrivial normal subgroup of G/L. By (ii), CL/L is a radical group and hence $\mathbf{Ln}(CL/L) = K/L \neq 1$. Since $CL/L \lhd G/L$, it follows that $K/L \lhd G/L$ and hence K is normal in G. It follows easily that $K = L(K \cap C)$. Consider $K \cap C$. If H is a finitely generated subgroup of $K \cap C$, then $H/H \cap L \cong HL/L$ is nilpotent. Also $H \le C = C_G(L)$, so $H \cap L \le \zeta(H)$. It follows that H is nilpotent and hence $K \cap C$ is a locally nilpotent normal subgroup of G, from which we deduce that $K \cap C \le L$. Hence $K \le L$ and we obtain the contradiction desired. $\square$

We shall also need the following important and useful result.

1.2.13. PROPOSITION. *Let G be a finitely generated group and let C be a subgroup of G of finite index. Then C is also finitely generated.*

PROOF. Let M be a finite subset of G such that $G = \langle M \rangle$. We may assume that if $x \in M$, then $x^{-1} \in M$ also. Suppose that $k = |G : C|$ and let $\{g_1, \ldots, g_k\}$ be a left transversal to C in G. Without loss of generality we may assume that $g_1 = 1$. If x is an arbitrary element of G, then $x(g_j C)$ is a left coset so there is an integer $x(j)$ such that $x(g_j C) = g_{x(j)} C$, for $1 \le j \le k$. Clearly the mapping $j \longmapsto x(j)$ is a permutation of the set $\{1, \ldots, k\}$. Furthermore, $xg_j = g_{x(j)} d(j, x)$ for some element $d(j, x) \in C$.

If c is an arbitrary element of C, then $c = x_m \ldots x_1$ for certain $x_j \in M$, for $1 \le j \le m$. We have

$$c = cg_1 = x_m \ldots x_2 g_{x_1(1)} d(1, x_1)$$
$$= g_{c(1)} d(x_{m-1} \ldots x_1(1), x_m) \ldots d(1, x_1).$$

But $g_{c(1)} C = cg_1 C = C$, so that $g_{c(1)} = 1$. It follows that $C = \langle d(j, x) | 1 \le j \le k, x \in M \rangle$ and hence C is finitely generated. $\square$

An immediate deduction we can make is as follows.

1.2.14. COROLLARY. *Let G be a group and let H be a normal locally finite subgroup of G such that G/H is also locally finite. Then G is locally finite.*

Our next result can be regarded as analogous to Proposition 1.2.10 and is also due to Plotkin [**216**, Corollary to Theorem 1].

1.2.15. PROPOSITION. *Let G be a group.*

(i) *The product of the normal locally finite subgroups of G is a normal locally finite subgroup of G;*

(ii) *If H is an ascendant locally finite subgroup of G, then H^G is a normal locally finite subgroup of G;*

(iii) *The product of the normal radical subgroups of G is a normal radical subgroup of G;*

(iv) *If H is an ascendant radical subgroup of G, then H^G is a normal radical subgroup of G.*

PROOF. (i) As in the proof of Proposition 1.2.10 it suffices to prove this assertion for the product of two normal locally finite subgroups T_1, T_2. The intersection $T_1 \cap T_2$ is locally finite and

$$T_1 T_2/(T_1 \cap T_2) = T_1/(T_1 \cap T_2) \times T_2/(T_1 \cap T_2)$$

is clearly also locally finite. Then Corollary 1.2.14 can be applied to deduce the result that $T_1 T_2$ is locally finite.

The proof of (ii) is similar to the proof of Proposition 1.2.10(ii).

To complete the proof of (iii) and (iv) we note that a radical normal subgroup R of a group G has a characteristic series whose factors are locally nilpotent, since the Hirsch–Plotkin radical is always a characteristic subgroup. This observation now makes it easy to prove the result.   □

It is also easy to see that a product of normal soluble subgroups is locally soluble. However an example due to P. Hall (see [**225**, Theorem 8.19.1], for example) shows that a product of two normal locally soluble groups need not be locally soluble in general and the same example shows that the product of two normal hyperabelian groups need not be hyperabelian.

Proposition 1.2.15 shows that if G contains a non-trivial ascendant locally finite subgroup, then it has a unique largest characteristic locally finite subgroup, the *locally finite radical*, which we denote by $\mathbf{Lf}(G)$. This is of considerable importance for the class of generalized radical groups, which we define next.

A group G is called *generalized radical* if G has an ascending series whose factors are locally nilpotent or locally finite. It should be noted that in the past some authors have used a more restrictive definition where the factors are locally nilpotent or *finite*. It follows easily from the definition that a generalized radical group G either has an ascendant locally nilpotent subgroup or an ascendant locally finite subgroup. In the former case, the Hirsch–Plotkin radical of G is nontrivial. In the latter case, G contains a nontrivial normal locally finite subgroup, by Proposition 1.2.15, so the locally finite radical is nontrivial. Thus every generalized radical group has an ascending series of *normal*, indeed characteristic, subgroups with locally nilpotent or locally finite factors. Consequently, every generalized radical group is hyper (locally nilpotent or locally finite).

1.2.16. LEMMA. *Let G be a group.*

(i) *G is generalized radical if and only if every non-trivial homomorphic image of G contains a non-trivial ascendant subgroup which is either locally nilpotent or locally finite;*

(ii) *If G is a generalized radical group and H is a subgroup of G, then H is generalized radical;*

(iii) *If G is a generalized radical group and H is a normal subgroup of G, then G/H is generalized radical;*

(iv) *If G is a generalized radical group and H, K are subgroups of G such that H is normal in K, then K/H is generalized radical.*

EXERCISE 1.5. *Prove Lemma 1.2.16.*

The next result shows that periodic subgroups of generalized radical groups are locally finite. We shall be interested in the class of *locally generalized radical groups*. A group G is a locally generalized radical group if every finitely generated subgroup is a generalized radical group. The class of locally generalized radical groups will play a very prominent role in this book.

1.2.17. LEMMA. *Let G be a finitely generated periodic group. If G is generalized radical, then G is finite. Consequently, every periodic locally generalized radical group is locally finite.*

PROOF. Let

$$1 = H_0 \leq H_1 \leq \ldots H_\alpha \leq H_{\alpha+1} \leq \ldots H_\gamma = G$$

be an ascending series in G whose factors are locally nilpotent or locally finite. We proceed by transfinite induction on γ. If $\gamma = 1$, then either G is locally nilpotent or locally finite. Since a finitely generated periodic nilpotent group is finite, in either case G is finite and the result follows.

Consequently, let $\gamma > 1$ and suppose inductively that if H is a finitely generated subgroup of H_β, where $\beta < \gamma$, then H is finite. If γ

is a limit ordinal then, since G is finitely generated, $G \leq H_\beta$ for some $\beta < \gamma$, and hence G is finite, by the induction hypothesis. If $\gamma - 1$ exists we let $L = H_{\gamma-1}$. Then G/L is a periodic finitely generated group that is locally nilpotent or locally finite. In either case G/L is finite, as above, and so L is also finitely generated by Proposition 1.2.13. By the induction hypothesis L is finite and therefore G is finite, as required.   □

Next we discuss the chief factors of a group. Suppose that G is a group and U, V are normal subgroups of G. If $U \leq V$, then V/U is called a *chief factor* of G if V/U contains no proper nontrivial normal subgroups of G/U. If $U = 1$, then we call V a *minimal normal subgroup* of G. It is a consequence of Zorn's Lemma that every group has chief factors although not every group possesses minimal normal subgroups. The chief factors of locally nilpotent and locally soluble groups are of particular interest and we now discuss these. These results are due to A. I. Maltsev, but we give here the method of D. H. McLain [**192**].

1.2.18. PROPOSITION. *Let G be a locally soluble group. Then the chief factors of G are abelian. Furthermore the chief factors are either elementary abelian p-groups for some prime p or are torsion-free divisible groups.*

PROOF. First we show that if M is a minimal normal subgroup of G, then M is abelian. If M is not abelian choose elements $x, y \in M$ such that $z = [x, y] \neq 1$. Then $\langle z^G \rangle = M$, since M is minimal normal in G and hence $\langle x, y \rangle \leq \langle z^K \rangle$ for some finite subset K of G. Let $H = \langle z, K \rangle$. Since H is finitely generated, and therefore soluble, it follows that $Z = \langle z^H \rangle$ is also soluble. On the other hand, $z = [x, y] \in Z'$, so $Z = Z'$, which is a contradiction. Hence M is abelian.

Since M is abelian it has a torsion subgroup $\mathbf{Tor}\,(M)$, which is characteristic in M, so either $\mathbf{Tor}\,(M) = 1$ or $\mathbf{Tor}\,(M) = M$. In the former case, it follows that M is torsion-free. In this case, for each natural number n, the subgroup M^n is normal in G and hence $M^n = M$, so that M is divisible. In the case when M is periodic it is the direct product of its p-components, each of which is characteristic in M (and hence normal in G). It follows that M is a p-group for some prime p. Moreover, $S = \{x \in M | x^p = 1\}$ is a characteristic subgroup of M, and hence normal in G so $M = S$ and M is elementary abelian. This completes the proof.   □

The immediate corollary is that simple locally soluble groups are not complicated.

1.2.19. COROLLARY. *Each simple locally soluble group has prime order.*

There is a similar result concerning the chief factors of a locally nilpotent group.

1.2.20. PROPOSITION. *The chief factors of a locally nilpotent group are central of prime order.*

PROOF. It is sufficient to prove that a minimal normal subgroup, M, of the locally nilpotent group G is central of prime order. If $\zeta(G)$ does not contain M, then there exists $x \in M$ and $y \in G$ such that $[x, y] = z \neq 1$. Since $z \in M$ and M is a minimal normal subgroup of G we see that $M = \langle z^G \rangle$. Hence there exists a finite subset S of G such that $x \in \langle z^S \rangle$. Let $H = \langle x, y, S \rangle$ and $K = \langle x^H \rangle \leq H$. Then $z = [x, y] \in [K, H]$, so $\langle z^S \rangle \leq [K, H]$. Hence $x \in [K, H]$ and it follows that $\langle x^H \rangle = K = [K, H]$. Thus $[K,_r H] = K$ for each $r \in \mathbb{N}$. However, H is a finitely generated nilpotent group so, for some natural number c, we have $[K,_c H] = 1$. Consequently, $K = 1$, so $x = 1$ and $z = 1$, a contradiction which implies that M is central. Since every subgroup of the centre of a group G is normal in G, it follows that M has prime order. $\qquad\square$

For the next result we need some more notation and terminology. If $\mathfrak{X}$ is a class of groups then the $\mathfrak{X}$-*residual* of a group G is defined to be

$$G^{\mathfrak{X}} = \bigcap \{N \lhd G | G/N \in \mathfrak{X}\}.$$

If the set $\mathbf{Res}_{\mathfrak{X}}(G) = \{N \lhd G | G/N \in \mathfrak{X}\}$ has a least element L, then $L = G^{\mathfrak{X}}$ and $G/G^{\mathfrak{X}} \in \mathfrak{X}$ but in general $G/G^{\mathfrak{X}}$ need not be an $\mathfrak{X}$-group.

When $\mathfrak{X} = \mathfrak{A}$, the class of abelian groups, then the $\mathfrak{A}$-residual of a group G is precisely the derived subgroup G' and $G/G^{\mathfrak{A}} \in \mathfrak{A}$, in this case.

More generally, if $\mathfrak{X} = \mathfrak{N}_c$, the class of nilpotent groups of nilpotency class at most c, then the $\mathfrak{N}_c$-residual is precisely $\gamma_{c+1}(G)$ and again $G/G^{\mathfrak{N}_c} \in \mathfrak{N}_c$.

However, when $\mathfrak{X} = \mathfrak{N}$, the class of all nilpotent groups then, in general, $G/G^{\mathfrak{N}} \notin \mathfrak{N}$. Indeed this factor group need not even be locally nilpotent. For example, as we remarked earlier, if G is a free group then, by the theorem of Magnus mentioned earlier, $\gamma_\omega(G) = 1$. It follows that $G^{\mathfrak{N}} = 1$ and that $G/G^{\mathfrak{N}}$ is a free group.

For the class of groups $\mathfrak{X}$, a group G is called a *residually $\mathfrak{X}$-group* if, for each non-trivial element $g \in G$, there is a normal subgroup H_g such that $g \notin H_g$ and $G/H_g \in \mathfrak{X}$.

EXERCISE 1.6. *Let G be a group and let $\mathfrak{X}$ be a class of groups. Prove that G is a residually $\mathfrak{X}$-group if and only if $\cap \mathbf{Res}_{\mathfrak{X}}(G) = 1$.*

We shall denote the class of residually $\mathfrak{X}$-groups by $\text{R}\mathfrak{X}$. If $\mathfrak{X} = \mathfrak{F}$, the class of all finite groups, then we obtain the familiar class $\text{R}\mathfrak{F}$ of *residually finite groups*. If p is a prime, we let $\mathfrak{F}_p$ denote the class of all finite p-groups and in this way we obtain the class $\text{R}\mathfrak{F}_p$ of residually $\mathfrak{F}_p$-groups. It is well known that every free group is residually $\mathfrak{F}_p$, for each prime p, a theorem originally due to Iwasawa [**119**].

It is easy to see that $G^{\mathfrak{X}}$ is always a characteristic subgroup of G and that $G/G^{\mathfrak{X}}$ is always a residually $\mathfrak{X}$-group. For example when $\mathfrak{X} = \mathfrak{F}$, the class of all finite groups, then $G^{\mathfrak{X}} = G^{\mathfrak{F}}$ is the finite residual of G and $G/G^{\mathfrak{F}}$ is residually finite.

For the group G, we let $\mathbf{Frat}(G)$ denote the Frattini subgroup of G, the intersection of all maximal subgroups of G, with the understanding that if G has no maximal subgroups then $\mathbf{Frat}(G) = G$. It is well known, since a maximal subgroup of a finite p-group G is a normal subgroup, that $\mathbf{Frat}(G) = G'G^p$ in this case. This observation is very useful in our next result.

1.2.21. PROPOSITION (Kurdachenko [**139**]). *Let p be a prime and let G be a locally finite and residually $\mathfrak{F}_p$-group. If $G/G'G^p$ is finite, then G is a finite p-group.*

PROOF. First we note that if K is a finite subgroup of G then, for each $g \in K$, there is a normal subgroup N_g of G such that $g \notin N_g$ and G/N_g is a finite p-group. If $N = \cap\{N_g | g \in K\}$, then G/N is also a finite p-group, by Remak's Theorem and $K \cap N = 1$, so that K too is a finite p-group. Consequently, G is a p-group.

Let $L = G'G^p$ so that G/L is a finite elementary abelian p-group. Then $G = KL$ for some finite subgroup K. Let

$$\mathcal{M} = \{H | H \lhd G \text{ such that } G/H \text{ is a finite } p\text{-group}\}.$$

If $H \in \mathcal{M}$, then $LH/H = \mathbf{Frat}(G/H)$ since

$$\mathbf{Frat}(G/H) = (G/H)'(G/H)^p = G'G^pH/H,$$

and hence $G/H = (KH/H)\mathbf{Frat}(G/H)$. It is well known that the Frattini subgroup of a group is the set of non-generators (see [**229**, 5.2.12]), so we deduce that $G/H = KH/H$. Since $KH/H \cong K/(K \cap H)$ we see that $|G/H| \leq |K|$ and this is true for every subgroup H of $\mathcal{M}$. Consequently, the family $\mathcal{M}$ must be finite and the embedding of G into $\text{Dr}_{H \in \mathcal{M}} G/H$ shows that G is also finite. $\square$

Finally in this section, we shall also need some information concerning *stability groups*. Suppose that

$$(1.4) \qquad 1 = G_0 \lhd G_1 \lhd \ldots \lhd G_j \lhd G_{j+1} \lhd \ldots \lhd G_n = G$$

is a finite chain of subgroups of the group G. An automorphism α of G is said to *stabilize* this chain if $\alpha(xG_j) = xG_j$ for every $x \in G_{j+1}$ and all j such that $0 \le j \le n - 1$. Thus α acts trivially on each of the factors of the series.

The set of all such automorphisms stabilizing the chain in Equation (1.4) is a subgroup of $\mathbf{Aut}(G)$ called the *stability group* of the chain.

1.2.22. THEOREM. *Let G be a group having a finite series*

$$1 = G_0 \le G_1 \le \cdots \le G_{n-1} \le G_n = G$$

of normal subgroups. Then the stability group of this series is nilpotent of class at most $n - 1$.

L. A. Kaloujnine first obtained this result in the paper [**122**]; it was further improved by P. Hall [**106**] where it is shown that the stability group of any chain of subgroups of length n (without any normality assumption) is nilpotent of class at most $\frac{1}{2}n(n - 1)$.

1.3. Chernikov Groups and the Minimum Condition

In this section we discuss the class $\mathcal{C}$ of Chernikov groups. Such groups have been characterized in many ways and have been discussed in great detail in a number of books. For more detail than we provide here the reader should consult one or more of the books written by S. N. Chernikov [**41**], M. R. Dixon [**49**], M. I. Kargapolov and Yu. I. Merzlyakov [**129**], O. H. Kegel and B. A. F. Wehrfritz [**134**], A. G. Kurosh [**172**], or D. J. S. Robinson [**223, 225, 229**], which contain the highlights of the theory.

A *subgroup theoretical property* $\mathcal{P}$ is a property of certain subgroups of a group G so that always the identity subgroup of G has $\mathcal{P}$, and whenever a subgroup H of G has the property $\mathcal{P}$, then $\theta(H)$ also has $\mathcal{P}$ for each isomorphism θ of G with some other group. Typical examples of $\mathcal{P}$ are the properties of being normal, subnormal, abelian, subnormal abelian, central and so on.

Let $\mathcal{M}$ be an ordered set, with ordering $\le$. Recall that $\mathcal{M}$ is said to satisfy the *minimum (or minimal) condition* if every non-empty subset of $\mathcal{M}$ has a minimal element.

We say that $\mathcal{M}$ has the *descending chain condition* if for every descending chain $a_1 \ge a_2 \ge \cdots \ge a_n \ge \ldots$ of elements of $\mathcal{M}$, there is some $k \in \mathbb{N}$ such that $a_k = a_{k+n}$ for all $n \in \mathbb{N}$. In this regard, we have the following well-known result.

1.3.1. PROPOSITION. *The ordered set $\mathcal{M}$ satisfies the minimal condition if and only if $\mathcal{M}$ satisfies the descending chain condition.*

EXERCISE 1.7. *Prove Proposition 1.3.1.*

Let $\mathcal{P}$ be a subgroup theoretical property. The group G is said to satisfy the *minimum (or minimal) condition on $\mathcal{P}$-subgroups* (min-$\mathcal{P}$ for short) if the set of $\mathcal{P}$-subgroups of G, ordered by inclusion, satisfies the minimal condition.

For example, if $\mathcal{P}$ is the property of being a subgroup, the condition min-$\mathcal{P}$ is called the minimum condition and often abbreviated to min. If $\mathcal{P}$ represents the property of being an abelian subgroup we obtain the condition min-ab, the minimum condition on abelian subgroups. The conditions min-p and min-sn denote the cases when $\mathcal{P}$ represents the property of being a p-subgroup (for the prime p) and, respectively, the condition of being a subnormal subgroup.

The minimal condition has played an important role both in ring theory and group theory for many years. However, the structure of groups with min is not as well understood as the corresponding structure of rings. Since groups with the minimal condition are highly relevant to the topic in which we are interested, we shall spend some time obtaining some of the positive results concerning the structure of groups with min. Of course, all finite groups have min. The following proposition lists some of the properties of groups with the minimal condition.

1.3.2. PROPOSITION. *Let G be a group.*

(i) *If G satisfies min, then every subgroup H of G satisfies min;*

(ii) *If G satisfies min and L is a normal subgroup of G, then G/L satisfies min;*

(iii) *If L is a normal subgroup of G such that L and G/L satisfy min, then G satisfies min;*

(iv) *If G has a finite series of subgroups*

$$1 = H_0 \lhd H_1 \lhd \ldots \lhd H_n = G,$$

where every factor H_j/H_{j-1} satisfies min, for $1 \leq j \leq n$, then G satisfies min;

(v) *If $G = H_1 \times H_2 \times \cdots \times H_n$, where each group H_j satisfies min, for $1 \leq j \leq n$, then G satisfies min;*

(vi) *Let H be a subgroup of G and $H = \underset{\lambda \in \Lambda}{Dr} H_\lambda$. If G satisfies min, then Λ is finite;*

(vii) G satisfies the minimum condition if and only if every countable subgroup of G satisfies the minimum condition.

PROOF. The assertions (i) and (ii) are clear so we proceed to part
(iii). Let
$$D_1 \geq D_2 \geq \cdots \geq D_n \geq \ldots$$
be a descending chain of subgroups of G. Since L and G/L satisfy min,
there exists a natural number m such that $D_m \cap L = D_{m+n} \cap L$ and
$D_m L = D_{m+n} L$ for all $n \in \mathbb{N}$. Therefore,
$$D_m = D_m \cap (D_{m+n}L) = D_{m+n}(D_m \cap L) = D_{m+n}(D_{m+n} \cap L) = D_{m+n},$$
for all $n \in \mathbb{N}$, and (iii) follows. Assertions (iv) and (v) are now immediate consequences of (iii). To prove (vi) we suppose that Λ is infinite and
note that, in this case, Λ contains a countably infinite subset Γ which
we denote by $\{\mu_n | n \in \mathbb{N}\}$. We let $\Delta = \Lambda \setminus \Gamma$, $\Lambda_n = \{\mu_k | k > n\} \cup \Delta$ and
$G_n = \underset{\lambda \in \Lambda_n}{\mathrm{Dr}} H_\lambda$, for each $n \in \mathbb{N}$. Then we obtain a strictly descending
chain
$$G_1 > G_2 > \cdots > G_n > \ldots,$$
which gives a contradiction. It follows that Λ is finite.

Finally we will prove (vii) and remark that necessity is clear. To
prove sufficiency, suppose that G has an infinite strictly descending
chain of subgroups
$$G_1 > G_2 > \cdots > G_n > \ldots.$$
For each $j \in \mathbb{N}$ choose $g_j \in G_j \setminus G_{j+1}$ and let $H = \langle g_j | j \in \mathbb{N} \rangle$. Then H
is a countable subgroup of G so satisfies min, by hypothesis. However
it is clear that
$$H \cap G_1 > H \cap G_2 > \cdots > H \cap G_n > \ldots$$
is a strictly descending chain of subgroups of H, which is a contradiction. Hence every descending chain of subgroups terminates so that
G has the minimum condition, by Proposition 1.3.1. The result follows. $\qquad\square$

The most important examples of infinite groups with the minimal
condition are the *Prüfer groups of type p^∞*. Initially such groups were
called *quasicyclic p-groups*, but this latter terminology is now often
used for groups whose proper subgroups are cyclic and among such
groups there are some which are very different from Prüfer groups. We
now describe the structure of Prüfer p-groups.

Let p be a prime. For each natural number $n \geq 1$ let $G_n = \langle a_n \rangle$ be
a cyclic group of order p^n and, for each $n \in \mathbb{N}$, let $\theta_n : G_n \longrightarrow G_{n+1}$

be the monomorphism defined by $\theta_n(a_n) = a_{n+1}^p$. In this way, we can think of G_n as a subgroup of G_{n+1} and hence we can form the group

$$G = \bigcup_{n \in \mathbb{N}} G_n,$$

which is the union of a chain of cyclic p-groups of orders $p, p^2, \ldots$.

The group G obtained above is called a *Prüfer p-group*, or a *Prüfer group of type p^∞*, and is denoted by C_{p^∞}.

There are numerous useful descriptions of this group. It can be realized in terms of generators and relations as

$$C_{p^\infty} = \langle a_n | a_1^p = 1, a_{n+1}^p = a_n, n \in \mathbb{N} \rangle.$$

Prüfer p-groups also arise somewhat more concretely; they can be thought of as the multiplicative group of complex pth roots of unity, or as the set of elements of p-power order in the additive abelian group $\mathbb{Q}/\mathbb{Z}$. Thus $C_{p^\infty} \cong \mathbf{Tor}_p(\mathbb{Q}/\mathbb{Z})$. The Prüfer groups play a very important role in group theory. They are the main examples of groups in which every proper subgroup is finite; such groups are called *quasifinite*.

1.3.3. LEMMA. *Let p be a prime. Every proper subgroup of C_{p^∞} is finite. In particular, C_{p^∞} has the minimum condition on subgroups.*

EXERCISE 1.8. *Prove Lemma 1.3.3.*

Consequently, the group C_{p^∞} is a fundamental example of a group which satisfies the minimal condition on subgroups.

We recall that a group G is called *divisible* (many authors use this word only if the group is abelian, and reserve the term *radicable* in the general case) if, for each $g \in G$ and each $n \in \mathbb{Z}$, the equation $x^n = g$ always has a solution $x \in G$. Let $G^n = \langle g^n | g \in G \rangle$. If G is an abelian group then $G^n = \{g^n | g \in G\}$ and it is then clear that an abelian group G is divisible if and only if $G = G^n$ for each $n \in \mathbb{N}$.

Thus C_{p^∞} and $\mathbb{Q}$ are examples of divisible groups. In fact the structure of divisible abelian groups is well known: For the prime p, divisible abelian p-groups are direct products of Prüfer p-groups, whilst torsion-free divisible abelian groups are direct products of copies of $\mathbb{Q}$ (see [**88**, Theorem 23.1], for example). Every abelian group G has a maximal divisible subgroup $D = \mathbf{Div}(G)$, *the divisible (or radicable) part of G.* Certainly, $\mathbf{Div}(G)$ is a characteristic subgroup of G. Since divisible subgroups of abelian groups are well known to be direct summands it follows that each abelian group can be written in the form $G = D \times R$, for some subgroup R which contains no nontrivial divisible subgroups (see [**88**, Theorem 21.3], for example). Such a subgroup R is said to be *reduced*.

The concept of divisibility can be generalized further. It is clear that no nontrivial finite group is divisible and since divisibility is inherited by factor groups a divisible group has no proper subgroups of finite index. Groups which contain no proper subgroups of finite index are called $\mathfrak{F}$-*perfect*, so divisible groups are $\mathfrak{F}$-perfect. Every abelian $\mathfrak{F}$-perfect group is divisible. Indeed if $G \neq G^n$ for some positive integer n, then the order of the elements of G/G^n divides n and, in particular, these orders are bounded. By the first Prüfer Theorem (see [**88**, Theorem 17.2], for example), G/G^n is a direct product of finite cyclic subgroups and hence has a proper subgroup of finite index. This contradiction shows that $G = G^n$, for each $n \in \mathbb{N}$, so that G is divisible.

By considering a minimal subgroup of finite index in a group with the minimal condition it is possible to deduce the following result.

1.3.4. LEMMA. *Every group satisfying the minimal condition on subgroups has a normal $\mathfrak{F}$-perfect subgroup of finite index.*

It follows from Proposition 1.3.2 that a finite direct product of Prüfer groups and finite cyclic groups has the minimum condition. This establishes the easy half of the following theorem due to Kurosh [**170**].

1.3.5. THEOREM. *Let G be an abelian group. Then G has the minimum condition if and only if G is a direct product of finitely many Prüfer p-groups and finite cyclic groups.*

PROOF. If G has min, then G must be periodic since the infinite cyclic group does not have min.

By Lemma 1.3.4, G contains an $\mathfrak{F}$-perfect subgroup D of finite index. As we saw above D is divisible, so $G = D \times R$, for some reduced subgroup R. Clearly R is a finite abelian group and hence is a direct product of finitely many cyclic subgroups. Also, as we mentioned above, D is a direct product of Prüfer subgroups, and Proposition 1.3.2 shows that the set of direct factors in this direct product is finite. The result follows. □

A group G is called a *Chernikov group* if G contains a normal subgroup D of finite index which is a direct product of finitely many Prüfer p-groups.

Such groups were named in honour of S. N. Chernikov, who made an extensive study of groups with the minimum condition. The subgroup D is easily seen to be the maximal divisible subgroup of G which we again call the *divisible (or radicable) part* of G, denoted by $\mathbf{Div}(G)$.

By Proposition 1.3.2 and Theorem 1.3.5 every Chernikov group satisfies the minimal condition. Using Theorem 1.3.5 and Proposition 1.3.2 we obtain the following properties of Chernikov groups. For an abelian p-group G and natural number n we let $\Omega_n(G) = \{x \in G | x^{p^n} = 1\}$.

1.3.6. COROLLARY. *Let G be a group.*

(i) *If G is a Chernikov group, then every subgroup of G is also Chernikov;*

(ii) *If G is a Chernikov group and L is a normal subgroup of G, then G/L is Chernikov;*

(iii) *If L is a normal subgroup of G such that L and G/L are divisible Chernikov groups, then G is a divisible abelian Chernikov group;*

(iv) *If L is a normal subgroup of G such that L and G/L are Chernikov groups, then G is a Chernikov group;*

(v) *If G has a finite series of subgroups $1 = H_0 \triangleleft H_1 \triangleleft \ldots \triangleleft H_n = G$, in which every factor H_j/H_{j-1} is Chernikov, for $1 \le j \le n$, then G is also a Chernikov group.*

PROOF. The assertions (i) and (ii) are clear so we prove (iii). Let $L = \operatorname*{Dr}_{p \in \Pi(L)} L_p$, where $L_p = \mathbf{Tor}_p(G)$, for $p \in \Pi(L)$. We also let $M_n = \operatorname*{Dr}_{p \in \Pi(L)} \Omega_n(L_p)$, for $n \in \mathbb{N}$. Then M_n is a finite G-invariant subgroup of G, for each $n \in \mathbb{N}$. It follows that $G/C_G(M_n)$ is finite, for each $n \in \mathbb{N}$. Since L is abelian, $L \le C_G(M_n)$. On the other hand, G/L is $\mathfrak{F}$-perfect and hence contains no proper subgroups of finite index. It follows that $G = C_G(M_n)$, for all $n \in \mathbb{N}$ and since $L = \bigcup_{n \in \mathbb{N}} M_n$ we deduce that $L \le \zeta(G)$. Let $G/L = \operatorname*{Dr}_{1 \le j \le t} P_j/L$, where P_j/L is a Prüfer subgroup for $1 \le j \le t$. Since $L \le \zeta(G)$ and P_j/L is locally cyclic, P_j is abelian. Clearly P_j is $\mathfrak{F}$-perfect, and hence divisible. Using the argument above, we obtain the inclusion $P_j \le \zeta(G)$, so $G = P_1 P_2 \ldots P_t$ is abelian. Clearly G is $\mathfrak{F}$-perfect and satisfies the minimal condition, by Proposition 1.3.2. Part (iii) now follows by Theorem 1.3.5.

(iv) Suppose first that L is finite. By Proposition 1.3.2, G satisfies min, and Lemma 1.3.4 shows that G contains a normal $\mathfrak{F}$-perfect subgroup K of finite index. Since L is finite, $G/C_G(L)$ is finite so $K/C_K(L)$ is finite. As K is $\mathfrak{F}$-perfect, we see that $K = C_K(L)$, so $K \cap L \le \zeta(K)$. Being $\mathfrak{F}$-perfect, KL/L coincides with the divisible part of the Chernikov group G/L. It follows that $K/(K \cap L) \cong KL/L$ is an abelian Chernikov group. As above $K/(K \cap L) = \bigcup_{n \in \mathbb{N}} K_n/(K \cap L)$, where the subgroups $K_n/(K \cap L)$ are finite for all $n \in \mathbb{N}$. It follows that $K = \bigcup_{n \in \mathbb{N}} K_n$, where the subgroups K_n are finite and normal for all $n \in \mathbb{N}$. Using the arguments above we see that K is abelian and hence divisible. Since G/K is finite, Theorem 1.3.5 now shows that G is Chernikov.

Suppose now that L is infinite and let $D = \mathbf{Div}(L)$. Then D is a G-invariant subgroup and L/D is finite. Our proof above shows that G/D is a Chernikov group. Let $K/D = \mathbf{Div}(G/D)$. Then K/D is a divisible Chernikov subgroup. By (iii) K is a divisible Chernikov subgroup and the finiteness of G/K shows that G is Chernikov.

The assertion (v) is an immediate corollary of (iv). □

S. N. Chernikov formulated the problem of whether every group satisfying the minimum condition is Chernikov (albeit using different terminology) and one of his first results in this direction was to show that a locally soluble group satisfies the minimal condition if and only if it is Chernikov (see [**41**, Theorem 1.1], for example). One of the highlights of the theory of locally finite groups is the theorem of V. P. Shunkov and, independently, O. H. Kegel and B. A. F. Wehrfritz who showed that every locally finite group satisfying min is Chernikov (see [**134**, Theorem 5.8]). We shall address some of these results in more detail in Chapter 3.

However, a negative answer to Chernikov's problem was obtained by A. Yu. Ol'shanskii (see [**209**, Chapters 28, 35, 38]) who constructed exotic examples of periodic quasifinite groups. These examples, termed Tarski monsters, so named after A. Tarski who first hypothesized their existence, can be constructed for any sufficiently large prime p. They are infinite 2-generator simple p-groups whose proper subgroups all have order p. Clearly such groups have min and are not even locally finite, let alone Chernikov. It is clear that Chernikov groups are countable. However there exist uncountable groups satisfying the minimum condition, by a result of V. N. Obraztsov [**207**]. We shall see how the construction of such groups proceeds in Chapter 6.

1.4. Linear Groups

In this book much of our work is concerned with generalized soluble groups; such groups have many abelian sections. If A, B are subgroups of a group G such that $A \triangleleft B$ and B/A is abelian then we can think of $K = N_G(B)/C_G(B/A)$ as a subgroup of the group of automorphisms of the abelian group B/A. Here, two fundamental cases arise, when $B_0 = B/A$ is torsion-free and also when $B_0 = B/A$ is periodic. In either case B_0 is a $\mathbb{Z}$-module. In the first case, the $\mathbb{Z}$-*injective hull* (or *divisible hull*) [**88**, Chapter IV, Section 24], a divisible abelian torsion-free group D, containing B_0, such that D/B_0 is periodic, plays an important role.

1.4.1. LEMMA. *Let A be a torsion-free abelian group and let G be a subgroup of $\mathbf{Aut}(A)$. If D is the divisible hull of A then G is isomorphic to a subgroup of $\mathbf{Aut}(D)$.*

EXERCISE 1.9. *Prove Lemma 1.4.1.*

We can think of D as a vector space over $\mathbb{Q}$ so that it is then possible to embed $K = N_G(B)/C_G(B/A)$ as a subgroup of the group $GL(\mathbb{Q}, D)$ of all non-singular linear transformations of D.

If B_0 is periodic, then our study often reduces to the case when B_0 is a p-group for some prime p. In this case, a very important role is played by the structure of the *lower layer,*

$$\Omega_1(B_0) = \{b \in B_0 | b^p = 1\}.$$

We may think of $\Omega_1(B_0)$ as a vector space over the prime field $\mathbb{F}_p$. Furthermore, suppose that $1 \neq z \in \zeta(K)$ and let $J = \mathbb{F}_p\langle x \rangle$ be the group ring of the infinite cyclic group $\langle x \rangle$ over the field $\mathbb{F}_p$. We make $\Omega_1(B_0)$ into a J-module by defining $b \cdot x = b^z$ for every $b \in \Omega_1(B_0)$. If this module is J-torsion-free, then we can construct $E = \Omega_1(B_0) \otimes_J F$, where F is the field of fractions of the integral domain J. Then E is a vector space over F and in this case $K = N_G(B)/C_G(B/A)$ can be considered as a subgroup of $GL(F, E)$.

Therefore, it very often happens that a generalized soluble group induces a group of automorphisms on its abelian factor groups that is isomorphic to a subgroup of $GL(F, V)$, for some field F and some vector space V. In many instances (and these are the ones usually of concern to us) $\dim_F(V)$ is finite. In this case $GL(F, V)$ is isomorphic to the group of all non-singular $n \times n$ matrices with coefficients in F, which we denote by $GL_n(F)$. The theory of matrix groups is well established; its scope is broad and not limited to mathematical disciplines alone. In our work, we have to use a variety of important results from this theory. For our considerations the most convenient reference is the excellent book of B. A. F. Wehrfritz [258].

We now list some of the results we shall need without proof, accompanied by the appropriate reference, so the reader has easy access to the proof. A group G is called *linear* (more precisely, finite dimensional linear) if it is isomorphic to a subgroup of $GL_n(F)$ for some natural number n and some field F. The first result we need is a result concerning the periodic subgroups of $GL_n(\mathbb{Q})$. A proof of this result can be found in [258, Theorem 9.33]. We shall let $\mathbb{Z}_p$ denote the field of p-adic numbers, for the prime p.

1.4.2. PROPOSITION. *Let T be a periodic subgroup of $GL_r(\mathbb{Q})$ or $GL_r(\mathbb{Z}_p)$, for some prime p. Then T is finite and there is a function $\rho : \mathbb{N} \longrightarrow \mathbb{N}$ such that $|T| \leq \rho(r)$.*

In the theory of finite groups an important role is played by minimal normal subgroups. In the study of torsion-free groups such minimal normal subgroups need not exist. It is often useful to look at normal abelian subgroups of minimal 0-rank (the meaning of which will be explained later) however. These are the so-called *rationally irreducible* groups. Much of the work we do concerning rank conditions is dependent upon such groups.

We mention next the following far-reaching theorem of J. Tits [252], which is known as the Tits alternative.

1.4.3. THEOREM. *Let G be a subgroup of $GL_n(F)$, for some field F. If G contains no non-abelian free group then G has a soluble normal subgroup D such that G/D is locally finite. Moreover if F has characteristic 0, then G/D is finite.*

1.4.4. COROLLARY. *Let G be a finitely generated subgroup of $GL_n(F)$ for some field F. If G contains no non-abelian free subgroups then G is soluble-by-finite.*

Since a non-abelian free group is certainly not locally generalized radical we deduce the following.

1.4.5. COROLLARY. *Let G be a locally generalized radical subgroup of $GL_n(F)$, for the field F. Then there is a soluble normal subgroup D of G such that G/D is locally finite. Moreover, if F has characterstic 0, then G/D is finite.,*

Let V be a vector space over a field F and let G be a subgroup of $GL_n(F)$. Suppose that C, D are G-invariant subspaces of V such that $D \leq C$. Then C/D is called a *G-chief factor* of V if the only G-invariant subspaces B such that $D \leq B \leq C$ are C and D. Since V is an FG-module, this means that C/D is a simple FG-module. The subgroup G is called *irreducible* if V is a simple FG-module.

If V is a finite dimensional vector space over F then the finiteness of $\dim_F(V)$ implies that V has a finite series

$$0 = B_0 \leq B_1 \leq \cdots \leq B_n = V$$

of G-invariant subspaces whose factors are G-chief factors. It then follows that the factor groups $G/C_G(B_j/B_{j-1})$ are irreducible, for $1 \leq j \leq n$. Let

$$Z = C_G(B_1) \cap C_G(B_2/B_1) \cap \cdots \cap C_G(B_n/B_{n-1}).$$

We can choose a basis for V in such a way that Z will be isomorphic to a subgroup of $UT_n(F)$, the group of unitriangular matrices with coefficients in F. Furthermore, by Remak's Theorem, G/Z is isomorphic to a subgroup of

$$G/C_G(B_1) \times G/C_G(B_2/B_1) \times \cdots \times G/C_G(B_n/B_{n-1})$$

which of course is a finite direct product of irreducible linear groups.

We require some information concerning the structure of the group $UT_n(F)$. For each integer k such that $1 \leq k \leq n$, we define subgroups $UT_n^k(F)$ as follows. If $k = 1$ we let $UT_n^1(F) = UT_n(F)$ and, if $1 < k < n$, we let $UT_n^k(F)$ be the subgroup of $UT_n(F)$ defined by

$$UT_n^k(F) = \{A = (\alpha_{i,j}) \in UT_n(F) | \alpha_{i,r} = 0 \text{ for } i+1 \leq r \leq i+k-1\}.$$

We set $UT_n^n(F) = E$, the identity matrix. Thus $UT_n^k(F)$ is precisely the subgroup of $UT_n(F)$ consisting of matrices in which the first $k-1$ superdiagonals are all 0. It is easy to show that if $A \in UT_n^k(F)$ and $B \in UT_n^l(F)$ then $[A, B] \in UT_n^{k+l}(F)$, so the identity $[UT_n^k(F), UT_n^l(F)] = UT_n^{k+l}$ holds (see [**129**, 3.2], for example). It then follows, again quite easily, that the series

$$UT_n(F) = UT_n^1(F) \geq UT_n^2(F) \geq \cdots \geq UT_n^{n-1}(F) \geq UT_n^n(F) = E$$

is simultaneously the upper and the lower central series of $UT_n(F)$. Thus the group $UT_n(F)$ is nilpotent of class at most $n-1$. Furthermore, if F_+ denotes the additive group of F then

$$UT_n^m(F)/UT_n^{m+1}(F) \cong \underbrace{F_+ \oplus \cdots \oplus F_+}_{n-m}$$

(see, [**129**, 4.2], for example).

Further, if $T_n(F)$ denotes the group of all upper triangular matrices over F with non-zero determinant, then it is clear that $T_n(F)$ is soluble of derived length at most n.

The structure of irreducible soluble linear groups has been described by A. I. Maltsev [**188**]. The reader can consult Wehrfritz [**258**, Theorem 3.6 and Lemma 3.5] for a proof. This structure is given in the next two theorems.

1.4.6. THEOREM. *Let G be a soluble subgroup of $GL_n(F)$ for the field F. If G is irreducible then there exists an integer valued function $\mu(n)$ such that G has an abelian normal subgroup of index dividing $\mu(n)$.*

It follows from the proof of Theorem 1.4.6 that $\mu(n) \leq n!(n^2(n^2)!)^n$. We shall call $\mu(n)$ the *Maltsev function*.

1.4.7. THEOREM. *Let G be a soluble subgroup of $GL_n(F)$ for some field F. Then there exists a finite field extension E of F such that G has a normal subgroup H of index dividing $\mu(n)$ which is conjugate in $GL_n(E)$ to a subgroup of $T_n(E)$. In particular, G is nilpotent-by-abelian-by-finite.*

If G is a soluble group then we let $\mathbf{dl}(G)$ denote the derived length of G. One crude estimate for the derived length of a finite soluble group is easy to establish.

If $n \in \mathbb{N}$ and $n = p_1^{k_1} \ldots p_m^{k_m}$ is the primary decomposition of n, we let $e(n) = k_1 + \cdots + k_m$ and clearly

$$e(n) = \log_{p_1}(p_1^{k_1}) + \cdots + \log_{p_m}(p_m^{k_m})$$

$$\leq \log_2(p_1^{k_1}) + \cdots + \log_2(p_m^{k_m}) = \log_2 n.$$

From this it follows that if G is a finite soluble group then $\mathbf{dl}(G) \leq e(|G|) \leq \log_2 |G|$.

A particular consequence of Theorem 1.4.7 is a theorem of Zassenhaus [278].

1.4.8. THEOREM. *Let G be a locally soluble linear group of degree n over a field F. Then G is soluble. Furthermore there is a function $\zeta : \mathbb{N} \longrightarrow \mathbb{N}$ such that $\mathbf{dl}(G) \leq \zeta(n)$.*

PROOF. We may assume that F is algebraically closed. If H is a finitely generated subgroup of G then H is soluble and Theorem 1.4.7 implies that H contains a normal subgroup $L(H)$ of index dividing $\mu(n)$, and such that $L(H)$ is triangularizable. We noted earlier that $\mathbf{dl}(T_n(F)) \leq n$, so $L(H)$ therefore has derived length at most n. Thus each finitely generated subgroup of G has derived length at most $\zeta(n) = e(\mu(n)) + n$ and the result follows. □

The function $\zeta(n)$ is called the *Zassenhaus function*. B. Huppert [117, Satz 9] proved that $\zeta(n) \leq 2n$ and furthermore if $n = 2$, then there is a soluble linear group of derived length 4 (see [117, p.#495]). The most precise bound for the function $\zeta(n)$ has been obtained by M. F. Newman [206]. His paper includes a table of values of $\zeta(n)$ for all $n \leq 74$ and he proved that for $r \geq 66$ the relation

$$5\log_9(r-1) + D \leq \zeta(r) \leq 5\log_9(r-2) + D + (3/2)$$

is valid, where $D = (17/2) - 15(\log 2)/(2\log 3)$. The lower bound is actually attained whenever $r = 24 \cdot 9^k + 1$ and the upper bound is attained whenever $r = 8 \cdot 9^k + 2$, for $k \in 0 \cup \mathbb{N}$.

The following corollary to Zassenhaus's theorem will also be very useful.

1.4.9. COROLLARY. *Let G be a locally radical linear group of degree n over a field F. Then G is soluble of derived length at most $\zeta(n)$.*

PROOF. Let K be an arbitrary finitely generated subgroup of G and let
$$1 = R_0 \le R_1 \le \ldots R_\alpha \le R_{\alpha+1} \le \ldots R_\gamma = K$$
be the radical series of K. We use transfinite induction on the length of the series to show, using Theorem 1.4.8, that K is soluble of derived length at most $\zeta(n)$.

If $\gamma = 1$ then K is nilpotent, hence soluble and we may apply Theorem 1.4.8 to deduce the result. Suppose that $\gamma > 1$ and that we have already proved that R_α is soluble of derived length at most $\zeta(n)$ for all $\alpha < \gamma$. If γ is a limit ordinal then for each $\alpha < \gamma$ the group R_α is soluble of derived length at most $\zeta(n)$ and hence K must be soluble of this derived length.

If $\gamma - 1$ exists then, inductively, $R_{\gamma-1}$ is soluble. Thus K is soluble-by-nilpotent, since it is finitely generated, so K is soluble and Theorem 1.4.8 implies that K is of derived length at most $\zeta(n)$. Consequently the group G is locally soluble and Theorem 1.4.8 implies the result. This completes the proof. □

We next describe one situation in which irreducible linear groups arise. Accordingly, let A be a torsion-free abelian group and let G be a group of automorphisms of A. We say that A is *rationally irreducible with respect to G* or that A is *G-rationally irreducible* if, for every non-trivial G-invariant subgroup B of A, the factor group A/B is periodic. We shall also say that G is *rationally irreducible on A*.

The reason for the terminology is as follows. Suppose that A is a torsion-free abelian group. As we have seen above, the action of G on A can be extended to an action of G on the vector space $V = A \otimes_{\mathbb{Z}} \mathbb{Q}$.

1.4.10. PROPOSITION. *Let G be a group of automorphisms of the torsion-free abelian group A. Then G is rationally irreducible on A if and only if G is irreducible as a group of linear transformations of the vector space $V = A \otimes_{\mathbb{Z}} \mathbb{Q}$.*

EXERCISE 1.10. *Prove Proposition 1.4.10.*

The following generalization of Theorem 1.4.6 will also be needed.

1.4.11. THEOREM. *Let G be a soluble-by-finite subgroup of $GL_n(F)$, for the field F. If G is irreducible, then G is abelian-by-finite.*

PROOF. Let V be a vector space of F-dimension n on which G acts irreducibly and let H be a soluble normal subgroup of finite index in G. Since $\dim_F(V)$ is finite, there is a non-zero H-invariant subspace, W, of minimal dimension over F. Then W is a simple FH-submodule. Theorem 1.4.6 shows that $H/C_H(W)$ is abelian-by-finite. If $W = V$, then $C_H(W) = C_H(V) = 1$, so that H and hence G is abelian-by-finite. Therefore we may suppose that $W \neq V$.

If $g \in G, a \in W, h \in H$ then we have $(ag)h = a(ghg^{-1})g \in Wg$, since $H \lhd G$, which shows that Wg is also H-invariant. Let C be a non-zero, H-invariant subspace of Wg. Then Cg^{-1} is a non-zero H-invariant subspace of W and the minimal choice of W implies that $Cg^{-1} = W$. It follows that $C = Wg$, so that Wg is a simple FH-submodule. Hence either $Wg = W$ or $Wg \cap W = 0$. Since $W \neq V$ it follows that there must be some element $g_1 \in G$ such that $Wg_1 \cap W = 0$. If $Wg \leq Wg_1 \oplus W$, for all $g \in G$, then $V = Wg_1 \oplus W$. Otherwise there is an element $g_2 \in G$ such that $Wg_2 \cap (Wg_1 \oplus W) = 0$ and we can then form $W \oplus Wg_1 \oplus Wg_2$. Proceeding in this fashion and using the fact that V has finite F-dimension, we deduce that there are elements $g_1, g_2, \ldots, g_s$ of G such that $V = W \oplus Wg_1 \oplus \cdots \oplus Wg_s$. Theorem 1.4.6 implies, as above, that $H/C_H(Wg_j)$ is abelian-by-finite for $1 \leq j \leq s$. Since $C_H(W) \cap C_H(Wg_1) \cap \cdots \cap C_H(Wg_s) = 0$, Remak's Theorem gives the embedding

$$H \longrightarrow H/C_H(W) \times H/C_H(Wg_1) \times \cdots \times H/C_H(Wg_s),$$

which in turn implies that H, and hence G, is abelian-by-finite.   □

Using [**31**, Theorem 2] and Corollary 1.4.5 we obtain the following consequence.

1.4.12. COROLLARY. *Let F be a finite field extension of $\mathbb{Q}$ and let G be a locally generalized radical subgroup of $GL_n(F)$. If G is irreducible, then G contains a countable free abelian normal subgroup of finite index.*

1.5. Some Relationships Between the Factors of the Upper and Lower Central Series

It is well known that the kth term, $\zeta_k(G)$, of the upper central series of a group G coincides with G if and only if the $(k + 1)$th term, $\gamma_{k+1}(G)$, of the lower central series of G is trivial. For infinite central series there is no such general result. For example, we noted above that, by Magnus's theorem [**172**, Chapter IX, §36], the lower central series of a non-abelian free group G has the property that $\gamma_\omega(G) = 1$, but of course $\zeta(G) = 1$. In a later chapter, we exhibit a hypercentral Chernikov p-group whose lower central series is finite and does not terminate in the identity. A very natural problem arises for groups with a finite central series as follows:

Let G be a group such that $\mathbf{zl}(G) = n$ is finite. For which classes of groups $\mathfrak{X}$ does the relation $G/\zeta_n(G) \in \mathfrak{X}$ imply that $\gamma_{n+1}(G) \in \mathfrak{X}$ or, more generally, when does the (locally) nilpotent residual belong to $\mathfrak{X}$?

An important special case arises here when $n = 1$. In this case our question asks for which classes $\mathfrak{X}$ is it the case that the relation $G/\zeta(G) \in \mathfrak{X}$ implies that $G' = \gamma_2(G) \in \mathfrak{X}$?

The case when $\mathfrak{X} = \mathfrak{F}$, the class of finite groups, is one that naturally springs to mind first. In this case, we obtain groups G whose central factor group $G/\zeta(G)$ is finite. Such groups are naturally called *centre-by-finite* groups and are closely connected with an important, interesting class of groups known as FC-groups.

A group G is called an *FC-group* if the conjugacy class $x^G = \{x^g | g \in G\}$ is finite for each element $x \in G$. Thus x has finitely many conjugates in G and it is well known that this implies that $G/C_G(x^G)$ is finite. In particular, a group G is an FC-group if and only if the index $|G : C_G(x)|$ is finite, for all $x \in G$.

1.5.1. LEMMA. *Let G be an FC-group. Then every finitely generated subgroup of G is centre-by-finite.*

EXERCISE 1.11. *Prove Lemma 1.5.1.*

An important role is played in the study of FC-groups by the following result, often called Dietzmann's Lemma, due to A. P. Dietzmann [**47**].

1.5.2. PROPOSITION. *Let G be a group and let M be a finite subset of G. Suppose that every element of M has finite order and that $x^G \subseteq M$, for every element $x \in M$. Then $\langle M \rangle = \langle M \rangle^G$ is finite.*

PROOF. Let $|M| = k$ and let $d = \text{lcm}\{|x| \,|\, x \in M\}$. Every element $y \in \langle M \rangle$ can be written in the form $y = x_1 x_2 \ldots x_m$, where $x_j \in M$ for $1 \leq j \leq m$. Suppose that $m > k(d-1)$. Then at least one of the elements of M occurs at least d times in this product and we denote this particular element by u. Let s be the first index for which $x_s = u$. We have

$$y = x_1 x_2 \ldots x_m = u(u^{-1}x_1 u)(u^{-1}x_2 u) \ldots (u^{-1}x_{s-1}u)(u^{-1}u)x_{s+1} \ldots x_m.$$

Let $w_j = u^{-1}x_j u$, so $w_j \in M$ for $1 \leq j \leq s-1$, by hypothesis, and $y = u w_1 \ldots w_{s-1} x_{s+1} \ldots x_m$. Let t be the next index for which $x_t = u$.

By the same argument we have

$$y = uw_1 \ldots w_{s-1}x_{s+1} \ldots x_m$$
$$= uu(u^{-1}w_1u) \ldots (u^{-1}w_{s-1}u)(u^{-1}x_{s+1}u)$$
$$\ldots (u^{-1}x_{t-1}u)(u^{-1}u)x_{t+1} \ldots x_m$$
$$= u^2 v_1 \ldots v_{s-1}v_{s+1} \ldots v_{t-1}x_{t+1} \ldots x_m,$$

suitably renaming the elements.

Continuing in this way we can write $y = u^d z_1 \ldots z_{m-d} = z_1 \ldots z_{m-d}$, for certain elements $z_1, \ldots, z_{m-d}$ of M. Repeating the argument as needed, we will eventually obtain the expression $y = y_1 \ldots y_n$, where $y_1, \ldots, y_n \in M$ and $n \leq k(d-1)$. Hence every element of $\langle M \rangle$ can be written as a product of at most $k(d-1)$ elements so $\langle M \rangle$ is finite. The hypotheses clearly imply that $\langle M \rangle = \langle M \rangle^G$. $\qquad \square$

The following result is immediate.

1.5.3. COROLLARY. *Let G be an FC-group. Then $\mathbf{Tor}\,(G)$ coincides with the set of elements of finite order. Furthermore $G/\mathbf{Tor}\,(G)$ is torsion-free.*

We present several results concerning torsion-free FC-groups. The first of the results we obtain, due to G. A. Miller and H. Moreno [**197**], seems totally unrelated.

1.5.4. PROPOSITION. *Let G be a finite group whose proper subgroups are abelian. Then G is soluble.*

PROOF. Suppose the contrary and among all finite insoluble groups whose proper subgroups are abelian choose one, G, of least order. If G contains a proper normal subgroup H then the minimal choice of $|G|$ implies that G/H is soluble-here we note that the proper subgroups of G/H are abelian. Since H is abelian, G is soluble and we obtain a contradiction. Hence G is simple.

Let M be a maximal subgroup of G and suppose that there is a proper subgroup K such that $K \nleq M$. Since M is maximal it follows that $G = \langle M, K \rangle$. Clearly, $[M \cap K, M] = [M \cap K, K] = 1$ and we deduce that $M \cap K \leq \zeta(G)$. Thus $M \cap K = 1$, since G is simple.

Let $p \in \Pi(M)$ and let M_p be a Sylow p-subgroup of M. Let P be a Sylow p-subgroup of G containing M_p. If $P \nleq M$ then $M \cap P = 1$, as above. However, $M_p \leq M \cap P$, a contradiction. Hence M_p is a Sylow p-subgroup of G and it follows that $m = |M|$ and $t = |G : M|$ are relatively prime. Since M is a maximal, non-normal subgroup of G we have $M = N_G(M)$. It follows that M has exactly t conjugates.

We note that M^g is a maximal subgroup of G, for each $g \in G$ and, as above, $M \cap M^g = 1$, whenever $g \notin M$. Each of the subgroups M^g contains exactly $m - 1$ nontrivial elements and hence M and all its conjugates contain exactly $(m - 1)t$ nontrivial elements. This leaves $t - 1$ nontrivial elements to account for.

Let q be a prime such that $q \notin \Pi(M)$ and let Q be a Sylow q-subgroup of G. Let V be a maximal subgroup of G containing Q. As above $M \cap V = 1$. Suppose that $\Pi(M) \cap \Pi(V) \neq \emptyset$ and let $r \in \Pi(M) \cap \Pi(V)$. Let V_r (respectively M_r) be a Sylow r-subgroup of V (respectively of M). As above, V_r and M_r are Sylow r-subgroups of G and hence there exists $y \in G$ such that $V_r^y = M_r$. It follows that $M_r \leq M \cap V^y$. Since V^y is a maximal subgroup of G, our work above proves that $M = V^y$. On the other hand, $q \notin \Pi(M)$ and $q \in \Pi(V) = \Pi(V^y)$, which is a contradiction. Hence $\Pi(M) \cap \Pi(V) = \emptyset$. Thus V and M have relatively prime orders, so $v = |V|$ divides $|G : M| = t$ and hence $t = vd$ for some natural number d. As above V has md conjugates each containing exactly $v - 1$ nontrivial elements, all distinct. None of these conjugates coincides with any of the conjugates of M so this gives $(v - 1)md$ elements not previously counted. However,

$$(v - 1)md = (vm - m)d \geq vd = t, \text{ since } v, m \geq 2,$$

which now yields a final contradiction.   □

1.5.5. LEMMA. *Let G be an infinite group whose nontrivial subgroups all have finite index. Then $\zeta(G) \neq 1$.*

EXERCISE 1.12. *Prove Lemma 1.5.5.*

We next obtain a result due to Fedorov [83].

1.5.6. THEOREM. *Let G be an infinite group whose nontrivial subgroups all have finite index. Then G is cyclic.*

PROOF. Clearly, G is torsion-free and, by Lemma 1.5.5, $\zeta(G) \neq 1$, so $G/\zeta(G)$ is finite. We prove the result by induction on $n = |G/\zeta(G)|$. If $n = 2$ then $G/\zeta(G)$ is cyclic and it is well known that G is then abelian, hence clearly cyclic. Suppose now that $n > 2$ and that we assume the obvious induction hypothesis. If H is a proper nontrivial subgroup containing $\zeta(G)$ then $\zeta(G) \leq \zeta(H)$ and hence $|H/\zeta(H)| < |G/\zeta(G)|$. By the induction hypothesis, H is infinite cyclic. Hence every proper subgroup of $G/\zeta(G)$ is cyclic and Proposition 1.5.4 implies that $G/\zeta(G)$ is soluble. Then $G/\zeta(G)$ contains a normal subgroup $K/\zeta(G)$ of prime index p. Again by the induction hypothesis, $K = \langle g \rangle$ is cyclic. If we suppose that $K \nleq \zeta(G)$ then there is an element y

such that $g^y = g^{-1}$. On the other hand, $G/\zeta(G)$ is finite, so $\langle g^k \rangle = K \cap \zeta(G) \neq 1$, for some k. Then $y^{-1}g^k y = g^k$ and also $y^{-1}g^k y = g^{-k}$, which is a contradiction. Consequently, $K \leq \zeta(G)$ and since G/K is cyclic, G is abelian and hence cyclic, as required. $\qquad\square$

1.5.7. COROLLARY. *Let G be a torsion-free group. If G contains a cyclic normal subgroup of finite index, then G is cyclic.*

PROOF. Let C be a cyclic normal subgroup of G and suppose that $|G : C| = k$. If $1 \neq g \in G$, then $g^k \in C$ and, since G is torsion-free, $g^k \neq 1$. Since every nontrivial subgroup of a cyclic group has finite index we have $|C : \langle g^k \rangle|$ is finite. Hence $|G : \langle g^k \rangle|$ is finite, so that $|G : \langle g \rangle|$ is also finite and Theorem 1.5.6 applies to give the result. $\quad\square$

The following result occurs in a paper of R. Baer [6].

1.5.8. LEMMA. *Let G be a torsion-free group. If every cyclic subgroup of G is normal, then G is abelian.*

EXERCISE 1.13. *Prove Lemma 1.5.8.*

1.5.9. COROLLARY. *Let G be a finitely generated torsion-free FC-group. Then G is abelian.*

PROOF. By Lemma 1.5.1, $G/\zeta(G)$ is finite so if $g \in G$, there is a natural number k such that $g^k \in \zeta(G)$. Hence $C = \langle g^k \rangle$ is a normal subgroup of G. The element gC is of finite order in G/C and Proposition 1.5.2 shows that $K/C = \langle gC \rangle^{G/C}$ is finite. The subgroup K is torsion-free and contains a cyclic subgroup of finite index, so by Corollary 1.5.7 K is cyclic. Since every subgroup of a cyclic group is characteristic, $\langle g \rangle$ is normal in G and Lemma 1.5.8 gives the result. $\quad\square$

The following corollary can be immediately deduced.

1.5.10. COROLLARY. *Every torsion-free FC-group is abelian.*

EXERCISE 1.14. *Let G be a finitely generated abelian group. Prove that the torsion subgroup of G is finite and deduce that every periodic finitely generated abelian group is finite.*

1.5.11. COROLLARY. *Let G be a finitely generated centre-by-finite group. Then G' is finite.*

PROOF. Proposition 1.2.13 shows that $\zeta(G)$ is also finitely generated. By Corollary 1.5.3, the subgroup $T = \mathbf{Tor}(G)$ consists of the set of elements of finite order, so Exercise 1.14 implies that $R = \mathbf{Tor}(\zeta(G)) = T \cap \zeta(G)$ is finite. Since $T\zeta(G)/\zeta(G)$ is finite it follows that T is finite. The group G/T is torsion-free and hence is abelian, by Corollary 1.5.10. Therefore $G' \leq T$, so G' is finite. $\quad\square$

1.5.12. THEOREM. *Let G be a centre-by-finite group. Then G' is finite.*

PROOF. There is a finitely generated subgroup K of G such that $G = K\zeta(G)$ and clearly $G' = K'$. By Corollary 1.5.11, K' is finite. $\square$

This theorem plays an important role in infinite groups where it is used in the proofs of many important, interesting results. There are already a number of alternative proofs of this theorem based on different approaches. We have presented here a proof that is self-contained. Furthermore, Theorem 1.5.6 is seen to be equivalent to Theorem 1.5.12. The history of Theorem 1.5.12 is quite interesting. In the form given here it first appeared in an article of B. H. Neumann [202]. However, at the end of this paper Neumann writes that R. Baer had informed him that the result is a consequence of a more general result, proved by Baer in [7]. In fact in [7, Theorem 3] it was proved that if the normal subgroup H of the group G has finite index then $(G' \cap H)/[H, G]$ is also finite. In a later article, Baer [8] gives the statement of Theorem 1.5.12 in its usual form and gave a new proof. In his famous lectures on nilpotent groups, P. Hall [109, Theorem 8.7] obtained a further generalization of Theorem 1.5.12. In these lectures Hall called this theorem "Schur's Theorem" but gave no specific references. Many algebraists then started calling Theorem 1.5.12 "Schur's Theorem" often citing the paper [234] as the source. However, Schur's paper is not concerned with infinite group theory. In this classic old paper, [234], Schur introduced (only for finite groups!) the idea of what is now called the Schur multiplicator (or multiplier) and studied its properties. In modern terminology, the Schur multiplier $M(G)$ of a group G is exactly the second cohomology group $H^2(G, U(\mathbb{C}))$. The following property of the Schur multiplier holds and appears in [262, Lemma 4.1].

1.5.13. THEOREM. *Let G be a group and let C be a subgroup of $\zeta(G)$. If G/C is finite, then $G' \cap C$ is an epimorphic image of $M(G/C)$.*

From this theorem, we can deduce the original result of Schur [234].

1.5.14. COROLLARY. *Let G be a finite group and let C be a subgroup of $\zeta(G)$. Then $G' \cap C$ is an epimorphic image of $M(G/C)$.*

The following important property of the Schur multiplier also occurs in [262, Corollaries 4.3 and 4.4].

1.5.15. THEOREM. *Let G be a finite group and let p be a prime.*

(i) *If G is a p-group then $M(G)$ is a p-group;*

(ii) *If $p \in \Pi(G)$ and P is a Sylow p-subgroup of G, then the Sylow p-subgroup of $M(G)$ is isomorphic to a subgroup of $M(P)$. In particular, $\Pi(M(G)) \subseteq \Pi(G)$.*

An interesting consequence is the following result.

1.5.16. COROLLARY. *Let G be a group and suppose that C is a subgroup of $\zeta(G)$ such that G/C is finite. Then $\Pi(G') \subseteq \Pi(G/C)$. In particular, if G/C is a p-group for some prime p, then G' is a p-subgroup of G.*

PROOF. Let $\Pi(G/C) = \pi$. Then G/C is a π-group and, by Theorem 1.5.15, $M(G/C)$ is also a π-group. Theorem 1.5.13 implies that $G' \cap C$ is also a π-group and, since $G'C/C \cong G'/(G' \cap C)$, it follows that G' is a π-group, as required. □

1.5.17. COROLLARY. *Let G be a group and let C be a subgroup of $\zeta(G)$. If G/C is locally finite, then G' is also locally finite and $\Pi(G') \subseteq \Pi(G/C)$.*

PROOF. Let F be an arbitrary finitely generated subgroup of G'. Then there exists a finitely generated subgroup K such that $F \leq K'$. Since G/C is locally finite, KC/C is finite, so that $K/(K \cap C)$ is finite. Clearly $K \cap C \leq \zeta(K)$, so Theorem 1.5.12 shows that K' is finite. Hence F is finite, so that G' is locally finite. Moreover, using Corollary 1.5.16 we have

$$\Pi(F) \subseteq \Pi(K') \subseteq \Pi(K/K \cap C) \subseteq \Pi(G/C)$$

and the second part also follows. □

Theorem 1.5.12 suggests a natural question as to whether there is a relationship between $t = |G/\zeta(G)|$ and $|G'|$. This question was posed in the paper [**202**] of B. H. Neumann, who obtained the first estimates for $|G'|$. The best estimate has been obtained by J. Wiegold [**261**], as follows.

1.5.18. THEOREM. *Let G be a group such that $G/\zeta(G)$ is finite of order t. Then*

(i) *$|G'| \leq w(t)$, where $w(t) = t^m$ and $m = (1/2)(\log_2 t - 1)$;*

(ii) *If p is a prime and $t = p^n$, then G' is a p-group of order at most $p^{\frac{n(n-1)}{2}}$;*

(iii) *For each prime p and each integer $n > 1$ there is a group G such that $|G/\zeta(G)| = p^n$ and $|G'| = p^{\frac{n(n-1)}{2}}$.*

When $G/\zeta(G)$ has more than one prime divisor, the full picture is still unclear here. Theorem 1.5.12 admits the following natural generalization, known as Baer's Theorem.

1.5.19. THEOREM. *Let G be a group and suppose that there is a natural number k such that $G/\zeta_k(G)$ is finite. Then $\gamma_{k+1}(G)$ is finite.*

Paradoxically, R. Baer did not explictly state and prove this result. In his article [8], mentioned above, Baer notes that it can be obtained from Zusatz zum Endlichkeitssatz of that paper. The title "Baer's Theorem" again appears for the first time in the lectures of P. Hall. Hall gave a generalization of Theorem 1.5.12 which has as a corollary Theorem 1.5.19, but again Hall gave no references. Later D. J. S. Robinson [229] used the results of Baer's paper to give a direct proof of Theorem 1.5.19. We give a proof of this result, but also indicate the connection between the orders of $G/\zeta_k(G)$ and $\gamma_{k+1}(G)$.

The next lemma has several variations which we shall obtain in Chapter 7.

1.5.20. LEMMA. *Let G be a group and let A be a normal abelian subgroup of G. Suppose that G satisfies the following conditions:*

(i) $G/C_G(A) = \langle x_1 C_G(A), \ldots, x_m C_G(A) \rangle$, *for certain elements $x_1, \ldots, x_m \in G$;*

(ii) $A/(\zeta(G) \cap A)$ *is finite of order t.*

Then $[A, G]$ is finite and $|[A, G]| \leq t^m$.

EXERCISE 1.15. *Prove Lemma 1.5.20.*

We now prove the quantitive version of Theorem 1.5.19 which appears in [168].

1.5.21. THEOREM. *Let G be a group and suppose that there is a natural number k such that $G/\zeta_k(G)$ is finite of order t. Then there is a function β_1 of k, t only such that $\gamma_{k+1}(G)$ is finite of order at most $\beta_1(t, k)$.*

PROOF. Let

$$1 = Z_0 \leq Z_1 \leq \cdots \leq Z_{k-1} \leq Z_k = Z$$

be the upper central series of G. We use induction on k to obtain the result. If $k = 1$ then G/Z_1 has order at most t and an application of Theorem 1.5.18 proves that $\gamma_2(G) = G'$ has order at most $w(t)$.

Assume now that $k > 1$. Then G/Z_1 has a shorter upper central series than G, so we may suppose inductively that we have found a function β_1 such that $|\gamma_k(G/Z_1)| \leq \beta_1(t, k-1)$. We note that $K/Z_1 = \gamma_k(G/Z_1) = \gamma_k(G)Z_1/Z_1$ and, letting $L = \gamma_k(G)$, we observe that $K = LZ_1$, so $L \leq K$ and $K' = L'$. Furthermore, $|L/(L \cap Z_1)| = |K/Z_1| \leq \beta_1(t, k-1)$. By Proposition 1.1.6, $[L, Z_k] = 1$, so that $G/C_G(L)$ is finite of order at most t. The subgroup K is centre-by-finite and an application of Theorem 1.5.18 shows that K' is finite of order at most $w(\beta_1(t, k-1))$. Of course L/K' is abelian. We have

$$(L/K')/(L/K' \cap Z_1K'/K') \cong L/(L \cap Z_1K'),$$

which shows that $(L/K')/(L/K' \cap Z_1K'/K')$ is an epimorphic image of $L/(L \cap Z_1)$ and hence $|(L/K')/(L/K' \cap Z_1K'/K')| \leq \beta_1(t, k-1)$. We apply Lemma 1.5.20 to G/K' to deduce that

$$V/K' = [L/K', G/K'] = [\gamma_k(G)/K', G/K'] = \gamma_{k+1}(G)K'/K'$$

is finite of order at most $t^{\beta_1(t,k-1)}$. Thus,

$$|\gamma_{k+1}(G)| \leq |K'| \cdot |V/K'| \leq w(\beta_1(t, k-1)) \cdot t^{\beta_1(t,k-1)},$$

so we define $\beta_1(t, k) = w(\beta_1(t, k-1)) \cdot t^{\beta_1(t,k-1)}$ and this completes the proof. $\qquad\qquad\square$

The proof gives us the bound, defined recursively as follows:

$$\beta_1(t, 1) = w(t),$$

$$\beta_1(t, k) = w(\beta_1(t, k-1)) \cdot t^{\beta_1(t,k-1)}.$$

Theorem 1.5.19 shows that if the upper hypercentre of a group G has finite index and $\mathbf{zl}(G)$ is finite, then G contains a finite normal subgroup K such that G/K is nilpotent. The question then arises: What can be said about a group G when the upper hypercentre has finite index but $\mathbf{zl}(G)$ is arbitrary? This question was discussed in [**77**] and we end this section with a short, simple proof of their main theorem, which appeared in [**157**]. We remind the reader that $\zeta_\infty(G)$ denotes the upper hypercentre of G.

1.5.22. THEOREM. *Let G be a group and suppose that $G/\zeta_\infty(G)$ is finite. Then G contains a finite normal subgroup L such that G/L is hypercentral.*

PROOF. We may suppose that G is not hypercentral. Let $Z = \zeta_\infty(G)$ and choose a finitely generated subgroup K with the property that $G = ZK$. Since $K/K \cap Z$ is finite, Proposition 1.2.13 implies that $K \cap Z$ is also finitely generated. It follows from Proposition 1.2.5 that $K \cap Z$ is nilpotent, and the proof of Proposition 1.2.5 shows that $t = \mathbf{zl}(K)$ is finite. Since G/Z is not nilpotent, neither is K and we let $C = \zeta_\infty(K) = \zeta_t(K)$. If $C \cap Z \neq C$ then CZ/Z is nontrivial which implies that the upper hypercentre of G/Z is nontrivial, a contradiction. Hence $C \cap Z = C$ and Theorem 1.5.19 shows that $\gamma_{t+1}(K)$ is finite. Hence the nilpotent residual, $L = K^{\mathfrak{N}}$, of K is finite.

It remains to show that L is normal in G and that G/L is hypercentral. To this end, we let $\mathcal{L}$ denote the family of finitely generated subgroups of G each member of which contains K. Let $V \in \mathcal{L}$. Since $G = ZV$ we note that, as above, $V \cap Z = \zeta_\infty(V) = \zeta_n(V)$, for some n. Since $V \le KZ$ and $K \le V$ we have $V = K(V \cap Z) = K\zeta_n(V)$. Corollary 1.1.8 shows that $\gamma_{n+1}(V) = \gamma_{n+1}(K)$ and hence $\gamma_{n+1}(K)$ is normal in V. Since L is a characteristic subgroup of $\gamma_{n+1}(K)$ it follows that $L \lhd V$. Since G is the union of all the subgroups $V \in \mathcal{L}$ we deduce that L is a normal subgroup of G. Moreover,

$$G/ZL \cong (G/L)/(ZL/L) = (K/L)(ZL/L)/(ZL/L)$$
$$\cong (K/L)/((K/L) \cap (ZL/L)),$$

which is nilpotent since K/L is nilpotent. However, the hypercentre of G/L contains ZL/L, so G/L is also hypercentral. $\qquad\square$

1.6. Some Direct Decompositions in Abelian Normal Subgroups

The topic in this section arises naturally as follows. Let G be a group and let B, C be normal subgroups of G such that $B \le C$. The factor group C/B is called G-central if $C_G(C/B) = G$ and G-eccentric if $C_G(C/B) \ne G$. Now suppose that A is a finite normal abelian subgroup of G. Then A has a finite G-chief series,

$$1 = A_0 \le A_1 \le \cdots \le A_n = A.$$

In this series, the G-central and G-eccentric factors may be arranged arbitrarily and the following question, roughly posed, arises: Under what conditions can all the G-central factors be gathered together and when can all the G-eccentric factors be gathered together? This question, in a more general form (for factors associated with formations of groups) was raised and studied in the works of L. A. Shemetkov [239] and R. Baer [16]. We will not go into the details and intricacies of this subject, but here limit ourselves only to essential results which we shall require for future use. A more detailed discussion is available in the book [153].

If G is a group and A is a normal subgroup of G then we define the *upper G-central series* of A,

$$1 = A_0 \le A_1 \le \ldots A_\alpha \le A_{\alpha+1} \le \ldots A_\gamma,$$

where

$$A_1 = \zeta_G(A) = \{a \in A | [a, g] = 1, \text{ for all } g \in G\},$$
$$A_{\alpha+1}/A_\alpha = \zeta_G(A/A_\alpha), \text{ for all ordinals } \alpha < \gamma \text{ and}$$
$$A_\lambda = \bigcup_{\beta < \lambda} A_\beta, \text{ for all limit ordinals } \lambda < \gamma.$$

Furthermore, $\zeta_G(A/A_\gamma) = 1$. We note that every subgroup in this series is G-invariant. Of course, $A_1 = C_A(G) = \zeta(G) \cap A$ and it is easy to see that $A_\alpha = A \cap \zeta_\alpha(G)$, for all ordinals α.

The last term A_γ of this series is called the *upper G-hypercentre* of A and is denoted by $\zeta_G^\infty(A)$. If $A = A_\gamma$, then A is called *G-hypercentral*; when γ is finite, A is called *G-nilpotent*. The normal subgroup A is called *G-hypereccentric* if it has an ascending series

$$1 = C_0 \leq C_1 \leq \cdots \leq C_\alpha \leq C_{\alpha+1} \leq \cdots \leq C_\gamma,$$

of G-invariant subgroups such that each factor $C_{\alpha+1}/C_\alpha$ is G-eccentric and G-chief, for every $\alpha < \gamma$.

A normal abelian subgroup A of G is said to have the $Z(G)$-*decomposition* if

$$A = \zeta_G^\infty(A) \oplus \eta_G^\infty(A),$$

where $\eta_G^\infty(A)$ is the maximal G-hypereccentric G-invariant subgroup of A. This concept was introduced by D. I. Zaitsev [**272**] (note that here we do not use the language of modules used by Zaitsev).

We note that in this case $\eta_G^\infty(A)$ contains every G-hypereccentric, G-invariant subgroup of A and hence is unique. To see this, let B be a G-hypereccentric, G-invariant subgroup of A and let $E = \eta_G^\infty(A)$. If $(BE)/E$ is nontrivial it contains a nontrivial minimal G-invariant subgroup U/E. Since $(BE)/E \cong B/(B \cap E)$ it follows that U/E is G-isomorphic to some G-chief factor of B and hence $G/C_G(U/E) \neq G$. On the other hand, $(BE)/E \leq A/E \cong \zeta_G^\infty(A)$ so that $G/C_G(U/E) = G$, which gives us a contradiction. Hence $B \leq E$ and our claim concerning $\eta_G^\infty(A)$ follows.

We now consider some results that give conditions for the existence of a $Z(G)$-decomposition in abelian normal subgroups. These results have been obtained in the article [**150**] and will be very useful.

1.6.1. LEMMA. *Let G be a group and let A be an abelian normal subgroup of G. Suppose that A contains a G-invariant subgroup C such that $C \leq \zeta_G(A)$ and A/C is a G-eccentric chief factor of A. If $G/C_G(A)$ is hypercentral, then A contains a minimal G-invariant subgroup D such that $A = C \times D$.*

EXERCISE 1.16. *Prove Lemma 1.6.1.*

By using a suitable inductive argument we deduce the next result.

1.6.2. COROLLARY. *Let G be a group and let A be an abelian normal subgroup of G. Suppose that A contains a G-invariant subgroup C such that C is G-nilpotent and A/C is a G-eccentric chief factor of A. If $G/C_G(A)$ is hypercentral, then A contains a minimal G-invariant subgroup D such that $A = C \times D$.*

1.6.3. LEMMA. *Let G be a group and let A be an abelian normal subgroup of G. Suppose that A contains a minimal G-invariant eccentric subgroup C such that $A/C \leq \zeta_G(G/C)$. If $G/C_G(A)$ is hypercentral, then A contains a G-invariant subgroup D such that $A = C \times D$.*

PROOF. Let $Z = C_G(A)$ and suppose that $Z \neq zZ \in \zeta(G/Z)$. The mapping $\xi_z : A \longrightarrow A$ defined by $\xi_z(a) = [a, z]$ is an endomorphism of A. Since $\xi_z(a)^g = \xi_z(a^g)$, for all $g \in G$, it follows that $\ker(\xi_z) = C_A(z)$ and $\text{Im}(\xi_z) = [A, z]$ are G-invariant subgroups of A. Since $A/C \leq \zeta_G(G/C)$ we have $[A, z] \leq C$ and since C is a minimal G-invariant subgroup of A it follows that $[A, z] = C$ or $[A, z] = 1$. The latter case cannot arise since it implies that $z \in C_G(A) = Z$, contrary to the choice of z. Hence $\text{Im}(\xi_z) = C$.

Assume, for a contradiction, that $[C, z] = 1$, so that $C \leq \ker(\xi_z)$. Since $G \neq C_G(C)$ there exist $c \in C, g \in G$ such that $c^g \neq c$ and since $C = \text{Im}(\xi_z)$ there exists $b \in A$ such that $c = \xi_z(b)$. Since $[b, g] \in C$ it follows that $[b, g] \in \ker(\xi_z)$ and we have

$$c^g = \xi_z(b^g) = \xi_z(b[b, g]) = \xi_z(b)\xi_z([b, g]) = \xi_z(b) = c,$$

which is the contradiction sought. Hence $[C, z] \neq 1$. However, $[C, z]$ is a G-invariant subgroup of C, since $zZ \in \zeta(G/Z)$, and hence $[C, z] = C$, from which it follows that $[A, z] = [C, z]$.

Let $a \in A \setminus C$ and choose $c \in C$ such that $[a, z] = [c, z]$. Then $[ac^{-1}, z] = 1$, so $ac^{-1} \in C_A(z)$. Hence $A = C \cdot C_A(z)$, the product of two G-invariant subgroups of A. It remains to show that $C \cap C_A(z) = 1$, at which point we set $D = C_A(z)$. However, $C \cap C_A(z)$ is also a G-invariant subgroup of A and the minimal choice of C implies that either $C \cap C_A(z) = 1$ or $C \leq C_A(z)$. As we observed above, the latter possibility cannot happen. Hence $A = C \times C_A(z) = C \times D$. The result follows.   □

A straightforward argument gives us the following result.

1.6.4. COROLLARY. *Let G be a group and let A be an abelian normal subgroup of G. Suppose that A contains a G-invariant subgroup C*

such that $A/C \leq \zeta_G(G/C)$ and suppose that C has a finite series of G-invariant subgroups whose factors are G-chief and G-eccentric. If $G/C_G(A)$ is hypercentral, then A contains a G-invariant subgroup D such that $A = C \times D$.

Let A be a normal subgroup of a group G having a finite G-chief series. By the Jordan–Hölder Theorem, every pair of such series have the same length and isomorphic factors. In particular, the length of such a G-chief series is an invariant of A, which we denote by $\mathbf{cl}_G(A)$.

We next give some sufficient conditions for an abelian normal subgroup of a group G to have a $Z(G)$-decomposition.

1.6.5. COROLLARY. *Let G be a group and let A be an abelian normal subgroup of G with a finite G-chief series. If $G/C_G(A)$ is hypercentral, then A has a $Z(G)$-decomposition.*

PROOF. We use induction on $\mathbf{cl}_G(A)$. If $\mathbf{cl}_G(A) = 1$ then A is a minimal normal subgroup of G and the result follows, so suppose that $\mathbf{cl}_G(A) > 1$. Since A has a finite G-chief series, it contains a G-invariant subgroup B such that A/B is a G-chief factor. Clearly $\mathbf{cl}_G(A) = \mathbf{cl}_G(B) + \mathbf{cl}_G(A/B)$ and $\mathbf{cl}(_G B) < \mathbf{cl}_G(A)$. By the induction hypothesis, B has a $Z(G)$-decomposition, say $B = E \times C$, where C is the upper G-hypercentre of B and E is a G-hypereccentric G-invariant subgroup.

Suppose first that A/B is G-eccentric and consider A/E. Since $B/E = CE/E \cong_G C$ we may apply Corollary 1.6.2 to A/E and deduce that it contains a minimal G-invariant subgroup D/E such that $A/E = B/E \times D/E$. Since $A/B \cong_G D/E$ it follows that D/E is G-eccentric and hence D is G-hypereccentric. Then $A = BD = ECD = CD$. Also $C \cap D \leq E$ so

$$C \cap D = (C \cap D) \cap E = C \cap E = 1.$$

Hence $A = C \times D$ and A has a $Z(G)$-decomposition.

Next, suppose that A/B is G-central and consider A/C. Since $B/C = EC/C \cong_G E$, we may apply Corollary 1.6.4 to A/C and deduce that there is a G-invariant, G-central subgroup Z/C such that $A/C = B/C \times Z/C$. Then $A = BZ = ECZ = EZ$. Also $E \cap Z \leq B \cap Z \leq C$ so

$$E \cap Z = (E \cap Z) \cap C = E \cap C = 1.$$

Since C is G-hypercentral and Z/C is G-central, Z is also G-hypercentral, so $A = E \times Z$ has a $Z(G)$-decomposition in this case as well. □

1.6.6. PROPOSITION. *Let G be a group and let A be an abelian normal subgroup of G. Suppose that A contains a G-invariant subgroup*

C such that C is G-nilpotent and A/C has a finite G-chief series. If $G/C_G(A)$ is hypercentral, then A has a $Z(G)$-decomposition. Furthermore $\mathbf{cl}_G(\eta_G^\infty(A)) \leq \mathbf{cl}_G(A/C)$. In particular, if A/C is finite then $\eta_G^\infty(A)$ is finite and $|\eta_G^\infty(A)| \leq |A/C|$.

PROOF. Let B be the upper G-hypercentre of A so that $C \leq B$. Since $\mathbf{cl}_G(A/C)$ is finite, $\mathbf{zl}_G(B)$ and $\mathbf{cl}_G(A/B)$ are both finite. If $B = 1$, then the result follows from Corollary 1.6.5. Hence we may assume that $B \neq 1$ and use induction on $\mathbf{cl}_G(A/B)$. If $\mathbf{cl}_G(A/B) = 0$, then $A = B$, so A is G-nilpotent and the result follows. Suppose then that $A \neq B$. Since $\mathbf{cl}_G(A/B)$ is finite it follows that A/B has a minimal G-invariant subgroup D/B. Since B is the upper G-hypercentre, D/B is G-eccentric and it follows from Corollary 1.6.2 that D contains a minimal G-invariant subgroup E such that E is G-eccentric and $D = BE$.

Consider A/E and let K/E be the upper G-hypercentre of A/E. Clearly $D/E = BE/E \leq K/E$ and it follows that $\mathbf{cl}_G((A/E)/(K/E)) < \mathbf{cl}_G(A/B)$. By the induction hypothesis A/E has a $Z(G)$-composition, say $A/E = K/E \times L/E$, where $L/E = \eta_G^\infty(A/E)$. It is easy to see, using Corollary 1.6.4 repeatedly, that K contains a G-invariant subgroup Y such that $K = Y \times E$. Clearly, $Y \cap L = 1$ and we have $A = KL = (YE)L = Y \times L$, as required.   □

We now show how to apply the results of this section to the problems of Section 1.5 and obtain further information concerning them. We recall that in that section we made use of a function which we denoted by $w(t)$.

1.6.7. PROPOSITION. *Let G be a finitely generated group and suppose that for some natural number k, $G/\zeta_k(G)$ is finite of order t. Then the nilpotent residual $G^{\mathfrak{N}}$ is finite of order at most $tw(t)$. Furthermore, $G/G^{\mathfrak{N}}$ is nilpotent.*

PROOF. Let

$$1 = Z_0 \leq Z_1 \leq \cdots \leq Z_{k-1} \leq Z_k = Z$$

be the upper central series of G, so that each of the subgroups Z_j is G-invariant and each factor Z_j/Z_{j-1} is G-central, for $1 \leq j \leq n$. By Theorem 1.2.22 $G/C_G(Z)$ is nilpotent of class at most $k - 1$. Let $C = C_G(Z)$. Then $Z \leq C_G(C)$ and hence $G/C_G(C)$ is finite of order at most t. Of course $C \cap Z \leq \zeta(C)$ so that $C/(C \cap Z) \cong CZ/Z$ is finite of order at most t. Theorem 1.5.18 implies that C' has order at most $w(t)$. Of course, C' is G-invariant and C/C' is abelian. Furthermore,

$(G/C')/C_{G/C'}(C/C')$ is a finite nilpotent group. We have
$$(C \cap Z)C'/C' \leq \zeta^\infty_{G/C'}(C/C') \text{ and}$$
$$(C/C')/((C \cap Z)C'/C') \cong C/(C \cap Z)C'$$
is a finite group of order at most t. By Proposition 1.6.6, it follows that C/C' contains a finite G-invariant subgroup V/C' of order at most t such that $(C/C')/(V/C')$ is G-nilpotent. Since G/C is nilpotent, G/V is also nilpotent, so $G^\mathfrak{N} \leq V$. Hence $|V| = |C'| \cdot |V/C'| \leq tw(t)$, as required.

Finally, since $G^\mathfrak{N} = \cap\{W \lhd G | G/W \in \mathfrak{N}\}$ and V is finite, there exist finitely many normal subgroups $V_1, \ldots, V_k$ so that $G^\mathfrak{N} = \cap^k_{i=1}\{V_i | G/V_i \in \mathfrak{N}\}$. Then, by Remak's Theorem,

$$G/G^\mathfrak{N} \longrightarrow \operatorname*{Dr}_{i=1}^{k} G/V_i$$

so $G/G^\mathfrak{N}$ is nilpotent. $\square$

The technique of using the $Z(G)$-decomposition allows us, then, to obtain the following generalization of Theorem 1.5.22, which appears in [**157**].

1.6.8. THEOREM. *Let G be a group and suppose that $G/\zeta_\infty(G)$ is finite of order t. Then G contains a finite normal subgroup L of order at most $tw(t)$ such that G/L is hypercentral.*

PROOF. Let $Z = \zeta_\infty(G)$ and let $\mathcal{L}$ be the family of all finitely generated subgroups of G. If $U \in \mathcal{L}$ then $U \cap Z \leq \zeta_\infty(U)$ and, since $U/(Z \cap U)$ is finite, $U/\zeta_\infty(U)$ is also finite. Proposition 1.6.7 shows that the nilpotent residual, $U^\mathfrak{N}$, of U is finite of order at most $tw(t)$. In particular, the orders of the subgroups $U^\mathfrak{N}$ are bounded, for $U \in \mathcal{L}$. It follows that there is a finitely generated subgroup V such that the order of $V^\mathfrak{N}$ is largest. Let $g \in G$ be arbitrary and consider $H = \langle g, V \rangle$. Now $V/(V \cap H^\mathfrak{N}) \cong VH^\mathfrak{N}/H^\mathfrak{N}$ is nilpotent, by Proposition 1.6.7, and it follows that $V^\mathfrak{N} \leq H^\mathfrak{N}$. By the choice of V we have $V^\mathfrak{N} = H^\mathfrak{N}$ and hence $V^\mathfrak{N} \lhd H$. Consequently, $g \in N_G(V^\mathfrak{N})$ which implies that $V^\mathfrak{N} \lhd G$.

Now let $K/V^\mathfrak{N}$ be an arbitrary finitely generated subgroup of $G/V^\mathfrak{N}$. Then K is finitely generated and, as above, $V^\mathfrak{N} = K^\mathfrak{N}$, which shows that $K/V^\mathfrak{N}$ is nilpotent. Hence $G/V^\mathfrak{N}$ is locally nilpotent. However, $ZV^\mathfrak{N}/V^\mathfrak{N}$ has finite index in $G/V^\mathfrak{N}$ and hence $(G/V^\mathfrak{N})/(ZV^\mathfrak{N}/V^\mathfrak{N})$ is nilpotent. It follows that $G/V^\mathfrak{N}$ is hypercentral. $\square$

CHAPTER 2

Groups of Finite 0-Rank

In this chapter, we consider an important numerical invariant suitable for non-periodic groups and which is a non-commutative analogue of the $\mathbb{Z}$-rank. There are a number of integer invariants that can be defined on groups and which are used in the study of their structure. Many of these invariants have their origin in the concept of the dimension of a vector space. One of the first generalizations of this concept was that of the dimension of an R-module over a commutative ring R. Since every abelian group is a module over the ring $\mathbb{Z}$ of integers, this led naturally to the idea of the $\mathbb{Z}$-rank of a module (typically in the theory of abelian groups, the term 0-rank is used) which has been an effective and fruitful tool in the study of abelian groups.

This new invariant first appeared naturally in polycyclic groups, which are a non-commutative analogue of finitely generated abelian groups. This invariant was first discussed in the paper [**115**] of K. A. Hirsch, and was later called the *Hirsch length* or *Hirsch number* of the polycyclic-by-finite group. On the other hand, it was logical to find an analogue of the $\mathbb{Z}$-rank in groups close to abelian ones, such as nilpotent or hypercentral groups. During his investigations of hypercentral $\mathfrak{F}$-perfect groups, S. N. Chernikov [**38**] noted that a torsion-free hypercentral group has an ascending central series with torsion-free locally cyclic factors and that the length of this series is an invariant of such a group. Later he observed that in a group having a finite subnormal series with locally cyclic torsion-free factors, a so-called rational series, the length of this series is an invariant of such a group. Later this concept was studied in the works of V. M. Glushkov [**91, 92, 93**], a student of Chernikov's at that time.

In particular, Glushkov studied the relationships between the *rational rank* and *special rank*, which had been introduced earlier by A. I. Maltsev in his paper [**186**]. In the paper [**188**], Maltsev introduced the class of soluble A_1-groups. This class consists of those groups having a finite subnormal series whose factors are periodic abelian or torsion-free and locally cyclic. These groups also possess finite subnormal series, the factors of which are periodic or infinite cyclic. It was observed that

Ranks of Groups: The Tools, Characteristics, and Restrictions
By Martyn R. Dixon, Leonid A. Kurdachenko and Igor Ya Subbotin
Copyright © 2017 John Wiley & Sons, Inc.

the number of infinite cyclic factors is a group invariant. It seemed to be a useful invariant, and was named the 0-*rank* or *torsion-free rank*. This concept proved to be useful and effective, and it immediately found applications in solving problems concerning the theory of generalized soluble groups. D. I. Zaitsev very actively used this concept and popularized it. In particular, in the papers [**270, 274, 273**] he demonstrated its effectiveness in the solution of certain group theoretic problems. In this chapter, we shall consider the structure of important classes of infinite groups having finite 0-rank.

2.1. The Z-Rank in Abelian Groups

The theory of abelian groups is a well-established, some might say separate, branch of group theory. In this short section, we remind the reader of some of the results and concepts that often prove useful in this particular branch of mathematics. A good general reference discussing abelian groups is the book of L. Fuchs [**88**]. It will be useful to recall some simple basic properties, so we present them here in the form of exercises.

As we saw in Proposition 1.2.11 the set $\mathbf{Tor}\,(G)$ consisting of all elements of finite order in the abelian group G, is a characteristic subgroup of G and the factor group $G/\mathbf{Tor}\,(G)$ is then torsion-free. Furthermore, for the prime p, the set $\mathbf{Tor}_p(G)$, consisting of all the elements of p-power order is also a characteristic subgroup of G and $\mathbf{Tor}\,(G) = \underset{p \in \Pi(G)}{\mathrm{Dr}}\ \mathbf{Tor}_p(G)$. Hence the study of abelian groups often reduces to the study of abelian p-groups, for the different primes p, and torsion-free abelian groups.

It is well known that the torsion subgroup of a finitely generated abelian group is finite and that a finitely generated torsion-free abelian group is isomorphic to a finite direct sum of copies of $\mathbb{Z}$. Our first exercises lead to a proof of this fact.

EXERCISE 2.1. *Let G be an abelian group. Prove that if H is a subgroup of G such that G/H is an infinite cyclic group, then $G = H \times C$, where C is an infinite cyclic group.*

EXERCISE 2.2. *Let G be an abelian group. Prove that if H is a subgroup of G such that G/H is a direct product of finitely many cyclic groups, then $G = H \times C$, for some subgroup C.*

EXERCISE 2.3. *Let G be an abelian group and let n be a natural number. Prove that the mapping $\xi : G \longrightarrow G$ defined by $\xi(g) = g^n$ is an endomorphism of G, for all $g \in G$. Prove, in particular, that*

(i) $\boldsymbol{ker}(\xi) = \{g \in G|$ the order of g divides $n\}$ and $\boldsymbol{Im}(\xi) = \{g^n | g \in G\} = G^n$ are subgroups of G;

(ii) If G is torsion-free, then G is isomorphic to G^n, for each such n.

EXERCISE 2.4. *Prove that if G is a finitely generated, torsion-free abelian group, then G is the direct product of finitely many infinite cyclic groups. Furthermore, prove that if G is a finitely generated abelian group, then $G = \boldsymbol{Tor}(G) \times C$, where $\boldsymbol{Tor}(G)$ is finite and C is a direct product of finitely many infinite cyclic groups.*

In many results concerned with abelian groups, it is often worthwhile to think of the analogy with vector spaces. The following definition is useful.

2.1.1. DEFINITION. Let G be an abelian group. A subset X of G, consisting of elements of infinite order, is said to be $\mathbb{Z}$-*independent* or simply *independent* if, given distinct elements $x_1, \ldots, x_n$ in X and integers $k_1, \ldots, k_n$, the relation $x_1^{k_1} \ldots x_n^{k_n} = 1$ implies that $x_j = 1$ for $1 \leq j \leq n$.

Using this definition it is possible to show that *an abelian group is a direct product of infinite cyclic groups if and only if it is generated by a $\mathbb{Z}$-independent set of elements.* We state this as the next proposition and leave the proof as an exercise.

2.1.2. PROPOSITION. *Let G be an abelian group. If $X = \{x_\lambda | \lambda \in \Lambda\}$ is a $\mathbb{Z}$-independent subset of G, then $\langle X \rangle = \underset{\lambda \in \Lambda}{Dr} \langle x_\lambda \rangle$ is a direct product of infinite cyclic groups. Conversely, if $H = \underset{\lambda \in \Lambda}{Dr} \langle c_\lambda \rangle$, where c_λ is an element of infinite order for every $\lambda \in \Lambda$, then the subset $\{c_\lambda | \lambda \in \Lambda\}$ is $\mathbb{Z}$-independent.*

EXERCISE 2.5. *Prove Proposition 2.1.2.*

We recall that a group G that is isomorphic to a direct product of infinite cyclic groups is called a *free abelian* group. Zorn's Lemma implies that a $\mathbb{Z}$-independent subset of an abelian group is always contained in some maximal $\mathbb{Z}$-independent subset. By analogy with vector space theory, we have the following fundamental result. The reader can find a proof in [**88**, Chapter III], or any one of a number of other excellent books.

2.1.3. PROPOSITION. *Let G be an abelian group.*

(i) *If G has an infinite $\mathbb{Z}$-independent subset, then all maximal $\mathbb{Z}$-independent subsets of G have the same cardinality;*

(ii) *If G has a finite maximal ℤ-independent subset M, then each maximal ℤ-independent subset S of G is finite and $|S| = |M|$;*

(iii) *If X is a maximal ℤ-independent subset of G, then the factor group $G/\langle X \rangle$ is periodic. Conversely, if Y is a ℤ-independent subset of G such that $G/\langle Y \rangle$ is periodic, then Y is a maximal ℤ-independent subset of G.*

We shall require the following generalization of Exercise 2.2.

2.1.4. LEMMA. *Let G be an abelian group and let H be a subgroup of G such that G/H is free abelian. Then $G = H \times C$ for some free abelian subgroup C.*

EXERCISE 2.6. *Prove Lemma 2.1.4.*

The next definition introduces the initial ideas of a concept with which a large part of this book will be concerned.

2.1.5. DEFINITION. Let G be an abelian group. The cardinality of a maximal ℤ-independent subset of G is called the ℤ-*rank* or *torsion-free rank* of G, denoted by $\mathbf{r}_ℤ(G)$.

Although the following result is very easy to prove it seems appropriate to include a proof here.

2.1.6. PROPOSITION. *Let G be an abelian group. Then $\mathbf{r}_ℤ(G) = \mathbf{r}_ℤ(G/\boldsymbol{Tor}(G))$.*

PROOF. Let $T = \mathbf{Tor}\,(G)$ and let $\{x_\lambda T | \lambda \in \Lambda\}$ denote a maximal ℤ-independent subset of G/T. By Proposition 2.1.2, the subgroup

$$X/T = \langle x_\lambda T | \lambda \in \Lambda \rangle = \underset{\lambda \in \Lambda}{\mathrm{Dr}}\, \langle x_\lambda T \rangle$$

is free abelian. Using Lemma 2.1.4 we deduce that $X = T \times Y$, for some free abelian group Y that is isomorphic to X/T. Hence $Y = \underset{\lambda \in \Lambda}{\mathrm{Dr}}\, \langle y_\lambda \rangle$, for certain elements $y_\lambda \in G$. Proposition 2.1.2 shows that the set $\{y_\lambda | \lambda \in \Lambda\}$ is ℤ-independent.

Now, by Proposition 2.1.3, the factor group $(G/T)/(X/T) \cong G/X$ is periodic. Also $X/Y \cong T$ is periodic and hence G/Y is periodic. A further application of Proposition 2.1.3 implies that $\{y_\lambda | \lambda \in \Lambda\}$ is a maximal ℤ-independent subset of G. This proves the result. □

A group G is said to be *locally cyclic* if every finitely generated subgroup of G is cyclic. We let $ℚ_+$ denote the additive group of the field $ℚ$ of rational numbers.

2.1.7. PROPOSITION.
  (i) $\mathbb{Q}_+$ has $\mathbb{Z}$-rank 1;
  (ii) Let G be a torsion-free abelian group. Then $\mathbf{r}_{\mathbb{Z}}(G) = 1$ if and only if G is locally cyclic;
  (iii) Let G be a torsion-free abelian group. Then $\mathbf{r}_{\mathbb{Z}}(G) = 1$ if and only if G is isomorphic to a subgroup of the additive group of $\mathbb{Q}$.

EXERCISE 2.7. Prove Proposition 2.1.7.

We remark that the subgroups of the additive group of rational numbers are well known and we refer the reader to [**88**, Chapter XIII] for a description.

2.1.8. COROLLARY. Let G be a torsion-free abelian group and let $\mathbf{r}_{\mathbb{Z}}(G) = k$ for some natural number k. Then G is isomorphic to some subgroup of $A_1 \times \cdots \times A_k$, for certain groups A_i such that $A_i \cong \mathbb{Q}_+$, for $1 \leq i \leq k$. Conversely if G is isomorphic to a subgroup of $A_1 \times \cdots \times A_k$, where $A_i \cong \mathbb{Q}_+$, for each i and if k is the least natural number with this property, then $\mathbf{r}_{\mathbb{Z}}(G) = k$.

PROOF. We first suppose that $\mathbf{r}_{\mathbb{Z}}(G) = k$. By Proposition 2.1.3, every maximal $\mathbb{Z}$-independent subset of G has precisely k elements. Let $S = \{a_1, a_2, \ldots, a_k\}$ denote one such $\mathbb{Z}$-independent subset of G. Since S is $\mathbb{Z}$-independent it follows, from Proposition 2.1.2, that $\langle S \rangle$ is a free abelian subgroup. For each j such that $1 \leq j \leq k$, let

$$B_j = \langle a_1, \ldots, a_{j-1}, a_{j+1}, \ldots a_k \rangle.$$

Then we have $\langle a_j \rangle \cap B_j = 1$. Let

$$\mathcal{M}_j = \{Y | B_j \leq Y \leq G, \text{ and } \langle a_j \rangle \cap Y = 1\}.$$

By Zorn's Lemma, $\mathcal{M}_j$ contains a maximal element D_j, say. Since $\langle a_j \rangle \cap D_j = 1$, the element $a_j D_j$ clearly has infinite order in the group G/D_j. Suppose, for a contradiction, that $T/D_j = \mathbf{Tor}\,(G/D_j)$ is non-trivial. Then $D_j \lneqq T$. If $T \cap \langle a_j \rangle \neq 1$, then there exists $n \in \mathbb{N}$ such that $a_j^n \in T$ and since T/D_j is periodic it follows that $a_j^l \in D_j$ for some natural number l, which contradicts the fact that $a_j D_j$ has infinite order. Hence $T \cap \langle a_j \rangle = 1$, which contradicts the maximal choice of D_j and we obtain the desired contradiction. Hence G/D_j is torsion-free. Clearly $\{a_j D_j\}$ is a maximal $\mathbb{Z}$-independent subset of G/D_j so $\mathbf{r}_{\mathbb{Z}}(G/D_j) = 1$.

Let $C = \bigcap_{j=1}^k D_j$, so that $C \cap \langle S \rangle = 1$. By Proposition 2.1.3, the factor group $G/\langle S \rangle$ is periodic so that the isomorphisms

$$C \cong C/(C \cap \langle S \rangle) \cong C\langle S \rangle / \langle S \rangle$$

show that C is periodic. However, G is torsion-free, so we must have $C = 1$. Using Remak's Theorem (see [**129**, Theorem 4.3.9]) we obtain the embedding

$$G \longrightarrow G/D_1 \times G/D_2 \times \cdots \times G/D_k.$$

Finally, by Proposition 2.1.7, each of the factor groups G/D_j is isomorphic to the additive group of $\mathbb{Q}$. This completes the proof since the converse is clear. □

From this we see that an abelian group G has finite $\mathbb{Z}$-rank at most r if and only if $G/\mathbf{Tor}\,(G)$ is isomorphic to a subgroup of the additive group

$$\underbrace{\mathbb{Q} \oplus \cdots \oplus \mathbb{Q}}_{r}.$$

A. I. Maltsev [**188**] called an abelian group G an abelian A_1-*group* if $\mathbf{r}_{\mathbb{Z}}(G)$ is finite, and we shall also sometimes use this notation and terminology.

2.2. The 0-Rank of a Group

In this section, we shall define the 0-rank and obtain some elementary results concerning it. We begin with the following fundamental result.

2.2.1. LEMMA. *Let G be a group and suppose that*

(a) $1 = G_0 \lhd G_1 \lhd \ldots G_\alpha \lhd G_{\alpha+1} \lhd \ldots G_\gamma = G$ *and,*
(b) $1 = H_0 \lhd H_1 \lhd \ldots H_\beta \lhd H_{\beta+1} \lhd \ldots H_\delta = G$

are two ascending series of subgroups whose factors are either infinite cyclic or periodic. If the number of infinite cyclic factors of the series (a) is exactly r, then the number of infinite cyclic factors in series (b) is also r.

PROOF. By the generalized Schreier and Jordan–Hölder theorems (see [**172**, §56], for example) there exist isomorphic series

(c) $1 = K_0 \lhd K_1 \lhd \ldots K_\alpha \lhd K_{\alpha+1} \lhd \ldots K_\rho = G$ and
(d) $1 = L_0 \lhd L_1 \lhd \ldots L_\beta \lhd L_{\beta+1} \lhd \ldots L_\sigma = G$

that are refinements of the series (a) and (b) respectively. If $G_{\alpha+1}/G_\alpha$ is an infinite cyclic factor and if $G_\alpha \lhd U \lhd V \lhd G_{\alpha+1}$ then only one of the factors $U/G_\alpha, V/U, G_{\alpha+1}/V$ is also infinite cyclic whereas the others are finite. It follows that series (c) has exactly r infinite cyclic factors and the other factors are periodic. Since the series (c) and (d) are isomorphic, it follows that series (d) also has precisely r infinite cyclic factors. Hence series (b) also has r infinite cyclic factors and the result follows. □

This result shows that the number of infinite cyclic factors in the type of series of interest in Lemma 2.2.1 is an invariant of the group.

2.2.2. DEFINITION. Let G be a group which has an ascending series whose factors are either infinite cyclic or periodic. If the number of infinite cyclic factors is finite, then the group G is said to have *finite 0-rank*. The 0-rank of the group G is the number of infinite cyclic factor groups in the series and is denoted by $\mathbf{r}_0(G)$. Thus if the number of infinite cyclic factors is exactly r, then $\mathbf{r}_0(G) = r$. If no such integer r exists, then we shall say that G has infinite 0-rank. If G has no such ascending series the 0-rank is undefined.

Of course, the 0-rank of a periodic group is 0. It will be implicit in much of this chapter that the groups we consider have reasonably well-behaved ascending series. We remark that this definition is a slight generalization of the traditional definition where groups with a *finite* subnormal series whose factors are either infinite cyclic or periodic are considered. The 0-rank has also been called the *torsion-free rank* of the group G in some papers. We note that when G is abelian then the $\mathbb{Z}$-rank and the 0-rank are equal, so $\mathbf{r}_0(G) = \mathbf{r}_{\mathbb{Z}}(G)$. Thus in future we shall use the notation $\mathbf{r}_0(G)$ for abelian groups also and use the terminology 0-rank instead of $\mathbb{Z}$-rank.

The following elementary lemma shows that the 0-rank is well behaved.

2.2.3. LEMMA. *Let G be a group. Suppose that H is a subgroup of G and L is a normal subgroup of G.*

   (i) *If G has finite 0-rank, then H has finite 0-rank and $\mathbf{r}_0(H) \leq \mathbf{r}_0(G)$;*
   (ii) *G has finite 0-rank if and only if both L and G/L have finite 0-rank. In this case, $\mathbf{r}_0(G) = \mathbf{r}_0(L) + \mathbf{r}_0(G/L)$.*

EXERCISE 2.8. *Prove Lemma 2.2.3.*

There are many natural examples of groups of finite 0-rank and we here describe one such example. In Section 1.4, we defined the group $UT_n(F)$, of $n \times n$ unitriangular matrices over a field F and some of its very natural subgroups which we denoted by $UT_n^k(F)$, for $1 \leq k \leq n$.

Suppose now that F is a finite field extension of degree k over the field $\mathbb{Q}$ of rational numbers. Writing F_+ for the additive group of F, it is clear that

$$F_+ \cong \underbrace{\mathbb{Q}_+ \oplus \cdots \oplus \mathbb{Q}_+}_{k}.$$

In Section 1.4, we saw that

$$UT_n^m(F)/UT_n^{m+1}(F) \cong \underbrace{F_+ \oplus \cdots \oplus F_+}_{n-m}.$$

It follows from Lemma 2.2.3 that

$$\mathbf{r}_0(UT_n(F)) = \mathbf{r}_0(UT_n^1(F)/UT_n^2(F)) + \cdots + \mathbf{r}_0(UT_n^{n-1}(F))$$
$$= k(n-1) + k(n-2) + \cdots + k$$
$$= kn(n-1)/2$$

and that, in particular, $\mathbf{r}_0(UT_n(\mathbb{Q})) = n(n-1)/2$.

We observed in Section 2.1 that a torsion-free abelian group of 0-rank 1 is a locally cyclic group and is isomorphic to a subgroup of $\mathbb{Q}_+$. Therefore a torsion-free abelian group of 0-rank 1 is also called a *rational group*. This makes the following definition entirely natural.

2.2.4. DEFINITION. A group G is called *polyrational* if it has a finite subnormal series whose factors are torsion-free locally cyclic groups.

For example, our work above shows that the group $UT_n(F)$ is an example of a polyrational group when F is a finite field extension of $\mathbb{Q}$. The following observation gives us a number of examples of groups with finite 0-rank.

2.2.5. LEMMA. *Let G be a polyrational group. Then G has finite 0-rank.*

EXERCISE 2.9. *Prove Lemma 2.2.5.*

2.3. Locally Nilpotent Groups of Finite 0-Rank

In this section, we obtain the basic properties of locally nilpotent groups of finite 0-rank. Of course, we shall assume that our groups are non-periodic, so that the 0-rank is non-zero. The following result is basic to our study and was obtained by N. F. Sesekin [**238**, Lemma 6].

2.3.1. LEMMA. *Let G be a torsion-free locally nilpotent group and suppose that G contains a normal abelian subgroup A such that $\mathbf{r}_0(A) = k$. Then the kth term of the upper central series of G contains A.*

PROOF. We first prove the result in the case when G is nilpotent. To this end, we let $C = A \cap \zeta(G)$ and observe that $C \neq 1$, by Lemma 1.2.2. We proceed by induction on k and first let $k = 1$. Since C is torsion-free and $\mathbf{r}_0(A) = 1$, A/C is periodic. Hence, for each element $g \in G$ and $a \in A$, there exists $m \in \mathbb{N}$ such that $ga^m = a^m g$. By Corollary 1.2.8, this implies that $ga = ag$ and hence $A \leq \zeta(G)$.

Suppose now that $k > 1$. If $A \leq \zeta(G)$ the result follows and we may therefore suppose that $\zeta(G)$ does not contain A. Then $A\zeta(G)/\zeta(G)$ is nontrivial. By Corollary 1.2.9, $G/\zeta(G)$ is torsion-free, as is $A/C \cong A\zeta(G)/\zeta(G)$. Since C is nontrivial it follows that $\mathbf{r}_0(A/C) < k$ and hence $\mathbf{r}_0(A\zeta(G)/\zeta(G)) < k$ also. By the induction hypothesis, we have $A\zeta(G)/\zeta(G) \leq \zeta_{k-1}(G/\zeta(G))$ and hence $A \leq \zeta_k(G)$. This proves the result when G is nilpotent.

In the general case, let Y be an arbitrary finitely generated subgroup of G and let $a \in A$. Then $H = \langle a, Y \rangle$ is finitely generated and hence nilpotent. By the first part of the proof we have $D = A \cap H \leq \zeta_k(H)$ and hence $[D,_k H] = 1$. In particular, $[a,_k Y] = 1$ and this holds for each $a \in A$. Hence $[A,_k Y] = 1$ and this is valid for all finitely generated subgroups Y. It follows that $[A,_k G] = 1$ and hence $A \leq \zeta_k(G)$, as required. $\quad\square$

Lemma 2.3.1 has been attributed to V. S. Charin by other authors, but it does not seem to appear in his work.

As is often the case, when dealing with soluble groups the structure of the abelian subgroups can be decisive. This is illustrated very well in the next couple of results, the first of which is again due to N. F. Sesekin [**238**, Theorem 7]. We denote the nilpotency class of a group G by $\mathbf{ncl}(G)$.

2.3.2. LEMMA. *Let G be a torsion-free hypercentral group and suppose that G contains a maximal normal abelian subgroup A such that $r_0(A) = k$. Then G is nilpotent and has finite 0-rank at most $k(k+1)/2$. Furthermore, $\mathbf{ncl}(G) \leq 2k$.*

PROOF. The subgroup A is torsion-free abelian of 0-rank k so there is a monomorphism $\phi : G/C_G(A) \longrightarrow GL_k(\mathbb{Q})$. By Lemma 2.3.1, $A \leq \zeta_k(G)$. Consequently $[A, \underbrace{G, \ldots, G}_{k}] = 1$. Written additively this gives that $a(g-1)^k = 0$ for all $a \in A$ and it follows that $\operatorname{Im}\phi \leq UT_k(\mathbb{Q})$, which is nilpotent of class k and of 0-rank precisely $k(k-1)/2$. Since G is hypercentral $A = C_G(A)$, by Lemma 1.2.2, and since $A \leq \zeta_k(G)$ it follows that G is nilpotent. Furthermore, $\mathbf{ncl}(G) \leq 2k$. By Lemma 2.2.3, we see that

$$\mathbf{r}_0(G) \leq \mathbf{r}_0(A) + \mathbf{r}_0(G/A) \leq k + k(k-1)/2 = k(k+1)/2$$

and the result follows. $\quad\square$

The following result is a theorem of Maltsev [**188**, Theorem 5]. It is fundamental to the study of groups of finite rank, as we shall see.

2.3.3. THEOREM. *Let G be a torsion-free locally nilpotent group and suppose that every abelian subgroup of G has finite 0-rank. Then G is a nilpotent group of finite 0-rank. Furthermore, if A is a maximal normal abelian subgroup of G such that $r_0(A) = k$, then $r_0(G) \leq k(k+1)/2$ and $ncl(G) \leq 2k$.*

PROOF. We suppose, for a contradiction, that G is not nilpotent. Then, for every finitely generated subgroup U, there is a finitely generated subgroup V such that $U \leq V$ and $ncl(U) < ncl(V)$. Hence we can construct an ascending chain of finitely generated subgroups

$$K_1 < K_2 < \cdots < K_n < K_{n+1} < \ldots$$

such that $ncl(K_n) < ncl(K_{n+1})$ for all $n \in \mathbb{N}$. Then $K = \bigcup_{n \in \mathbb{N}} K_n$ is a countable, non-nilpotent group and we let $C_n = \zeta(K_n)$. Thus $[C_n, C_{n+1}] = 1$ for each $n \in \mathbb{N}$. Consequently, the subgroup $C = \langle C_n | n \in \mathbb{N} \rangle$ is abelian and hence $r_0(C)$ is finite, by hypothesis. Choose a free abelian subgroup D of C such that C/D is periodic and note that, since D is finitely generated, $D \leq C_1 \ldots C_m \leq K_m$, for some $m \in \mathbb{N}$. We observe also that, since C/D is periodic, $C_k D/D \cong C_k/(D \cap C_k)$ is also periodic and, since C_k is torsion-free, it follows that $D \cap C_k$ is nontrivial, for each $k \in \mathbb{N}$. If $j \geq m$ we have

$$D \cap C_{j+1} \leq D \leq C_1 \ldots C_m \leq K_m \leq K_j \leq K_{j+1}$$

and it follows that

$$D \cap C_{j+1} \leq K_j \cap C_{j+1} = K_j \cap \zeta(K_{j+1}) \leq \zeta(K_j).$$

Thus $D \cap C_{j+1} \leq D \cap \zeta(K_j) = D \cap C_j$. Furthermore

$$(D \cap C_j)/(D \cap C_{j+1}) \cong (D \cap C_j)C_{j+1}/C_{j+1} \leq K_{j+1}/C_{j+1}.$$

Since $K_{j+1}/C_{j+1} = K_{j+1}/\zeta(K_{j+1})$ is torsion-free, by Corollary 1.2.9, $D \cap C_j/D \cap C_{j+1}$ is torsion-free and, if it is nontrivial, we deduce that $r_0(D \cap C_{j+1}) < r_0(D \cap C_j)$. By considering the descending chain

$$D \cap C_m \geq D \cap C_{m+1} \geq \cdots \geq D \cap C_{m+k} \geq \ldots$$

and observing that $r_0(D \cap C_m)$ is finite it follows that there exists a natural number $t \geq m$ such that $D \cap C_j = D \cap C_{j+1}$ for all $j \geq t$. In turn, it follows that $1 \neq D \cap C_t = D \cap C_j \leq C_j = \zeta(K_j)$ for all $j \geq t$. Hence $1 \neq D \cap C_t \leq \zeta(K)$ and, in particular, $\zeta(K) \neq 1$.

We now show that $K/\zeta(K)$ inherits the properties of K and to this end we let $U/\zeta(K)$ be an arbitrary abelian subgroup of $K/\zeta(K)$. Then U is nilpotent and has finite 0-rank, by Lemma 2.3.2. Lemma 2.2.3 implies that $r_0(U/\zeta(K))$ is also finite. By Corollary 1.2.9, $K/\zeta(K)$ is torsion-free, so $U/\zeta(K)$ is also torsion-free and in this case $r_{\mathbb{Z}}(U/\zeta(K)) = r_0(U/\zeta(K))$. Thus every abelian subgroup of $K/\zeta(K)$ has finite 0-rank and we may now repeat the above argument to see that $\zeta(K/\zeta(K)) \neq 1$. Consequently, $\zeta(K) = \zeta_1(K) \neq \zeta_2(K)$ and it follows by induction that $\zeta_n(K) \neq \zeta_{n+1}(K)$ for all $n \in \mathbb{N}$. Since K is not nilpotent, $K \neq \zeta_n(K)$, for all $n \in \mathbb{N}$, so we obtain a strictly ascending series

$$1 < \zeta(K) = \zeta_1(K) < \zeta_2(K) < \cdots < \zeta_n(K) < \zeta_{n+1}(K) < \dots.$$

By Corollary 1.2.9, $K/\zeta_n(K)$ is torsion-free, for each natural number n, so $\zeta_{n+1}(K)/\zeta_n(K)$ is torsion-free. It follows that $Z = \bigcup_{n \in \mathbb{N}} \zeta_n(K)$ has infinite 0-rank. On the other hand, Z is hypercentral and Lemma 2.3.2 shows that $r_0(Z)$ is finite, a contradiction which shows that G is nilpotent and a further application of Lemma 2.3.2 completes the proof. $\square$

Certainly, if a group G has finite 0-rank, then so does each of its subgroups, so the following corollary is immediate from the previous theorem.

2.3.4. COROLLARY. *Let G be a torsion-free locally nilpotent group. If G has finite 0-rank, then G is a nilpotent, polyrational group.*

We need some facts concerning periodic groups next but leave the details as a couple of exercises.

EXERCISE 2.10. *Let G be an infinite periodic locally cyclic group. Prove that if G is a p-group, for some prime p, then G is a Prüfer group of type p^∞.*

For the following exercise, we first note that a *Sylow p-subgroup* of a group G is defined to be a maximal p-subgroup of G.

EXERCISE 2.11. *Let G be an infinite periodic abelian group. Prove that G is locally cyclic if and only if, for every prime p, its Sylow p-subgroup is cyclic or is a Prüfer group of type p^∞.*

2.3.5. COROLLARY. *Let G be a torsion-free locally nilpotent group. If G has finite 0-rank, then G has a finite subnormal series whose factors are infinite cyclic or periodic abelian groups with Chernikov Sylow p-subgroups for all primes p.*

PROOF. By Corollary 2.3.4, G has a subnormal series

$$1 = K_0 \lhd K_1 \lhd \dots K_r \lhd K_{r+1} \lhd \dots K_n = G$$

such that each K_{j+1}/K_j is locally cyclic and torsion-free, for $0 \leq j \leq n - 1$. It follows that K_{j+1}/K_j contains an infinite cyclic subgroup L_{j+1}/K_j such that K_{j+1}/L_{j+1} is a periodic locally cyclic group. Thus the Sylow p-subgroup of K_{j+1}/L_{j+1} is cyclic or a Prüfer p-subgroup for every prime p, by Exercise 2.11. $\square$

2.3.6. COROLLARY. *Let G be a torsion-free locally nilpotent group of finite 0-rank. Then the Sylow p-subgroups of every periodic section of G are Chernikov, for all primes p.*

2.4. Groups of Finite 0-Rank in General

The definition we originally gave for a group to have finite 0-rank included the possibility that such a group may have an infinite series whose factors are either periodic or infinite cyclic, with only finitely many of them being infinite cyclic. Our first theorem in this section shows that our original definition and the one often given are equivalent.

2.4.1. THEOREM. *Let G be a group of finite 0-rank. Then G has a finite series of normal subgroups, every factor of which either is a periodic group or a torsion-free nilpotent polyrational group.*

PROOF. We proceed by induction on $r = r_0(G)$. If $r = 0$, then G is periodic, and the result is clear. Suppose now that $r > 0$ and that we have already proved this result for groups whose 0-rank is less than r. By definition, G has an ascending series of subgroups, each factor of which is infinite cyclic or periodic. If G contains an ascendant periodic subgroup, then by Proposition 1.2.10, $T_0 = \mathbf{Tor}\,(G) \neq 1$ and G/T_0 contains no periodic ascendant subgroups. If $\mathbf{Tor}\,(G) = 1$, then let $T_0 = 1$. In either case, G/T_0 contains no periodic ascendant subgroups, and therefore G/T_0 contains an ascendant infinite cyclic subgroup. In this case, the Hirsch–Plotkin radical L/T_0 of G/T_0 is nontrivial, by Proposition 1.2.15, and since $\mathbf{Tor}\,(G/T_0) = 1$, L/T_0 is torsion-free. By Corollary 2.3.4, L/T_0 is nilpotent and polyrational. For G/L we have $r_0(G/L) < r$, and we can apply the induction hypothesis to G/L. This proves the result. □

This result allows us to obtain the following corollary, which is the usual definition of a group of finite 0-rank.

2.4.2. COROLLARY. *Let G be a group of finite 0-rank. Then G has a finite subnormal series whose factors either are periodic or infinite cyclic.*

2.4.3. LEMMA. *Let A be a torsion-free abelian group of finite 0-rank r. Then, for each prime p, the order of every elementary abelian p-section of A is at most p^r. Furthermore, if B is a subgroup of A such that $A^m \leq B$, for some positive integer m, then $|A/B| \leq m^r$.*

EXERCISE 2.12. *Prove Lemma 2.4.3.*

If $n \in \mathbb{N}$, then there is a function $a(n)$ such that if G is a finite group of order n then $\mathbf{Aut}\,(G)$ has order at most $a(n)$. It is clear that $a(n) \le n!$. Our next result involves a very typical factor shifting argument.

2.4.4. LEMMA. *Let G be a group and suppose that G contains a finite normal subgroup T such that G/T is a torsion-free abelian group of finite 0-rank r. Then G contains a characteristic torsion-free abelian subgroup A of 0-rank r such that G/A is finite. Moreover, if $|T| = t$, then $|G/A| \le a(t)t^{2r+1}$.*

PROOF. Clearly T is the torsion subgroup of G and hence is characteristic in G. Consequently the subgroup $C = C_G(T)$ is also characteristic in G and G/C is finite of order at most $d = a(t)$. It is clear that $C \cap T \le \zeta(C)$. Also, since $G' \le T$, C is nilpotent of class at most 2 and hence, by the commutator laws, we have, for all $b, c \in C$,

$$[b^t, c] = [b, c]^t = 1.$$

Hence $C^t \le \zeta(C)$. Certainly, $C/(C \cap T) \cong CT/T$ is torsion-free abelian of rank at most r. Applying Lemma 2.4.3 to $C/(C \cap T)$ and $\zeta(C)/(C \cap T)$ we deduce that $|C/\zeta(C)| \le t^r$. Since $\zeta(C)$ has finite torsion subgroup $C \cap T$, there is a torsion-free subgroup Z such that $\zeta(C) = (C \cap T) \times Z$ (see [**88**, Theorem 27.5], for example). Thus $A = \zeta(C)^t$ is a characteristic torsion-free abelian subgroup of G and

$$|G/A| \le a(t)t^r t t^r = a(t)t^{2r+1},$$

as required.   □

A straightforward induction argument allows us to obtain the following corollary, due to D. Zaitsev [**269**, Lemma 3], concerned with finite-by-polyrational groups.

2.4.5. COROLLARY. *Let G be a group and suppose that G has a series of normal subgroups*

$$1 = T_0 \le T_1 = A_0 \le A_1 \le \cdots \le A_n \le A_{n+1} = G$$

such that T_1 is finite and A_{j+1}/A_j is torsion-free abelian of finite 0-rank, for $0 \le j \le n$. Then G has a series of normal subgroups

$$1 = B_0 \le B_1 \le \cdots \le B_n \le B_{n+1} = H \le G$$

such that B_{j+1}/B_j is torsion-free abelian of 0-rank precisely $r_0(A_{j+1}/A_j)$, for $0 \le j \le n$ and G/H is finite. Furthermore, if $|T_1| = t$ and $r = \max\{r_0(A_{j+1}/A_j)|0 \le j \le n\}$, then there exists an integer valued function z such that $|G/H| \le z(t, r, n)$.

It is easy to obtain a recursion formula for the function $z(t, r, n)$. From Lemma 2.4.4 we see that $z(t, r, 0) = a(t)t^{2r+1}$ and hence, for all integers $n \geq 0$,

$$z(t, r, n+1) = a(z(t, r, n))z(t, r, n)^{2r+1}.$$

The type of situation occurring in Corollary 2.4.5 is very natural when we compare it with the next result, where we use the function $\rho(r)$, introduced in Proposition 1.4.2.

2.4.6. LEMMA. *Let G be a group and suppose that $\mathbf{Tor}\,(G) = 1$. If G contains a normal torsion-free abelian subgroup A of finite 0-rank r such that G/A is locally finite, then G contains a normal torsion-free abelian subgroup B of finite index and 0-rank at most r such that $|G/B| \leq \rho(r)$. Furthermore, G contains a characteristic torsion-free abelian subgroup C of finite index at most $\rho(r)^{r+1}$.*

EXERCISE 2.13. *Prove Lemma 2.4.6.*

We note that the group G in Lemma 2.4.6 has finite 0-rank. However the result itself no longer holds if G/A is not locally finite. In the paper [1] an example is constructed of a finitely generated non-abelian torsion-free group in which every pair of cyclic subgroups has nontrivial intersection. This group is a central extension of an infinite cyclic group by an infinite bounded periodic group so has 0-rank 1, but certainly is not torsion-free abelian-by-finite. More recently, A. Yu. Ol'shanskii [209, Theorem 31.4] constructed an example of a non-abelian torsion-free group in which every pair of cyclic subgroups has nontrivial intersection and every proper subgroup is cyclic. Again, this group has finite 0-rank. Thus the study of groups with finite 0-rank can be rather complicated, because of the presence of infinite Burnside groups, and at this stage it is appropriate to place some restrictions on the groups we consider. One restriction which we impose here follows very naturally from Lemma 2.4.6 and concerns the study of those groups G that have an ascending series whose factors are either infinite cyclic or locally finite, and in which the number of infinite cyclic factors is finite.

2.4.7. COROLLARY. *Let G be a group and suppose that $\mathbf{Tor}\,(G) = 1$. If G has an ascending series*

$$1 = H_0 \lhd H_1 \lhd \ldots H_\alpha \lhd H_{\alpha+1} \lhd \ldots H_\gamma = G$$

such that r factors of this series are infinite cyclic and the remaining factors are locally finite, then G has a finite series

$$1 = K_0 \leq L_1 \leq K_1 \leq L_2 \leq K_2 \leq \cdots \leq L_n \leq K_n = G$$

of normal subgroups such that the factors L_{j+1}/K_j are torsion-free abelian of finite 0-rank, for $0 \le j \le n - 1$ and the factors K_j/L_j are finite of order at most $\rho(r)^{r+1}$. Furthermore $n \le r$.

PROOF. Clearly the group G has finite 0-rank exactly r. It is also clear that the periodic sections of G are locally finite. We proceed by induction on r. Since $\mathbf{Tor}\,(G) = 1$, Proposition 1.2.10 implies that H_1 is infinite cyclic. Thus G contains an ascendant abelian subgroup, so the Hirsch–Plotkin radical L of G is nontrivial and indeed $H_1 \le L$. Since the set of elements having finite order in a locally nilpotent group is a characteristic subgroup and since $\mathbf{Tor}\,(G) = 1$, it follows that L is torsion-free. By Corollary 2.3.4, L is nilpotent and hence $Z = \zeta(L) \ne 1$. When $r = 1$ this argument gives the result since $G = K_1$ in this case.

Let $K_1/Z = \mathbf{Tor}\,(G/Z)$ so K_1/Z is locally finite. By Lemma 2.4.6 K_1 contains a characteristic torsion-free abelian subgroup L_1 of finite 0-rank such that K_1/L_1 is finite and $|K_1/L_1| \le \rho(r)^{r+1}$. By the choice of K_1 we have $\mathbf{Tor}\,(G/K_1) = 1$ and hence G/K_1 has an ascending series in which at most $r - 1$ of the factors are infinite cyclic and the remaining factors are locally finite. Hence we may apply our induction hypothesis to G/K_1 to deduce the result.   □

2.4.8. COROLLARY. *Let G be a group and suppose that $\mathbf{Tor}\,(G) = 1$. Suppose that G has an ascending series*

$$1 = H_0 \triangleleft H_1 \triangleleft \ldots H_\alpha \triangleleft H_{\alpha+1} \triangleleft \ldots H_\gamma = G$$

such that r factors of this series are infinite cyclic and the remaining factors are locally finite. Then G has a soluble normal subgroup S of finite index and $|G/S| \le a(\rho(r)^{r+1})^r$.

PROOF. By Corollary 2.4.7, G has a finite series

$$1 = K_0 \le L_1 \le K_1 \le L_2 \le K_2 \le \cdots \le L_n \le K_n = G$$

of normal subgroups such that L_{j+1}/K_j is torsion-free abelian of finite 0-rank for $0 \le j \le n - 1$, and the factors K_j/L_j are finite of order at most $\rho(r)^{r+1}$. Clearly $n \le r$.

Let $S = \bigcap_{1 \le j \le n} C_G(K_j/L_j)$. It is easy to see that S is soluble and, by Remak's Theorem, there is an embedding

$$G/S \longrightarrow G/C_G(K_1/L_1) \times \cdots \times G/C_G(K_n/L_n).$$

Since $|G/C_G(K_j/L_j)| \le a(|K_j/L_j|) \le a(\rho(r)^{r+1})$, it follows that

$$|G/S| \le a(\rho(r)^{r+1})^n \le a(\rho(r)^{r+1})^r$$

and this proves the result.   □

For groups of finite 0-rank we now have the following easily proved lemma which appeared in [**67**].

2.4.9. LEMMA. *Let G be a group. The following are equivalent.*

(i) *G is generalized radical of finite 0-rank;*
(ii) *G is locally generalized radical of finite 0-rank;*
(iii) *G has an ascending series*

$$1 = H_0 \lhd H_1 \lhd \ldots H_\alpha \lhd H_{\alpha+1} \lhd \ldots H_\gamma = G,$$

where finitely many factors of this series are infinite cyclic and the remaining factors are locally finite;
(iv) *G has a finite subnormal series each of whose factors is locally finite or infinite cyclic.*

PROOF. Every generalized radical group is locally generalized radical so the assertion that (i) implies (ii) is clear. It is also clear that the assertion (iv) implies assertion (i). Furthermore, that (iii) implies (iv) follows from Corollary 2.4.7, since **Tor** $(G/\textbf{Tor}\,(G))$ is trivial. Thus we assume that G is a locally generalized radical group of finite 0-rank and prove that (iii) holds. In this case, G has an ascending series whose factors are either infinite cyclic or periodic. The periodic factors are locally finite, by Lemma 1.2.17. Furthermore, the number of infinite cyclic factors is the 0-rank and hence (iii) follows. □

For the next result we shall require the following easily proved preliminary lemma.

2.4.10. LEMMA. *Let G be a soluble group of finite 0-rank and suppose that*

$$1 = G_0 \leq G_1 \leq \cdots \leq G_n = G$$

is a finite series of normal subgroups of G in which each factor G_{j+1}/G_j is torsion-free abelian, for $0 \leq j \leq n-1$. Then G has a finite series of normal subgroups whose factors are torsion-free abelian, G-rationally irreducible groups. Furthermore, if the original series is a central series then the new series is also a central series.

PROOF. Clearly $n \leq r$ and if $\textbf{r}_0(G) = 1$ then the result is clear. We prove the result by induction on $\textbf{r}_0(G)$, assuming the result for all groups of the type under consideration with torsion-free rank less than $\textbf{r}_0(G)$. We let U_1 be a nontrivial G-invariant subgroup of G_1 of minimal 0-rank and note that U_1 is G-rationally irreducible. Otherwise, if W is a G-invariant subgroup of U_1 such that U_1/W is not periodic then by Lemma 2.2.3 we have the equation

$$\textbf{r}_0(U_1) = \textbf{r}_0(W) + \textbf{r}_0(U_1/W),$$

which implies that $r_0(W) < r_0(U_1)$ contrary to the choice of U_1.

If $V_1/U_1 = \mathbf{Tor}\,(G_1/U_1)$, then V_1 is also G-rationally irreducible, for the same reason, and G_1/V_1 is torsion-free. Clearly $r_0(G/V_1) < r_0(G)$ so, by induction, we construct a series $\{V_j/V_1 | 2 \le j \le k\}$ of the desired type in G/V_1. Then

$$1 = V_0 \le V_1 \le \cdots \le V_k = G$$

is a series with the required properties.

Clearly, if the original series is central, then V_1 is central and G/V_1 also has a central series so the final claim also follows. □

2.4.11. THEOREM. *Let G be a soluble group such that $\mathbf{Tor}\,(G) = 1$. If G has finite 0-rank r, then G contains normal subgroups $L \le K \le G$ such that L is a characteristic torsion-free nilpotent subgroup, K/L is a finitely generated torsion-free abelian group and G/K is finite. Furthermore, there are functions $f_1, f_2 : \mathbb{N} \longrightarrow \mathbb{N}$ such that $|G/K| \le f_1(r)$ and $\mathbf{dl}(G) \le f_2(r)$.*

PROOF. (i) Let L be the Hirsch–Plotkin radical of G and note that L is a characteristic subgroup of G. Since G is soluble it is clear that $L \ne 1$. We note that the set of elements of finite order in L forms a characteristic subgroup of L and hence, since $\mathbf{Tor}\,(G) = 1$, it follows that L is torsion-free. Consequently, L is nilpotent, by Corollary 2.3.4. Also, by an obvious slight extension of Lemma 2.4.10, L has a finite series of G-invariant subgroups

$$(2.1) \qquad\qquad 1 = A_0 \le A_1 \le \cdots \le A_n = L,$$

each factor of which is torsion-free, central in L and G-rationally irreducible. Furthermore, it is clear that $n \le r$ and, since a centre-by-cyclic group is abelian, L is nilpotent of class at most $r - 1$.

Let $C_j = C_G(A_{j+1}/A_j)$, for $0 \le j \le n-1$. We can view $G_j = G/C_j$ as a soluble irreducible subgroup of $GL_k(\mathbb{Q})$ where $k = r_0(A_{j+1}/A_j)$. By Theorem 1.4.6, G_j is abelian-by-finite and Corollary 1.4.12 implies that G_j contains a normal free abelian subgroup U_j such that G_j/U_j is finite. Let $T_j = \mathbf{Tor}\,(G_j)$ and note that, by Proposition 1.4.2, T_j is finite of order at most $\rho(r)$. Lemma 2.4.6 implies that G_j/T_j contains a normal torsion-free abelian subgroup W_j/T_j such that $|G_j/W_j| \le \rho(r)$. By Lemma 2.4.4, W_j contains a characteristic torsion-free abelian subgroup K_j such that W_j/K_j is finite of order at most $a(t)t^{2r+1}$, where $t = \rho(r)$. Hence G_j/K_j is finite of order at most $a(t)t^{2r+2}$.

Let $C = \bigcap_{j=0}^{n-1} C_j$. Certainly, $L \le C$ since the series (2.1) is central and we suppose, for a contradiction, that $L \lneq C$. Since C/L is soluble it contains a nontrivial G-invariant abelian subgroup P/L. Of course

$P' \leq L$. Since $P \leq C$ we have $[P, A_{j+1}] \leq A_j$, for $j = 0, \ldots, n - 1$ and this makes it easy to see that

$$[L, \underbrace{P, \ldots, P}_{n}] = 1.$$

Consequently, $\gamma_{n+2}(P) = 1$ so that P is nilpotent and hence $P \leq L$. This contradiction shows that $C = L$.

The embedding $G/L \hookrightarrow \mathrm{Dr}_{j=0}^{n-1} G/C_j$ implies that G/L contains a finitely generated torsion-free abelian subgroup K/L such that G/K is finite. Furthermore,

$$|G/K| \leq |G_j/K_j|^n \leq (a(t)t^{2r+2})^n \leq (a(t)t^{2r+2})^r = f_1(r).$$

Hence

$$\mathbf{dl}(G) \leq \mathbf{dl}(K) + \mathbf{dl}(G/K) \leq (r - 1) + 1 + \mathbf{dl}(G/K) = r + \mathbf{dl}(G/K).$$

The group G/K is finite and soluble so $\mathbf{dl}(G/K) \leq e(|G/K|)$, where e is the function introduced following Theorem 1.4.7. By the choice of K we see that $\mathbf{dl}(G/K) \leq e((a(t)t^{2r+2})^r)$ and we set

$$f_2(r) = r + e((a(t)t^{2r+2})^r) \leq r + r \log_2(a(t)t^{2r+2})$$

This completes the proof of the theorem.   □

There are alternative possibilities for the function $f_2(r)$ which the proof suggests. The function μ that appeared in Theorem 1.4.6 implies that $\mathbf{dl}(G) \leq r + e(\mu(r))$. As we saw above, $\mu(r) \leq r!(r^2 a(r^2))^r$ so we may let

$$f_3(r) = r + e(r!(r^2 a(r^2))^r) \leq r + \log_2(r!(r^2 a(r^2))^r)$$

and in this case we have $\mathbf{dl}(G) \leq f_3(r)$. In any case, these functions grow rather rapidly and the bounds obtained are rather large even for small values of r. It would be interesting to see what the best possible bounds are, but we offer no speculation concerning this.

Another possible method for computing the function appearing in Theorem 2.4.11 is by means of the function ζ that appeared in Theorem 1.4.8 in which case we find that $\mathbf{dl}(G) \leq r + \zeta(r)$.

We highlight the following special case of this theorem.

2.4.12. COROLLARY. *Let G be a polyrational group. Then G contains normal subgroups $L \leq K \leq G$ such that L is torsion-free nilpotent, K/L is finitely generated torsion-free abelian and G/K is finite. Furthermore, if $r_0(G) = r$, then $|G/K| \leq f_1(r)$ and $\mathbf{dl}(G) \leq f_2(r)$.*

We now obtain the main result of this chapter, which describes the structure of locally generalized radical groups of finite 0-rank. This result appeared in [67].

2.4.13. THEOREM. *Let G be a locally generalized radical group of finite 0-rank. Then G has normal subgroups $T \leq L \leq K \leq S \leq G$ such that*

(i) *T is locally finite and G/T is soluble-by-finite;*

(ii) *L/T is a torsion-free nilpotent group;*

(iii) *K/L is a finitely generated torsion-free abelian group;*

(iv) *G/K is finite and S/T is the soluble radical of G/T.*

Moreover, if $r_0(G) = r$, then there are functions $f_2, f_4 : \mathbb{N} \longrightarrow \mathbb{N}$ such that $|G/K| \leq f_4(r)$ and $\mathbf{dl}(S/T) \leq f_2(r)$.

PROOF. Let $T = \mathbf{Tor}\,(G)$. Using Lemma 2.4.9 and Corollary 2.4.8 we see that G contains a normal subgroup S such that S/T is soluble, G/S is finite and $|G/S| \leq a(\rho(r)^{r+1})^r$. An application of Theorem 2.4.11 shows that S has S-invariant subgroups $L \leq R \leq S$ such that $T \leq L$ and L/T is torsion-free nilpotent, R/L is finitely generated torsion-free abelian and S/R is finite. The group L/T is characteristic in S/T, so L is G-invariant. Also, $|S/R| \leq f_1(r)$ and $\mathbf{dl}(S/T) \leq f_2(r)$, using the notation of Corollary 2.4.12. Then

$$|G : R| = |G : S| \cdot |S : R| \leq a(\rho(r)^{r+1})^r \cdot f_1(r).$$

Thus if $K = \mathbf{core}_G\, R$, then G/K is finite of order at most $f_4(r) = (f_1(r) \cdot a(\rho(r)^{r+1})^r)!$. The result follows. $\qquad\square$

We also observe the following fact, which can be deduced using Corollary 2.3.4.

2.4.14. COROLLARY. *Let G be a locally radical group of finite 0-rank. Then $G/\mathbf{Tor}\,(G)$ contains a normal polyrational subgroup of finite index.*

2.5. Local Properties of Groups of Finite 0-Rank

In this section, we investigate how the 0-rank of the subgroups of a group G affects the 0-rank of the entire group. We begin with the following result, which appeared in [67].

2.5.1. LEMMA. *Let G be a soluble group. Suppose that there is a positive integer r such that every finitely generated subgroup of G has finite 0-rank at most r. Then G has finite 0-rank at most r.*

PROOF. Let $\mathcal{L} = \{H_\lambda | \lambda \in \Lambda\}$ be the local system consisting of all the finitely generated subgroups of G and let

$$1 = D_0 \lhd D_1 \lhd \ldots \lhd D_n \lhd D_{n+1} = G$$

be the derived series of G. We proceed by induction on n and first consider the case when G is abelian. Choose $\nu \in \Lambda$ such that $r_0(H_\nu)$

is maximal. This choice implies that G/H_ν is periodic, so $\mathbf{r}_0(G) = \mathbf{r}_0(H_\nu) \le r$ and the result follows in this case.

Suppose now that $n > 1$. Since the hypotheses are clearly inherited by factor groups we may assume inductively that $\mathbf{r}_0(G/D_1)$ is finite, at most r, and that there is an index $\mu \in \Lambda$ such that $\mathbf{r}_0(G/D_1) = \mathbf{r}_0(H_\mu D_1/D_1)$. Since $\mathcal{L}$ is a local system for G, $\{H_\lambda \cap D_1 | \lambda \in \Lambda\}$ is a local system for D_1 and there exists $\kappa \in \Lambda$ such that $\mathbf{r}_0(H_\kappa \cap D_1)$ is maximal. Since $\mathcal{L}$ is a local system for G, there is an index $\sigma \in \Lambda$ such that $\langle H_\mu, H_\kappa \rangle \le H_\sigma$. The maximality of $\mathbf{r}_0(H_\mu D_1/D_1)$ implies that $\mathbf{r}_0(G/D_1) = \mathbf{r}_0(H_\sigma D_1/D_1)$, and since $H_\sigma D_1/D_1 \cong H_\sigma/H_\sigma \cap D_1$ we have, by Lemma 2.2.3,

$$\mathbf{r}_0(G) = \mathbf{r}_0(G/D_1) + \mathbf{r}_0(D_1) = \mathbf{r}_0(H_\sigma/H_\sigma \cap D_1) + \mathbf{r}_0(H_\sigma \cap D_1)$$
$$= \mathbf{r}_0(H_\sigma) \le r.$$

This completes the proof.   □

Now we extend this result to locally generalized radical groups. If G is a group, then we let $\delta_k(G)$ denote the kth term of the derived series of the group G.

2.5.2. PROPOSITION ([67]). *Let G be a locally generalized radical group and let r be a positive integer such that every finitely generated subgroup of G has finite 0-rank at most r. Then G has 0-rank at most r. In particular, G is a generalized radical group.*

PROOF. Let $\mathcal{L} = \{H_\lambda | \lambda \in \Lambda\}$ be the local system consisting of all finitely generated subgroups of G. Then each H_λ is a generalized radical group. By Theorem 2.4.13, each subgroup H_λ has normal subgroups $T_\lambda \le S_\lambda \le H_\lambda$ such that H_λ/S_λ is finite, S_λ/T_λ is soluble and T_λ is locally finite. Moreover, $|H_\lambda/S_\lambda| \le f_4(r) = d$ and $\mathbf{dl}(S_\lambda/T_\lambda) \le f_2(r) = k$. Clearly $K_\lambda = H_\lambda^d \le S_\lambda$ and therefore $\delta_k(K_\lambda) \le T_\lambda$. Thus $\delta_k(K_\lambda)$ is locally finite.

Observe that since H_λ is finitely generated and H_λ/K_λ is locally finite, it follows that H_λ/K_λ is finite. Let λ, μ be indices such that $H_\lambda \le H_\mu$. Then $K_\lambda = H_\lambda^d \le H_\mu^d = K_\mu$ so $K_\lambda \le K_\mu$ and hence $\{K_\lambda | \lambda \in \Lambda\}$ is a local system for $K = \bigcup_{\lambda \in \Lambda} K_\lambda$. Furthermore, K_λ is normal in H_λ, for all $\lambda \in \Lambda$, so K is a normal subgroup of G. If g is an arbitrary element of G then there is an index $\lambda \in \Lambda$ such that $g \in H_\lambda$. Hence $g^d \in K_\lambda$, so G/K is of exponent d and hence is locally finite. Next we note that $\delta_k(K) = \bigcup_{\lambda \in \Lambda} \delta_k(K_\lambda)$, so $\delta_k(K)$ is locally finite. Since every finitely generated subgroup of the soluble group $K/\delta_k(K)$ has finite 0-rank at most r, Lemma 2.5.1 implies that $K/\delta_k(K)$ has

0-rank at most r. It follows that G has 0-rank at most r, since G/K and $\delta_k(K)$ are locally finite. Clearly G is generalized radical. $\square$

We can obtain a further generalization of this local theorem, but first we need the following useful results, which appear in a paper of B. H. Neumann [**204**, Lemma 5.4*].

2.5.3. LEMMA. *Let H be a finite group and let $G = Cr_{\lambda \in \Lambda} G_\lambda$, where $G_\lambda \cong H$, for each $\lambda \in \Lambda$. Then G is a locally finite group.*

PROOF. Clearly we may suppose that Λ is infinite. Let $g_1, \ldots, g_k \in G$, and suppose that $g_j = (g_{j,\lambda})_{\lambda \in \Lambda}$, for $1 \leq j \leq k$.

We first consider the case in which $g_{j,\lambda} = g_{j,\mu}$ for each pair $\lambda, \mu \in \Lambda$. Let $L = \langle g_1, \ldots, g_k \rangle$. If $x = (x_\lambda)_{\lambda \in \Lambda} \in L$, it is clear that all the components of x are equal, that is $x_\lambda = x_\mu$ for each pair $\lambda, \mu \in \Lambda$. For each $\nu \in \Lambda$, let $pr_\nu : G \longrightarrow G_\nu$ be the canonical projection mapping, and set $K_\nu = \ker(pr_\nu)$. Then $G_\nu \cong G/K_\nu$ and $LK_\nu/K_\nu \cong L/(L \cap K_\nu)$. Therefore $L/(L \cap K_\nu)$ is finite. If $x = (x_\lambda)_{\lambda \in \Lambda} \in L \cap K_\nu$, then $x_\nu = 1$ and hence $x_\lambda = 1$ for every $\lambda \in \Lambda$. This means that $L \cap K_\nu = 1$ and then $L = L/(L \cap K_\nu) \cong LK_\nu/K_\nu$ is finite. Hence, in this case, the result has been proved.

Next consider the general case. Since H is finite, for the element g_1, there exist finitely many subsets $\Gamma_1, \ldots, \Gamma_m$ of Λ such that $\Lambda = \Gamma_1 \cup \cdots \cup \Gamma_m$ and $g_{1,\lambda} = g_{1,\mu}$ for each pair $\lambda, \mu \in \Gamma_j$, $1 \leq j \leq m$. For the element g_2 there are integers s_j and subsets $\Delta_{1,1}, \Delta_{1,2} \ldots, \Delta_{1,s_j}$ such that $\Gamma_j = \Delta_{1,1} \cup \cdots \cup \Delta_{1,s_j}$, where $g_{2,\lambda} = g_{2,\mu}$, for each pair $\lambda, \mu \in \Delta_{1,k}$, $1 \leq k \leq s_j, 1 \leq j \leq m$. In other words, there is an integer n and subsets, $\Delta_1, \Delta_2, \ldots, \Delta_n$ of Λ with the property that $\Lambda = \Delta_1 \cup \cdots \cup \Delta_n$ and $g_{1,\lambda} = g_{1,\mu}, g_{2,\lambda} = g_{2,\mu}$, whenever $\lambda, \mu \in \Delta_j$, for $1 \leq j \leq n$. The same argument can be applied to each element $g_3, \ldots, g_k$ in turn so that, proceeding in this way, we find subsets $\Sigma(1), \ldots, \Sigma(s)$ of Λ such that $\Lambda = \Sigma(1) \cup \cdots \cup \Sigma(s)$ and $g_{j,\lambda} = g_{j,\mu}$ for each pair $\lambda, \mu \in \Sigma(t)$, $1 \leq t \leq s$, $1 \leq j \leq k$. For every $1 \leq j \leq s$, we put $G_j = Cr_{\lambda \notin \Delta(j)} G_\lambda$. Therefore $G/G_j \cong Cr_{\lambda \in \Delta(j)} G_\lambda$ and $G_1 \cap \cdots \cap G_s = 1$. By Remak's Theorem we obtain the embedding

$$G \hookrightarrow G/G_1 \times \cdots \times G/G_s.$$

It follows that

$$L \hookrightarrow LG_1/G_1 \times \cdots LG_s/G_s.$$

Applying the argument given in the first paragraph, LG_j/G_j is finite for $1 \leq j \leq s$ and hence L is finite. Therefore G is locally finite. $\square$

2.5.4. PROPOSITION. *Let $G = Cr_{\lambda \in \Lambda} G_\lambda$, where G_λ is a finite group for each $\lambda \in \Lambda$. If there exists a positive integer t such that $|G_\lambda| \leq t$, then G is a locally finite group.*

PROOF. Let $|G_\lambda| = m \leq t$. Then G_λ is isomorphic to a subgroup of S_m, the symmetric group on m symbols, and since $m \leq t$ we have $S_m \leq S_t$. Hence $G \leq R = \underset{\lambda \in \Lambda}{Cr} R_\lambda$, where $R_\lambda \cong S_t$, for each $\lambda \in \Lambda$. By Lemma 2.5.3, R is locally finite and hence G is also locally finite. The result follows.   □

2.5.5. COROLLARY. *Let G be a finitely generated group and let*

$$\mathcal{B}(m) = \{H | H \text{ is a subgroup of finite index and } |G : H| \leq m\}.$$

If B is the intersection of all subgroups in the family $\mathcal{B}(m)$ then B has finite index in G.

PROOF. Clearly B is a characteristic subgroup of G. If $H \in \mathcal{B}(m)$ then $G/\mathbf{core}_G H$ is finite and $G/\mathbf{core}_G H \leq m!$. It is also clear that $B = \cap_{H \in \mathcal{B}(m)} \mathbf{core}_G H$. By Remak's Theorem we obtain an embedding

$$G/B \longrightarrow Cr_{H \in \mathcal{B}(m)} G/\mathbf{core}_G H.$$

By Proposition 2.5.4, G/B is locally finite. On the other hand, G/B is finitely generated and hence G/B must be finite.   □

Our next result appears as [**134**, 1.K.2 Proposition] and has many useful applications. We first require a little information concerning inverse (or projective) limits. Suppose that Λ is a non-empty partially ordered set and let $\{S_\lambda : \lambda \in \Lambda\}$ be a family of non-empty sets indexed by Λ. Suppose that if $\lambda, \mu \in \Lambda$ and $\mu \leq \lambda$ then there is a map $\pi_{\mu\lambda} : S_\lambda \longrightarrow S_\mu$ such that

    (a) $\pi_{\lambda\lambda} = 1_{S_\lambda}$, the identity map on S_λ.

    (b) If $\kappa, \lambda, \mu \in \Lambda$ and $\kappa \leq \mu \leq \lambda$ then $\pi_{\kappa\mu}\pi_{\mu\lambda} = \pi_{\kappa\lambda}$.

The *inverse limit* or *projective limit* of the system $\{S_\lambda, \pi_{\mu\lambda} : \lambda, \mu \in \Lambda, \mu \leq \lambda\}$ of sets and mappings is a subset S of the Cartesian product $C = \underset{\lambda \in \Lambda}{Cr} S_\lambda$ given by

$$S = \{(s_\lambda)_{\lambda \in \Lambda} \in C : \text{ if } \mu \leq \lambda \text{ then } \pi_{\mu\lambda}(s_\lambda) = s_\mu\}.$$

We denote this inverse limit by $\underleftarrow{\lim}\{S_\lambda, \pi_{\lambda\mu} | \lambda, \mu \in \Lambda\}$ or $\underleftarrow{\lim} S_\lambda$. If Λ is also *directed*, that is, given $\mu, \kappa \in \Lambda$ there exists $\lambda \in \Lambda$ such that $\mu, \kappa \leq \lambda$, then we call $\{S_\lambda, \pi_{\mu\lambda} : \lambda, \mu \in \Lambda, \mu \leq \lambda\}$ an *inverse system* of sets and mappings. Unfortunately, the inverse limit of an inverse system may be empty. It is, however, a well-known fact due to A. G. Kurosh (see [**172**, §55] or [**171**])that the inverse limit of an inverse system of finite sets is non-empty.

Further information concerning inverse limits can be found in [**172**, §55].

2.5.6. PROPOSITION. *Let $\mathfrak{X}$ be a class of groups such that for groups $X \geq S \geq Y$ with $X, Y \in \mathfrak{X}$ the finiteness of either of the indices $|X : S|$ and $|S : Y|$ implies that S belongs to $\mathfrak{X}$. If the group G has a local system $\mathcal{L}$ consisting of finitely generated subgroups, such that each subgroup $L \in \mathcal{L}$ has a subgroup of index at most n which belongs to $\mathfrak{X}$, then G has a subgroup of index at most n which has a local system consisting of finitely generated $\mathfrak{X}$-subgroups.*

PROOF. For each $L \in \mathcal{L}$ let

$$\mathcal{X}_L = \{K | K \text{ is an } \mathfrak{X}\text{-subgroup of } L \text{ such that } |L : K| \leq n\}.$$

By hypothesis, $\mathcal{X}_L \neq \emptyset$. Let Z be the intersection of all subgroups of L having finite index at most n in L. Since the subgroup L is finitely generated, L/Z is finite by Corollary 2.5.5. In particular, L has only finitely many subgroups of index at most n, and every one of these subgroups is also finitely generated, by Proposition 1.2.13. Hence the family $\mathcal{X}_L$ is finite. If $Y \in \mathcal{X}_L$, then Y contains Z and Z contains the $\mathfrak{X}$-subgroup 1. By our hypotheses $Z \in \mathfrak{X}$.

Let $M \in \mathcal{L}$, suppose $L \leq M$, and let $X \in \mathcal{X}_M$. Then $|L : L \cap X| \leq n$. It follows that $Z \leq L \cap X$. Since L/Z is finite, $L \cap X \in \mathfrak{X}$ and hence $L \cap X \in \mathcal{X}_L$. Next define the map $\pi_{L,M} : \mathcal{X}_M \longrightarrow \mathcal{X}_L$ where, for each $X \in \mathcal{X}_M$, we let $\pi_{L,M}(X) = L \cap X$. The system $\{\mathcal{X}_M, \pi_{L,M} | L, M \in \mathcal{L}, L \leq M\}$ is a projective system of finite sets. By the result due to A. G. Kurosh [**172**, §55] the projective limit $\varprojlim\{\mathcal{X}_M, \pi_{L,M} | L, M \in \mathcal{L}, L \leq M\}$ of this system is non-empty. Let

$$(W_L)_{L \in \mathcal{L}} \in \varprojlim\{\mathcal{X}_M, \pi_{L,M} | L, M \in \mathcal{L}, L \leq M\},$$

where $W_L \in \mathcal{X}_L$ and set $W = \bigcup_{L \in \mathcal{L}} W_L$. Clearly $W \cap L = W_L$ for $L \in \mathcal{L}$, and the family $\{W_L | L \in \mathcal{L}\}$ is a local system of finitely generated $\mathfrak{X}$-subgroups of W. We now show that $|G : W| \leq n$.

If this were not so, then there would exist at least $n + 1$ distinct left cosets of W in G. Let $g_1, \ldots, g_{n+1}$ be representatives of these cosets. There exists a subgroup $L \in \mathcal{L}$ with $\{g_1, \ldots, g_{n+1}\} \subseteq L$ and these elements represent $n + 1$ distinct left cosets of $W \cap L = W_L$. However, this is a contradiction, since $|L : W_L| \leq n$, by assumption. This contradiction shows that $|G : W| \leq n$, and this proves the proposition. $\square$

Let $\mathfrak{X}$ be a class of groups. A group G is called an *almost $\mathfrak{X}$-group* if G contains a normal subgroup $H \in \mathfrak{X}$ such that the index $|G : H|$ is finite. The class of all almost $\mathfrak{X}$-groups is denoted by $\mathfrak{X}\mathfrak{F}$. In particular, a group is *almost locally soluble* if it has a locally soluble subgroup of finite index.

Sometimes groups in the class $\mathfrak{X}\mathfrak{F}$ are called virtually $\mathfrak{X}$-groups, but here we shall use the term "almost $\mathfrak{X}$-group", which seems to be a more descriptive term.

2.5.7. THEOREM. *Let G be a group and suppose that G satisfies the following conditions:*

(i) *for every finitely generated subgroup L of G the factor group $L/\textbf{Tor}\,(L)$ is a generalized radical group;*

(ii) *there is a positive integer r such that $r_0(L) \leq r$ for every finitely generated subgroup L.*

Then $G/\textbf{Tor}\,(G)$ contains a soluble normal subgroup $D/\textbf{Tor}\,(G)$ of finite index. Moreover, G has finite 0-rank r and there is a function $f_5 : \mathbb{N} \longrightarrow \mathbb{N}$ such that $|G/D| \leq f_5(r)$.

PROOF. During the proof we shall use the functions introduced in Theorem 2.4.13. Without loss of generality we may suppose that $\textbf{Tor}\,(G) = 1$. If F is a finitely generated subgroup of G then, by Theorem 2.4.13, F contains a normal subgroup $K_F \geq \textbf{Tor}\,(F)$ such that F/K_F is finite of order at most n where $n = f_4(r)$ and $K_F/\textbf{Tor}\,(F)$ is soluble.

Let $\mathfrak{X}$ denote the class of all periodic-by-soluble groups. Clearly $\mathfrak{X}$ satisfies the conditions of Proposition 2.5.6. This proposition implies that G contains a subgroup H of index at most n such that H has a local system of periodic-by-soluble subgroups. Set $D = \text{core}_G H$; note that D is a normal subgroup of G and $|G/D| \leq n!$. Since D is normal we have $\textbf{Tor}\,(D) \leq \textbf{Tor}\,(G) = 1$. Next, let F be an arbitrary finitely generated subgroup of D and let $\mathcal{D}$ denote the family of all finitely generated subgroups of D containing F. Clearly $\mathcal{D}$ is a local system for D. It follows that $\bigcap_{U \in \mathcal{D}} \textbf{Tor}\,(U)$ is a periodic normal subgroup of D and hence $\bigcap_{U \in \mathcal{D}} \textbf{Tor}\,(U) = 1$. Note that , for each $U \in \mathcal{D}$, $U/\textbf{Tor}\,(U)$ is finitely generated and soluble and, by hypothesis, this factor group has finite 0-rank at most r. By Theorem 2.4.11, $\textbf{dl}(U/\textbf{Tor}\,(U)) \leq f_2(r)$. Using Remak's Theorem we obtain an embedding,

$$F \longrightarrow \text{Cr}_{U \in \mathcal{D}} F/(\textbf{Tor}\,(U) \cap F).$$

The isomorphism

$$F/(\textbf{Tor}\,(U) \cap F) \cong F\textbf{Tor}\,(U)/\textbf{Tor}\,(U) \leq U/\textbf{Tor}\,(U)$$

shows that $\textbf{dl}(F/(\textbf{Tor}\,(U) \cap F)) \leq f_2(r)$ for each $U \in \mathcal{D}$ and hence $\textbf{dl}(F) \leq f_2(r)$. Since this is true for an arbitrary finitely generated subgroup F of D, D is soluble and $\textbf{dl}(D) \leq f_2(r)$. Thus G is almost soluble. Lemma 2.5.1 implies that G has finite 0-rank and, furthermore, $\mathbf{r}_0(G) \leq r$. Finally, $|G/D| \leq f_5(r)$ where $f_5(r) = (f_4(r))!$ $\square$

CHAPTER 3

Section p-Rank of Groups

In the previous chapter, we discussed the 0-rank, which has its genesis in the concept of the $\mathbb{Z}$-rank of an abelian group and, as we have seen, it is an important numerical invariant of a group. Indeed, the 0-rank is so important that it allowed us to obtain a description of a very broad class of groups with finite 0-rank. However, this numerical invariant only gives an interesting theory in non-periodic groups, since $\mathbf{r}_0(G) = 0$ for each periodic group G. In this chapter, we shall consider another important numerical invariant which can be used in both periodic and non-periodic groups. It is associated with the 0-rank and in some ways looks a little more general. Its roots also lie in the concept of vector space dimension, but here the relation does not seem so obvious. As in Chapter 2, we will start with abelian groups and there define the concept of p-rank and show how this concept transforms into the concept of the section p-rank.

3.1. p-Rank in Abelian Groups

As we observed in Chapter 2, the 0-rank can be used very effectively in torsion-free abelian groups. For periodic abelian groups another numerical invariant, which is also based on the concept of dimension, can be introduced. We introduced the following notation earlier, but here use it in more depth, so we give a formal definition.

3.1.1. DEFINITION. Let p be a prime and let n be a non-negative integer. If P is an abelian p-group, then the n-layer of P is the subgroup $\Omega_n(P) = \{a \in P | a^{p^n} = 1\}$.

This notation will be used frequently in the remainder of this book. It is clear that $\Omega_{n+1}(P)/\Omega_n(P)$ is an elementary abelian p-group, for each $n \geq 0$, which can therefore be thought of as a vector space over the prime field $\mathbb{F}_p = \mathbb{Z}/p\mathbb{Z}$. In particular, $\Omega_1(P)$ is such a vector space.

3.1.2. DEFINITION. Let p be a prime and P an abelian p-group. The p-rank of P is defined to be the dimension of $\Omega_1(P)$ over $\mathbb{F}_p$.

Ranks of Groups: The Tools, Characteristics, and Restrictions
By Martyn R. Dixon, Leonid A. Kurdachenko and Igor Ya Subbotin

More generally, if G is an arbitrary abelian group then we define the p-rank of G to be the the p-rank of $\mathbf{Tor}_p(G)$. The p-rank of G is denoted by $\mathbf{r}_p(G)$.

In particular, if $\mathbf{Tor}_p(G) = 1$, then $\mathbf{r}_p(G) = 0$. It is clear that the p-rank of $C_p \times \mathbb{Z}$ is 1 and that the p-rank of the factor group $C_p \times C_p$ is 2. Thus the p-rank can increase when we pass to factor groups. From the definition given above it follows that if G is a finite elementary abelian p-group, isomorphic to a direct product of n copies of C_p, the cyclic group of order p, then $\mathbf{r}_p(G) = n$.

EXERCISE 3.1. *Prove the following assertion:*
Let p be a prime and let G be an abelian p-group. Suppose that $\mathbf{r}_p(G) = r$ is finite and that $G^m = 1$, where $m = p^k$ for some positive integer k. Then G is finite and $|G| \leq p^{rk}$.

3.1.3. LEMMA. *Let P be an abelian p-group for some prime p. Then $\mathbf{r}_p(P) = r$ is finite if and only if every elementary abelian p-section U/V of P is finite, $\mathbf{r}_p(U/V) \leq \mathbf{r}_p(P)$ and there is an elementary abelian section A/B of P such that $\mathbf{r}_p(A/B) = r$.*

PROOF. First we suppose that $\mathbf{r}_p(P) = r$ is finite. By definition this implies that $\mathbf{r}_p(\mathbf{\Omega}_1(P)) = r$ and hence there is an elementary abelian p-section precisely of p-rank r. Let K be an arbitrary finite subgroup of P. Then $K = \mathrm{Dr}_{1 \leq j \leq k}\langle x_j \rangle$, for certain elements $x_j \in K$ and a natural number k. It is clear that $\mathbf{\Omega}_1(K) = \mathrm{Dr}_{1 \leq j \leq k}\mathbf{\Omega}_1(\langle x_j \rangle)$ and since $\mathbf{\Omega}_1(K) \leq \mathbf{\Omega}_1(P)$ it follows that $k \leq r$. Furthermore, $K^p = \mathrm{Dr}_{1 \leq j \leq k}\langle x_j^p \rangle$ and hence
$$K/K^p \cong \mathrm{Dr}_{1 \leq j \leq k}\langle x_j \rangle / \langle x_j^p \rangle.$$
Consequently $\mathbf{r}_p(K/K^p) = k \leq r$.

Now let U/V be an elementary abelian p-section of P. If K is an arbitrary subgroup of U we have $K/(K \cap V) \cong KV/V \leq U/V$ so $K/(K \cap V)$ is elementary abelian and hence $K^p \leq K \cap V$. From our earlier argument it follows that $\mathbf{r}_p(K/K \cap V) \leq r$. Hence the orders of the finite subgroups of U/V are bounded and we deduce that U/V is finite. Consequently, there is a finite subgroup K such that $U = KV$ and then $\mathbf{r}_p(K/K \cap V) = \mathbf{r}_p(U/V) \leq r$.

Conversely, $\mathbf{r}_p(A/B) = r \geq \mathbf{r}_p(\mathbf{\Omega}_1(P)) = \mathbf{r}_p(P)$ so $\mathbf{r}_p(P)$ is finite and it follows from the previous argument that $\mathbf{r}_p(P)$ is precisely r. $\square$

EXERCISE 3.2. *Prove that if p is a prime and G is an abelian p-group, then G is divisible if and only if $G = G^p$.*

Our next result characterizes abelian p-groups with finite p-rank.

3.1.4. PROPOSITION. *Let P be an abelian p-group for some prime p. Then $\mathbf{r}_p(P)$ is finite if and only if P is a Chernikov group.*

PROOF. Suppose first that $\mathbf{r}_p(P) = r$ is finite. Using Lemma 3.1.3, we observe that P/P^p is finite. Hence there is a finite subgroup K such that $P = KP^p$. Since K is finite, there is an integer d such that $K^d = 1$. If $D = P^d$ then

$$D = P^d = (KP^p)^d = K^d P^{pd} = (P^d)^p = D^p.$$

Consequently, D is a divisible p-group, by Exercise 3.2, and hence is a direct product of Prüfer p-groups (see [**88**, Theorem 23.1], for example). If $|\mathbf{\Omega}_1(D)| = p^m$, then $m \le \mathbf{r}_p(P)$ by Lemma 3.1.3 and it follows that D is a direct product of only finitely many Prüfer subgroups. However, P/D is finite and hence P is a Chernikov group.

Conversely, let P be a Chernikov group. Then $\mathbf{\Omega}_1(P)$ is a direct product of subgroups of order p. Being a Chernikov group, G satisfies the minimal condition on subgroups and Proposition 1.3.2 implies that this direct product has only finitely many direct factors. It follows that $\mathbf{\Omega}_1(P)$ is finite and hence $\mathbf{r}_p(P)$ is finite. $\quad\square$

One particular consequence of this is the following corollary.

3.1.5. COROLLARY. *Let P be an abelian p-group for some prime p. Then P is a Chernikov group if and only if $\mathbf{\Omega}_1(P)$ is finite.*

The theory of abelian groups can be delineated using the 0-rank and the p-rank. The 0-rank works well with torsion-free groups and the p-rank is very useful for periodic groups. However, D. J. S. Robinson [**223**, 6.1] united the concepts of 0-rank and p-rank by introducing the class of $\mathfrak{A}_0$-groups.

An abelian group A is called an $\mathfrak{A}_0$-group if and only if $\mathbf{r}_0(A)$ is finite and $\mathbf{r}_p(A)$ is finite for all primes p.

He also extended the definition to soluble groups in a traditional way by defining a class $\mathfrak{S}_0$.

A soluble group G is called an $\mathfrak{S}_0$-group if and only if G has a finite subnormal series, each factor of which is an abelian $\mathfrak{A}_0$-group.

3.2. Finite Section p-Rank

In this section, we introduce a concept that is more versatile since it is appropriate for both periodic and non-periodic groups, while incorporating many important properties of the 0-rank and the p-rank. It arises from the properties of p-rank which were obtained in Lemma 3.1.3. We call this new rank the *section p-rank* of a group.

3.2.1. DEFINITION. Let p be a prime. We say that a group G has *finite section p-rank* $\mathbf{sr}_p(G) = r$ if every elementary abelian p-section of G is finite of order at most p^r and *there is* an elementary abelian p-section A/B of G such that $|A/B| = p^r$.

It is always the case, for an abelian group G, that $\mathbf{r}_p(G) \leq \mathbf{sr}_p(G)$, as is easily seen. However, the p-rank and the section p-rank need not coincide for abelian groups in general, as we observed following Definition 3.1.2. It is clear that the p-rank of $C_p \times \mathbb{Z}$ is 1, but this group has the elementary abelian p-section $(C_p \times \mathbb{Z})/p\mathbb{Z}$ of order p^2. It is easy to see that the section p-rank of $C_p \times \mathbb{Z}$ is precisely 2. Nevertheless, Lemma 3.1.3 shows that at least for abelian p-groups, and hence for all periodic abelian groups, the two concepts amount to the same thing.

We observe that if a group G has an element g of infinite order, then G has a section $\langle g \rangle / \langle g^p \rangle$ of order p, for every prime p. Hence if G is not periodic, then $\mathbf{sr}_p(G) \geq 1$ for each prime p. If a group G has an element g of order p, for the prime p, then G has the section $\langle g \rangle / \langle 1 \rangle$ of order p. Hence if $p \in \Pi(G)$, then $\mathbf{sr}_p(G) \geq 1$. The equality $\mathbf{sr}_p(G) = 0$ implies that G is a periodic group containing no p-elements.

The following elementary lemma enables us to easily determine a number of examples of groups of finite section p-rank.

3.2.2. LEMMA. *Let p be a prime and let G be a group.*

(i) *Suppose that G has finite section p-rank. If K is a subgroup of G and H is a normal subgroup of K, then $\boldsymbol{sr}_p(K/H) \leq \boldsymbol{sr}_p(G)$;*

(ii) *Suppose that G has finite section p-rank. If H is a normal subgroup of G, then $\boldsymbol{sr}_p(G) \leq \boldsymbol{sr}_p(H) + \boldsymbol{sr}_p(G/H)$;*

(iii) *Suppose that G has finite section p-rank. If H is a normal periodic subgroup of G such that $p \notin \Pi(H)$, then $\boldsymbol{sr}_p(G/H) = \boldsymbol{sr}_p(G)$;*

(iv) *Let H be a normal subgroup of G. If H and G/H have finite section p-rank, then G has finite section p-rank.*

EXERCISE 3.3. *Prove Lemma 3.2.2.*

We need not have equality in part (ii) of this lemma as the cyclic group of order 4 readily shows. It follows from this lemma that the property of having finite section p-rank is closed under taking subgroups and homomorphic images. Furthermore, a slight modification of the proof of part (ii) of the lemma shows that the property of having finite section p-rank is closed under taking extensions. In particular, this makes it easy to see that a Chernikov group is necessarily a group of finite section p-rank for each prime p, although we shall see later that this is a consequence of a much stronger property enjoyed by Chernikov groups.

The following result allows us to characterize abelian groups with finite section p-rank in general. As we have already observed, the Sylow p-subgroup of an abelian group G is unique and is precisely the p-component $\mathbf{Tor}_p(G)$.

3.2.3. PROPOSITION. *Let A be an abelian group and let p be a prime. Then A has finite section p-rank if and only if the Sylow p-subgroup P of A is Chernikov and the 0-rank of A is finite. Moreover, $\boldsymbol{sr}_p(A) = \boldsymbol{sr}_p(P) + \boldsymbol{r}_0(A)$.*

PROOF. Suppose first that A has finite section p-rank and that $\mathbf{sr}_p(A) = r$. By Lemma 3.1.3 and Proposition 3.1.4, P is Chernikov. Let $\mathcal{M}$ be a maximal $\mathbb{Z}$-independent subset of A and let $C = \langle \mathcal{M} \rangle$. Then C is a free abelian subgroup of A. Since C/C^p is an elementary abelian p-section of A it has order at most p^r. This observation implies that $\mathbf{r}_0(A) = |\mathcal{M}| = s$ is finite and that $\mathbf{r}_0(A) \leq \mathbf{sr}_p(A)$. In addition,

$$\mathbf{sr}_p(PC) = \mathbf{sr}_p(P \times C) = \mathbf{sr}_p(P) + \mathbf{r}_0(C)$$
$$(3.1) \qquad = \mathbf{sr}_p(P) + \mathbf{r}_0(A) \leq \mathbf{sr}_p(A).$$

Conversely, suppose that the Sylow p-subgroup of A is Chernikov and that $\mathbf{r}_0(A)$ is finite. By Proposition 3.1.4, $\mathbf{sr}_p(P)$ is finite. Suppose that U/V is a finite elementary abelian p-section of A. Then there is a finitely generated subgroup X of U such that $U = XV$. From Exercise 2.4 we have $X = T \times F$, where T is finite and F is free abelian, say $F \cong \underbrace{\mathbb{Z} \times \cdots \times \mathbb{Z}}_{t}$.

Clearly $\mathbf{sr}_p(T) \leq \mathbf{sr}_p(P)$ and $\mathbf{r}_0(F) \leq \mathbf{r}_0(A)$. Furthermore $\mathbf{sr}_p(X) = \mathbf{sr}_p(T) + \mathbf{r}_0(F)$ and

$$\mathbf{sr}_p(U/V) = \mathbf{sr}_p(X/X \cap V) \leq \mathbf{sr}_p(T) + \mathbf{r}_0(F) \leq \mathbf{sr}_p(P) + \mathbf{r}_0(A)$$

so that U/V is of bounded order. It follows that A has finite section p-rank at most $\mathbf{sr}_p(P) + \mathbf{r}_0(A)$, that is

$$(3.2) \qquad \mathbf{sr}_p(A) \leq \mathbf{sr}_p(P) + \mathbf{r}_0(A).$$

The result follows using Equations (3.1) and (3.2). $\square$

Lemma 2.4.3 was a useful observation in a number of the proofs in Section 2.4 and it will also be useful in the present section, so here we rephrase it using our new terminology.

3.2.4. LEMMA. *Let A be a torsion-free abelian group of finite 0-rank r. Then, for each prime p, A has finite section p-rank at most r.*

Furthermore, if $B \leq A$ and $A^m \leq B$, for some natural number m, then $|A/B| \leq m^r$.

We now turn to the study of the structure of locally nilpotent groups having finite section p-rank. The end result is almost the same as that in the case of abelian groups, but it requires preparation taking much greater effort. The first natural case is the consideration of periodic locally nilpotent groups having finite section p-rank. It is clear that this case reduces to the consideration of locally nilpotent p-groups (equivalently, locally finite p-groups). The study of nilpotent groups of finite section p-rank requires the use of some general properties of nilpotent groups, which we note in passing. The following simple result, due to O. Grün [**100**], will be very useful.

3.2.5. LEMMA. *Let G be a group and suppose that $\zeta_2(G) \neq \zeta_1(G)$. If $g \in \zeta_2(G) \setminus \zeta_1(G)$, then the mapping $\xi_g : G \longrightarrow G$ defined by $\xi_g(x) = [g, x]$, for $x \in G$, is an endomorphism of G. Moreover, $\mathbf{Im}(\xi_g) = [g, G]$, $\mathbf{ker}(\xi_g) = C_G(g)$ and $[g, G] \cong G/C_G(g)$. Furthermore, if $g^k \in \zeta_1(G)$, then $[g, G]^k = 1$.*

EXERCISE 3.4. *Prove Lemma 3.2.5.*

3.2.6. PROPOSITION ([**141**]). *Let G be a hypercentral group and let A be a normal abelian p-subgroup of G. If $G/C_G(A)$ contains no subgroups of index p then $A \leq \zeta(G)$.*

PROOF. Suppose, to the contrary, that $\zeta(G)$ does not contain A. Since G is hypercentral, Lemma 1.2.2 yields $A\zeta(G)/\zeta(G) \cap \zeta_2(G)/\zeta(G) \neq 1$ and hence there is an element $d \in A$ such that $d \in \zeta_2(G) \setminus \zeta(G)$. Without loss of generality we may assume that $d^p \in \zeta(G)$. By Lemma 3.2.5, $C_G(d) \lhd G$. Since $d \in A$ we have $C_G(A) \leq C_G(d)$ and hence $G/C_G(d)$ contains no subgroups of index p. Lemma 3.2.5 shows that $[d, G]$ is an elementary abelian p-subgroup of $\zeta(G)$. However, $[d, G] \cong G/C_G(d)$, by Lemma 3.2.5, which gives us a contradiction. This proves that $A \leq \zeta(G)$. $\square$

3.2.7. COROLLARY ([**37**]). *Let G be a hypercentral group. If G is $\mathfrak{F}$-perfect, then $\mathbf{Tor}\,(G) \leq \zeta(G)$. In particular, every $\mathfrak{F}$-perfect periodic hypercentral group is abelian.*

PROOF. Assume, for a contradiction, that $T = \mathbf{Tor}\,(G) \cap \zeta(G) \neq \mathbf{Tor}\,(G)$. Since G is hypercentral, $\mathbf{Tor}\,(G)\zeta(G)/\zeta(G) \cap \zeta_2(G)/\zeta(G) \neq 1$, by Lemma 1.2.2 and hence there is an element $d \in \mathbf{Tor}\,(G)$ such that $d \in \zeta_2(G) \setminus \zeta(G)$. The subgroup $A = \langle d, T \rangle$ is clearly abelian and also normal in G. To see this we note that if $g \in G$ then the choice of d

implies that $[d, g] \in \zeta(G)$. On the other hand, $\mathbf{Tor}\,(G)$ is a normal subgroup of G so that $[d, g] \in \mathbf{Tor}\,(G)$ and hence $[d, g] \in \mathbf{Tor}\,(G) \cap \zeta(G) = T$. Hence $A \lhd G$, as claimed.

Let B be the p-component of A and note that $B \lhd G$. Since G is $\mathfrak{F}$-perfect, $G/C_G(B)$ contains no subgroup of index p. Then Proposition 3.2.6 shows that $B \leq \zeta(G)$ and since this holds for all primes p we have $A \leq \zeta(G)$, which gives the contradiction that $d \in \zeta(G)$. This proves the result. $\qquad\square$

3.2.8. LEMMA. *Let G be a nilpotent Chernikov group. Then G contains a finite normal subgroup F such that G/F is a divisible Chernikov group.*

PROOF. Let t denote the nilpotency class of G and use induction on t. The case $t = 1$ is trivial since G is then abelian. Let $t > 1$ and suppose inductively that we have already proved that $G/\zeta(G)$ contains a finite normal subgroup $K/\zeta(G)$ such that G/K is divisible. We have $\zeta(G) = D \times C$, for some divisible subgroup D and some finite subgroup C (see [88, Theorem 21.2], for example). By Theorem 1.5.12, K' is finite. Then, as above, the abelian group K/K' can be written as $K/K' = DK'/K' \times L/K'$, where L/K' is finite. Let m be a natural number such that $(L/K')^m = 1$. Since K/K' is abelian, $((L/K')^{G/K'})^m = 1$. However, K/K' is Chernikov and this implies that $F/K' = (L/K')^{G/K'}$ is finite. Since K' is finite it follows that F is finite. Of course $F \lhd G$ and the choice of F implies that K/F is divisible. Since G/K is also divisible, G/F is $\mathfrak{F}$-perfect. Using Corollary 3.2.7, we deduce that G/F is abelian and hence divisible. $\qquad\square$

3.2.9. LEMMA. *Let G be an FC-group. Then the centre of G contains every $\mathfrak{F}$-perfect subgroup.*

EXERCISE 3.5. *Prove Lemma 3.2.9.*

3.2.10. COROLLARY. *Let G be a nilpotent Chernikov group. Then G is centre-by-finite.*

PROOF. Let D be the divisible part of G. Then certainly D is $\mathfrak{F}$-perfect. By Lemma 3.2.8, G' is finite and hence G is an FC-group. By Lemma 3.2.9, we have $D \leq \zeta(G)$ and the result follows. $\qquad\square$

3.2.11. PROPOSITION. *Let G be a periodic nilpotent group. Then the centre of G contains the largest $\mathfrak{F}$-perfect subgroup of G.*

EXERCISE 3.6. *Prove Proposition 3.2.11.*

We now extend Proposition 3.1.4 to nilpotent p-groups in the following way.

3.2.12. PROPOSITION. *Let p be a prime and let P be a nilpotent p-group. Then $sr_p(P)$ is finite if and only if P is a centre-by-finite Chernikov group.*

PROOF. Suppose that $sr_p(P)$ is finite. Then $sr_p(\zeta_{m+1}(P)/\zeta_m(P))$ is finite for each factor of the upper central series so, by Proposition 3.2.3, each upper central factor is Chernikov. By Corollary 1.3.6, the class of Chernikov groups is closed under taking extensions, so P is a Chernikov group. By Corollary 3.2.10, G is centre-by-finite.

Conversely, if P is a Chernikov group then $sr_p(P)$ is finite by a repeated application of Lemma 3.2.2.

$\square$

Our goal now is to generalize Proposition 3.2.3 to locally finite p-groups and it is to this that we turn attention. Such groups are locally nilpotent and certainly there are locally nilpotent p-groups that are not hypercentral. However such groups cannot have finite section p-rank. We obtain a sequence of results in this direction.

3.2.13. LEMMA. *Let p be a prime and let G be a countable locally finite p-group. If every abelian subgroup of G has finite section p-rank, then $\zeta(G) \neq 1$.*

EXERCISE 3.7. *Prove Lemma 3.2.13.*

3.2.14. LEMMA. *Let G be a prime and let G be a p-group. Suppose that G contains a normal hypercentral subgroup L of finite index. Then G is hypercentral.*

EXERCISE 3.8. *Prove Lemma 3.2.14.*

Immediate corollaries are the following results.

3.2.15. COROLLARY. *Let p be a prime and let G be a p-group. If G is abelian-by-finite, then G is hypercentral.*

3.2.16. COROLLARY. *Let p be a prime and let G be a Chernikov p-group. Then G is hypercentral.*

We next state a classical result concerning the group of automorphisms of a Chernikov group. This result is due to R. Baer [10]. Proofs of this theorem have been given in a number of books. A standard proof of this result can be found, for example, in [49, Theorem 1.5.16].

3.2.17. THEOREM. *Let G be a Chernikov group and let A be a periodic group of automorphisms of G. Then A is Chernikov and if G is abelian, then A is finite.*

3.2.18. COROLLARY. *Let p be a prime and let G be a hypercentral p-group. If every abelian subgroup of G has finite section p-rank, then G is a Chernikov group.*

PROOF. Let A be a maximal normal abelian subgroup of G. By Lemma 1.2.2, $A = C_G(A)$. Proposition 3.2.3 shows that A is Chernikov and Theorem 3.2.17 implies that G/A is finite. Hence G is a Chernikov group also. □

3.2.19. PROPOSITION. *Let G be a periodic FC-group. If G is infinite, then every element of G is contained in an infinite abelian subgroup.*

EXERCISE 3.9. *Prove Proposition 3.2.19.*

We use the previous results to prove the following very important theorem of S. N. Chernikov [39], one of the first of a number of results of its kind.

3.2.20. THEOREM. *Let G be a locally finite p-group for some prime p. If every abelian subgroup of G is Chernikov, then G is Chernikov.*

PROOF. First we assume that G is countable and assume, for a contradiction, that G is not Chernikov. Then Corollary 3.2.18 implies that G is not hypercentral and hence $Z = \zeta_\infty(G) \neq G$. Again using Corollary 3.2.18 we deduce that Z is Chernikov. If every abelian subgroup of G/Z is Chernikov, then Corollaries 3.2.18 and 3.2.16 show that $\zeta(G/Z) \neq 1$, contrary to the choice of Z. Hence G/Z contains an abelian subgroup, B/Z, which is not Chernikov. An application of Proposition 3.1.4 shows that $C/Z = \Omega_1(B/Z)$ is infinite. Let D denote the divisible part of Z, so that Z/D is finite. Then C/D has finite derived subgroup and hence it is an FC-group. By Proposition 3.2.19, C/D contains an infinite abelian subgroup E/D. Since C/D has finite exponent, E/D is not Chernikov. By Theorem 3.2.17, $G/C_G(D)$ is finite and hence $E/D \cap C_G(D)/D$ is not Chernikov either. Again using Proposition 3.1.4 we deduce that $A/D = \Omega_1(E/D \cap C_G(D)/D)$ is infinite and we note that $D \leq \zeta(A)$.

By the choice of A we have that $a^p \in D$ for each element a of A. Then Lemma 3.2.5 shows that $[a, A]$ is an elementary abelian p-subgroup of D and it follows that A' is likewise an elementary abelian p-subgroup of D. Since D is Chernikov, we deduce that A' is finite.

For the abelian group A/A' we have $A/A' = D/A' \times V/A'$, for some subgroup V (by [88, Theorem 21.2], for example), and the isomorphism $V/A' \cong A/D$ shows that the orders of the elements of V are also bounded. The subgroup V also has finite derived subgroup, so V is an FC-group. By Proposition 3.2.19, V contains an infinite abelian subgroup U and since V has finite exponent, U is not Chernikov, contrary to our hypothesis. This contradiction implies that G is hypercentral and Corollary 3.2.18 allows us to deduce that G is a Chernikov group.

In the general case, the argument above shows that every countable subgroup of G is Chernikov and hence hypercentral, by Corollary 3.2.16. Then G is hypercentral, by Proposition 1.2.4 and a further application of Corollary 3.2.18 implies that G is also Chernikov.   □

3.2.21. COROLLARY. *Let p be a prime and let G be a locally finite p-group. Then G has finite section p-rank if and only if G is a hypercentral Chernikov group.*

PROOF. We observed, following the proof of Lemma 3.2.2, that a Chernikov group always has finite section p-rank for each prime p, so it suffices to prove sufficiency of the condition.

Suppose, then, that G has finite section p-rank. Using Proposition 3.2.3 we see that every abelian subgroup of G is Chernikov. Then Theorem 3.2.20 implies that G is also Chernikov. The fact that G is hypercentral follows from Corollary 3.2.16.   □

Unlike with the nilpotent case, the centre of a Chernikov p-group need not contain the divisible part. The simplest example of this is the locally dihedral group, constructed as follows. Let K be a Prüfer 2-group. Then K has an automorphism $\iota : x \mapsto x^{-1}$, for $x \in K$, and it is easy to see that $G = K \rtimes \langle \iota \rangle$ is a Chernikov 2-group with finite centre.

There are more complicated examples than this for each prime p. Here we consider the case $p = 3$ and let

$$K_j = \langle a_{j,n} | a_{j,1}^3 = 1, a_{j,n+1}^3 = a_{j,n}, n \in \mathbb{N} \rangle$$

be a Prüfer 3-group, for $j = 1, 2$. Let $A = K_1 \times K_2$. Then A has an automorphism χ of order 3 defined by:

$$\chi(a_{1,1}) = a_{1,1}, \qquad \chi(a_{2,1}) = a_{2,1}a_{1,1}$$

$$\chi(a_{1,n+1}) = a_{1,n+1}a_{2,n}^{-1}, \qquad \chi(a_{2,n+1}) = a_{2,n+1}a_{1,n+1}a_{2,n}^{-1}, \text{ for } n \geq 1.$$

The natural semidirect product $G = A \rtimes \langle \chi \rangle$ is a Chernikov 3-group whose upper central factors each have order 3. Other examples are given in [19].

EXERCISE 3.10. *Prove that χ defined above is an automorphism of A of order 3 and that the group G defined above is a Chernikov 3-group whose upper central factors are cyclic of order 3.*

Our next goal is the general description of the structure of nilpotent groups having finite section p-rank. As with the abelian case there is a very close connection with the 0-rank.

3.2.22. LEMMA ([18]). *Let G be a finitely generated nilpotent group. Suppose that $T = \mathbf{Tor}(G)$ is an elementary abelian p-subgroup for some prime p. Then G contains normal subgroups $U \leq V$ such that G/U is finite and V/U is an elementary abelian p-group of order p^k, where $k = \mathbf{sr}_p(T) + \mathbf{r}_0(G/T)$. In particular, $\mathbf{sr}_p(G) \geq \mathbf{sr}_p(T) + \mathbf{r}_0(G/T)$.*

PROOF. Since G/T is torsion-free, Corollary 1.2.9 implies that the factors of the upper central series of G/T are torsion-free also, so that G has a series of normal subgroups,

$$1 = L_0 \leq L_1 = T \leq L_2 \leq \cdots \leq L_m = G,$$

in which L_{j+1}/L_j is a torsion-free abelian group and $L_{j+1}/L_j \leq \zeta(G/L_j)$ for all $j > 0$. Since G is a finitely generated nilpotent group, T is finite and hence its centralizer $C = C_G(T)$ has finite index in G. If $x, y \in C \cap L_2$, we have $x^y = xa$ for some element $a \in T$ and since $T \leq \zeta(C \cap L_2)$ we deduce that $y^{-p}xy^p = xa^p = x$, using the hypotheses on T. This means that $y^p \in \zeta(C \cap L_2)$ and hence $(C \cap L_2)/\zeta(C \cap L_2)$ is periodic. Since L_2 is a finitely generated nilpotent group it follows that $(C \cap L_2)/\zeta(C \cap L_2)$ is finite. Since G/C is finite, $L_2/(C \cap L_2)$ is finite and then $L_2/\zeta(C \cap L_2)$ is also finite. It follows that $\mathbf{r}_0(L_2) = \mathbf{r}_0(\zeta(C \cap L_2))$. Let $V_2 = \zeta(C \cap L_2)$, a normal subgroup of G. Since T is the finite torsion subgroup of the abelian group V_2 it follows that $V_2 = T \times A$, for some torsion-free subgroup A. Then $U_2 = V_2^p = A^p$ and $|V_2/U_2| = |T| |A/A^p|$. Note that U_2 is G-invariant. Since $A \cong V_2/T$ we have $\mathbf{r}_0(A) = \mathbf{r}_0(V_2) = \mathbf{r}_0(L_2)$. Since A is free abelian, we deduce that $|A/A^p| = p^{\mathbf{r}_0(A)} = p^{\mathbf{r}_0(L_2)}$ and hence $|V_2/U_2| = p^{k_2}$, where $k_2 = \mathbf{sr}_p(T) + \mathbf{r}_0(L_2/T)$. We note that V_2/U_2 is elementary abelian.

Now L_2/U_2 is finite so $\mathbf{r}_0(L_3/U_2) = \mathbf{r}_0(L_3/L_2)$. Let $C_2 = C_G(V_2/U_2)$ and consider the subgroup $C_2 \cap L_3$. Arguing as above, we can show that $V_3/U_2 = \zeta((C_2 \cap L_3)/U_2)$ has finite index in L_3/U_2. We note that $V_3 \triangleleft G$. It follows that $\mathbf{r}_0(L/U_2) = \mathbf{r}_0(V_3/U_2)$ and then $\mathbf{r}_0(L_3/L_2) = \mathbf{r}_0(V_3/U_2)$.

Since V_2/U_2 is the finite torsion subgroup of the abelian group V_3/U_2, we have $V_3/U_2 = V_2/U_2 \times A_3/U_2$ for some subgroup A_3. It follows that $U_3/U_2 = (V_3/U_2)^p = (A_3/U_2)^p$, so that

$$|V_3/U_3| = |V_2/U_2| \, |(A_3/U_2)/(A_3/U_2)^p|.$$

Again note that U_3 is G-invariant and V_3/A_3 is finite. Therefore

$$\mathbf{r}_0(A_3/U_2) = \mathbf{r}_0(V_3/U_2) = \mathbf{r}_0(L_3/L_2).$$

Since A_3/U_2 is free abelian, $|(A_3/U_2)/(A_3/U_2)^p| = p^{\mathbf{r}_0(A_3/U_2)} = p^{\mathbf{r}_0(L_3/L_2)}$. Consequently,

$$|V_3/U_3| = p^{\mathbf{sr}_p(T)+\mathbf{r}_0(L_2/T)} p^{\mathbf{r}_0(L_3/L_2)} = p^{\mathbf{sr}_p(T)+\mathbf{r}_0(L_2/T)+\mathbf{r}_0(L_3/L_2)}$$

$$= p^{\mathbf{sr}_p(T)+\mathbf{r}_0(L_3/T)}.$$

Arguing as above, we can construct, after finitely many steps, two normal subgroups U and V such that G/U is finite and V/U is an elementary abelian p-group of order p^k where $k = \mathbf{sr}_p(T) + \mathbf{r}_0(G/T)$. This completes the proof. $\qquad\square$

3.2.23. COROLLARY ([18]). *Let G be a finitely generated torsion-free nilpotent group. Then $\mathbf{sr}_p(G) = \mathbf{r}_0(G)$ for all primes p. In particular, G is nilpotent of class at most $\mathbf{sr}_p(G)$.*

PROOF. Let

$$1 = C_0 \le C_1 \le \cdots \le C_k = G$$

be the upper central series of G, so that each factor is torsion-free abelian and $\mathbf{r}_0(C_{j+1}/C_j) = \mathbf{sr}_p(C_{j+1}/C_j)$, for $0 \le j \le k - 1$. It follows from Lemma 3.2.2 that

$$\mathbf{sr}_p(G) \le \sum_{0 \le j \le k-1} \mathbf{sr}_p(C_{j+1}/C_j) = \sum_{0 \le j \le k-1} \mathbf{r}_0(C_{j+1}/C_j) = \mathbf{r}_0(G).$$

However, by Lemma 3.2.22, $\mathbf{sr}_p(G) \ge \mathbf{r}_0(G)$ so $\mathbf{sr}_p(G) = \mathbf{r}_0(G)$. Finally, $\mathbf{ncl}\,(G) = k \le \sum_{0 \le j \le k-1} \mathbf{r}_0(C_{j+1}/C_j) = \mathbf{r}_0(G) = \mathbf{sr}_p(G).$ $\qquad\square$

3.2.24. COROLLARY ([18]). *Let G be a torsion-free locally nilpotent group with finite section p-rank, for some prime p. Then G is nilpotent of class at most $\mathbf{sr}_p(G)$ and $\mathbf{sr}_p(G) = \mathbf{r}_0(G)$.*

PROOF. Suppose that $\mathbf{sr}_p(G) = s$ and let K be an arbitrary finitely generated subgroup of G. Corollary 3.2.23 implies that K is nilpotent of class at most s and hence the same is true for G. Let

$$1 = C_0 \le C_1 \le \cdots \le C_k = G$$

be the upper central series of G. Since G is torsion-free, Corollary 1.2.9 implies that all factors of this series are torsion-free abelian and

$$\mathbf{r}_0(C_{j+1}/C_j) = \mathbf{sr}_p(C_{j+1}/C_j), \text{ for } 0 \le j \le k - 1.$$

In each factor C_{j+1}/C_j we choose a finitely generated subgroup L_{j+1}/C_j such that C_{j+1}/L_{j+1} is periodic. Let V_{j+1} be a finitely generated subgroup such that $L_{j+1} = V_{j+1}C_j$ and let V be the subgroup generated by the subgroups V_j, for $1 \leq j \leq k$. Then V is also finitely generated. Since G is nilpotent, V is a subnormal subgroup of G and there is a finite series

$$V = U_0 \lhd U_1 \lhd U_2 \lhd \ldots \lhd U_t = G.$$

The choice of V implies that the factors U_{j+1}/U_j are periodic, for $0 \leq j \leq t-1$ and hence $\mathbf{r}_0(G) = \mathbf{r}_0(V)$. By Corollary 3.2.23, $\mathbf{r}_0(V) = \mathbf{sr}_p(V) \leq \mathbf{sr}_p(G) = s$. However, G contains subgroups A, B such that $B \lhd A$ and A/B is an elementary abelian p-group of order p^s. Let D be a finitely generated subgroup of G such that $A = DB$. Then $D/(D \cap B) \cong DB/B = A/B$ so $D/(D \cap B)$ is an elementary abelian p-group of order p^s. Thus $\mathbf{r}_0(V) = \mathbf{sr}_p(V) \leq \mathbf{sr}_p(D) = s$. Set $E = \langle V, D \rangle$. Then $\mathbf{sr}_p(E) = \mathbf{sr}_p(D) = \mathbf{sr}_p(G)$ and $\mathbf{r}_0(E) = \mathbf{r}_0(V) = \mathbf{r}_0(G)$. Since E is finitely generated, we can apply Corollary 3.2.23 to deduce that $\mathbf{sr}_p(E) = \mathbf{r}_0(E)$. It now follows that $\mathbf{sr}_p(G) = \mathbf{r}_0(G)$, as required. $\square$

We note that in general $\mathbf{sr}_p(G) \neq \mathbf{r}_0(G)$ if G is a torsion-free soluble group. Indeed our next example shows that equality does not hold even for polyrational groups.

To see this, let p be a prime and let $A = \mathbb{Q}_p = \{m/p^n | m, n \in \mathbb{Z}\}$ be the additive subgroup of $\mathbb{Q}$, the group of rational numbers. Since $A = pA$ the mapping $\rho : A \longrightarrow A$ defined by $\rho(a) = a/p$, for $a \in A$, is an automorphism of A of infinite order. Let $G = A \rtimes \langle \rho \rangle$ denote the natural semidirect product, namely the set of ordered pairs (a, ρ^k), for $a \in A, k \in \mathbb{Z}$ with multiplication defined by

$$(a, \rho^k)(b, \rho^t) = (a + p^k b, \rho^{k+t}) \text{ for all } a, b \in A, k, t \in \mathbb{Z}.$$

The subgroup A is normal in G and G/A is infinite cyclic, so G is a polyrational group and it is clear that $\mathbf{r}_0(G) = 2$. The group G is soluble and finitely generated by the elements $(1, 1)$ and $(0, \rho)$. However, G has no elementary abelian sections of order p^2, as we ask the reader to verify in the next exercise.

EXERCISE 3.11. *Prove that the group G defined above has only elementary abelian p-sections of order p, so that $\mathbf{sr}_p(G) = 1$.*

3.2.25. COROLLARY ([18]). *Let G be a finitely generated nilpotent group. Suppose that $T = \mathbf{Tor}(G)$ is a p-subgroup for some prime p. Then $\mathbf{sr}_p(G) = \mathbf{sr}_p(T) + \mathbf{r}_0(G/T)$.*

PROOF. Corollary 3.2.23 shows that $\mathbf{sr}_p(G/T) = \mathbf{r}_0(G/T)$ so, using Lemma 3.2.2, we deduce that

$$\mathbf{sr}_p(G) \leq \mathbf{sr}_p(T) + \mathbf{sr}_p(G/T) = \mathbf{sr}_p(T) + \mathbf{r}_0(G/T).$$

Let $\mathbf{sr}_p(T) = s$. Then T has subgroups A, B such that $B \lhd A$ and A/B is an elementary abelian p-group of order p^s. Since G is a finitely generated nilpotent group, T is finite and hence $C = C_G(T)$ has finite index in G. Hence $\mathbf{r}_0(G/T) = \mathbf{r}_0(G) = \mathbf{r}_0(C)$. Let $D = CA$. Since every finitely generated nilpotent group is residually finite (for example, see [20, Corollary 1.21]) D has a normal subgroup K of finite index such that $K \cap \mathbf{Tor}\,(D) = A$. Then $\mathbf{r}_0(D) = \mathbf{r}_0(K)$ since D/K is finite. Applying Lemma 3.2.22 to K/B we conclude that K has normal subgroups U, V such that $U \leq V$, K/U is finite and V/U is an elementary abelian p-group of order p^k, where $k = \mathbf{sr}_p(A/B) + \mathbf{r}_0(K/A) = \mathbf{sr}_p(T) + \mathbf{r}_0(G/T)$. It follows that $\mathbf{sr}_p(G) \geq \mathbf{sr}_p(T) + \mathbf{r}_0(G/T)$. Using our earlier inequality we deduce that $\mathbf{sr}_p(G) = \mathbf{sr}_p(T) + \mathbf{r}_0(G/T)$, as required. $\square$

Now we are in a position to prove the main result of this section, namely the description of locally nilpotent groups with finite section p-rank.

3.2.26. THEOREM. *Let G be a locally nilpotent group. Then G has finite section p-rank if and only if the Sylow p-subgroup P of G is Chernikov and the 0-rank of G is finite. Moreover, $\boldsymbol{sr}_p(G) = \boldsymbol{sr}_p(P) + r_0(G)$.*

PROOF. Let $T = \mathbf{Tor}\,(G)$ and note that, since G is locally nilpotent, $T = P \times Q$, where Q is the Sylow p'-subgroup of G. Lemma 3.2.2 shows that $\mathbf{sr}_p(G) = \mathbf{sr}_p(G/Q)$, so without loss of generality we may suppose that $T = P$.

Using Lemma 3.2.2 and Corollary 3.2.21 we see that P is Chernikov. Lemma 3.2.2 also shows that $\mathbf{sr}_p(G/P)$ is finite. An application of Corollary 3.2.24 implies that $\mathbf{sr}_p(G/P) = \mathbf{r}_0(G/P) = \mathbf{r}_0(G)$.

By Lemma 3.2.2 again, $\mathbf{sr}_p(G) \leq \mathbf{sr}_p(P) + \mathbf{sr}_p(G/P) = \mathbf{sr}_p(P) + \mathbf{r}_0(G/P)$. Let $\mathbf{sr}_p(G) = s$ and choose subgroups A, B of G such that $B \lhd A$ and A/B is an elementary abelian p-group of order p^s. Then there is a finitely generated subgroup D_1 such that $A = D_1 B$. Since $A/B = D_1 B/B \cong D_1/(D_1 \cap B)$, the subgroup D_1 also has an elementary abelian p-section of order p^s so $\mathbf{sr}_p(G) = \mathbf{sr}_p(D_1)$. By the same argument, we may choose a finite subgroup D_2 of P such that $\mathbf{sr}_p(P) = \mathbf{sr}_p(D_2)$. As in the proof of Corollary 3.2.24, there is a subgroup D_3 such that $P \leq D_3$, D_3/P is finitely generated and $\mathbf{r}_0(G) = \mathbf{r}_0(D_3)$. Consequently, there is a finitely generated subgroup D_4 such that $D_3 = D_4 P$ and the isomorphism $D_3/P = D_4 P/P \cong D_4/(D_4 \cap P)$ shows that $\mathbf{r}_0(D_4) = \mathbf{r}_0(D_3) = \mathbf{r}_0(G)$.

Let $D = \langle D_1, D_2, D_4 \rangle$. By Lemma 3.2.2, $\mathbf{sr}_p(D) \geq \mathbf{sr}_p(D_1) = \mathbf{sr}_p(G)$ so that $\mathbf{sr}_p(D) = \mathbf{sr}_p(G)$. By Lemma 2.2.3, $\mathbf{r}_0(D) \geq \mathbf{r}_0(D_4) = \mathbf{r}_0(G)$ so that $\mathbf{r}_0(D) = \mathbf{r}_0(G)$. Since D is finitely generated we deduce from Corollary 3.2.25 that

$$\mathbf{sr}_p(D) = \mathbf{sr}_p(D \cap P) + \mathbf{r}_0(D/(D \cap P)) = \mathbf{sr}_p(D \cap P) + \mathbf{r}_0(D).$$

By Lemma 3.2.2, $\mathbf{sr}_p(D \cap P) \geq \mathbf{sr}_p(D_2) = \mathbf{sr}_p(P)$, so $\mathbf{sr}_p(D \cap P) = \mathbf{sr}_p(P)$ and we finally obtain

$$\mathbf{sr}_p(G) = \mathbf{sr}_p(D) = \mathbf{sr}_p(D \cap P) + \mathbf{r}_0(D) = \mathbf{sr}_p(P) + \mathbf{r}_0(G).$$

$\square$

The basic goal of this chapter is to consider locally generalized radical groups of finite section p-rank where p is a prime. The theory splits naturally into two parts. A very important role is played by the maximal normal periodic subgroup $\mathbf{Tor}\,(G)$ of a group G and one might expect this subgroup to have considerable influence on the structure of G. It is of course also necessary to consider the factor group $G/\mathbf{Tor}\,(G)$. For the case when G is a locally generalized radical group the subgroup $\mathbf{Tor}\,(G)$ is locally finite and $G/\mathbf{Tor}\,(G)$ is almost torsion-free, as we shall see. Of course the study of locally finite groups and the study of almost torsion free groups requires completely different approaches and different techniques. For the description of the structure of $G/\mathbf{Tor}\,(G)$ we will use the results from the previous chapters. However the study of the periodic part $\mathbf{Tor}\,(G)$ requires certain additional concepts and results and it is to these that we now turn.

3.3. Locally Finite Groups with Finite Section p-Rank

This section will be devoted to the study of locally finite groups of finite section p-rank for some prime p. If G is such a group then it follows from Corollary 3.2.21 that every p-subgroup (and hence, every Sylow p-subgroup) of G is Chernikov. Thus G will have the minimal condition on p-subgroups, for the prime p, which is sometimes referred to as the condition min-p. In infinite groups, there are many different variations of the term "Sylow theory", as can be seen in [49]. Corollary 3.2.21 shows that the locally finite groups of finite section p-rank are precisely the locally finite groups whose Sylow p-subgroups are Chernikov. We remark at once that the Sylow p-subgroups of such groups are not isomorphic in general. In fact the Sylow p-subgroups of locally finite groups with Chernikov p-subgroups can be quite badly behaved. However, in such groups it is possible to find certain "nice" Sylow p-subgroups. We name these special Sylow p-subgroups after B. A. F. Wehrfritz who first showed their existence in locally finite groups whose Sylow p-subgroups are Chernikov.

3.3.1. DEFINITION. Let G be a locally finite group and p a prime. A maximal p-subgroup P of G is called a *Wehrfritz p-subgroup* [**49**, Definition 2.5.2] if P contains an isomorphic copy of every p-subgroup of G.

As we noted above, locally finite groups whose Sylow p-subgroups are Chernikov always contain Wehrfritz p-subgroups. This was first shown by B. A. F. Wehrfritz [**256**]. In turn, Wehrfritz used this "Sylow theory" to obtain strong structural results, due to M. I. Kargapolov [**126**] and V. P. Shunkov [**241**], for locally finite groups whose Sylow p-subgroups are Chernikov. We give an account of Wehrfritz's work.

The theory of inverse limits is all that we now require in order to prove the following lemma.

3.3.2. LEMMA. *Let P and Q be Chernikov groups. If Q contains an isomorphic copy of every finite subgroup of P, then Q contains a subgroup isomorphic to P.*

PROOF. Let S and T be the respective divisible parts of P and Q. Then S is a direct product of finitely many Prüfer groups. Let k be a fixed but arbitrary positive integer and let

$$S_k = \underset{p \in \Pi(S)}{\mathrm{Dr}} \, \Omega_k(\mathbf{Tor}_p(S)),$$

so that S_k is a finite, characteristic subgroup of S. Let

$|Q : T| = q_1^{a_1} \dots q_t^{a_t}$ where $a_j \geq 0$ for $1 \leq j \leq t$ and $\Pi(Q) = \{q_1, \dots, q_t\}$.

Let m be the greatest element of the set $\{a_1, \dots, a_t\}$. Note that since $|P : S|$ is finite there exists a finite subgroup H of P such that $P = HS$.

By assumption HS_{k+m} can be embedded in Q, so there is a monomorphism

$$\rho : HS_{k+m} \longrightarrow Q.$$

Suppose that $x \in \Omega_k(\mathbf{Tor}_p(S))$ for some prime p and positive integer k. Since S is divisible, there exists $y \in S$ such that $y^{p^m} = x$ and clearly $y \in S_{k+m}$. Then

$$\rho(x) = (\rho(y))^{p^m} \in T,$$

by definition of m, and the fact that a p-element of Q/T has order at most p^m. It follows that ρ induces a monomorphism $\sigma = \rho|_{S_k}$:

$S_k \longrightarrow T$. Also $\rho|_H$ composed with the natural map from Q to Q/T is a mapping $\tau : H \longrightarrow Q/T$ and both σ and τ are induced by ρ. Let $\mathcal{A}_k$ be the set of all such pairs (σ, τ) induced by a monomorphism $\rho : HS_{k+m} \longrightarrow Q$. Then the above argument shows that $\mathcal{A}_k$ is non-empty. Also $\mathcal{A}_k$ is finite since each of the groups H, S_k and Q/T is finite and T is a direct product of finitely many Prüfer groups.

Suppose that $k, j \in \mathbb{N}$ and $j \leq k$. Then there exists a mapping

$$\alpha_{jk} : \mathcal{A}_k \longrightarrow \mathcal{A}_j$$

defined by

$$\alpha_{jk}(\sigma, \tau) = (\sigma^*, \tau), \text{ where } \sigma^* \text{ is the restriction of } \sigma \text{ to } S_j$$

whenever $(\sigma, \tau) \in \mathcal{A}_k$. It is clear that $\alpha_{mj}\alpha_{jk} = \alpha_{mk}$ and that α_{kk} is the identity map on $\mathcal{A}_k$ so the collection $\{\mathcal{A}_k, \alpha_{jk} | j \leq k; j, k \in \mathbb{N}\}$ is an inverse system of non-empty finite sets and mappings. It follows from the theorem of A. G. Kurosh [**172**, §55] that $\varprojlim\{\mathcal{A}_k, \alpha_{jk}\} \neq \emptyset$.

Let $((\sigma_k, \tau_k))_{k \in \mathbb{N}} \in \varprojlim\{\mathcal{A}_k, \alpha_{jk}\}$ and suppose that $\rho_k : HS_{k+m} \longrightarrow Q$ induces σ_k, τ_k. Define $U = \bigcup_{k \in \mathbb{N}} \sigma_k(S_k)$ and let $\phi : S \longrightarrow U$ be the map defined by $\phi|_{S_k} = \sigma_k$. Since $((\sigma_k, \tau_k)) \in \varprojlim\{\mathcal{A}_k, \alpha_{jk}\}$, it is clear that ϕ is an isomorphism. Suppose that the order of H is n. Then $H \cap S \leq S_n$. Let

$$V = \langle U, \rho_n(H) \rangle$$

and define $\Phi : P \longrightarrow V$ by $\Phi(x) = \rho_n(h) \cdot \phi(s)$, whenever $x = hs \in P$ for some elements $h \in H, s \in S,$. Then Φ is well defined. For if $x = hs = h_1 s_1$, for some $h, h_1 \in H, s, s_1 \in S$, then $s_1 s^{-1} = h_1^{-1} h \in H \cap S \leq S_n$. We have

$$\phi(s_1 s^{-1}) = \phi(h_1^{-1} h) = \sigma_n(h_1^{-1} h) = \rho_n(h_1^{-1} h).$$

Thus $\rho_n(h_1) \cdot \phi(s_1) = \rho_n(h) \cdot \phi(s)$, and that Φ is well defined follows.

Furthermore, Φ is a homomorphism. For, if $h, h_1 \in H, s, s_1 \in S$, then

$$
\begin{aligned}
\Phi(hs \cdot h_1 s_1) &= \Phi(hh_1(h_1^{-1}sh_1)s_1) = \rho_n(hh_1) \cdot \phi(h_1^{-1}sh_1 s_1) \\
&= \rho_n(h) \cdot \rho_n(h_1) \cdot \rho_k(h_1^{-1}sh_1) \cdot \rho_k(s_1) \text{ if } h_1^{-1}sh_1, s_1 \in S_k \\
&= \rho_n(h) \cdot \phi(s) \cdot \rho_n(h_1) \cdot \phi(s_1) \text{ since } \tau_k = \tau_n \\
&= \Phi(hs) \cdot \Phi(h_1 s_1).
\end{aligned}
$$

Finally, if $\Phi(hs) = 1$, then $\rho_n(h) = (\phi(s))^{-1}$. So $\phi(s)$ has order dividing $|H|$ and hence $s \in S_n$ by definition of n. Thus $\phi(s) = \rho_n(s)$ and $\Phi(hs) = \rho_n(hs) = 1$. Since ρ_n is a monomorphism, $hs = 1$ and hence Φ is also a monomorphism. This completes the proof. $\square$

This result is the main ingredient required for the existence of Wehrfritz p-subgroups in a group with Chernikov p-subgroups for the prime p.

3.3.3. THEOREM. *Let G be a locally finite group whose p-subgroups are Chernikov for the prime p. Then G contains Wehrfritz p-subgroups and every finite p-subgroup lies in at least one of these.*

PROOF. Let H be a finite p-subgroup of G, set $P_1 = H$ and let $\{P_n | n \in \mathbb{N}\}$ be a set of finite p-subgroups of G such that if Q is a finite p-subgroup of G then $Q \cong P_n$ for some n. Such a choice of subgroups is possible since there are only countably many non-isomorphic finite p-groups. Let $S_n = \langle P_j : 1 \leq j \leq n \rangle$, a finite subgroup of G. Suppose that R_n is a Sylow p-subgroup of S_n. Then there exists a Sylow p-subgroup R_{n+1} of S_{n+1} such that $R_n \leq R_{n+1}$, so we can construct a tower

$$R_1 \leq R_2 \leq \cdots \leq R_n \leq R_{n+1} \leq \cdots$$

of finite p-subgroups of G. Let R be a maximal p-subgroup of G containing $\bigcup_{n \in \mathbb{N}} R_n$. By Sylow's Theorem, $P_n^g \leq R_n$ for some $g \in S_n$, so R contains a copy of every finite p-subgroup of G. If P is an arbitrary p-subgroup of G, then P and R are Chernikov and it follows from Lemma 3.3.2 that R contains a subgroup isomorphic to P. Hence R is a Wehrfritz p-subgroup of G. Since $H = P_1 \leq R$ the result follows. □

In fact we can characterize the Wehrfritz p-subgroups of a locally finite group with Chernikov Sylow p-subgroups in terms of the finite p-subgroups alone, as the following result shows.

3.3.4. PROPOSITION. *Let G be a locally finite group. Suppose that the p-subgroups of G are Chernikov for some prime p. If P is a p-subgroup of G and Q is a Wehrfritz p-subgroup of G, then the following are equivalent.*

   (i) *P is a Wehrfritz p-subgroup of G;*

   (ii) *P contains an isomorphic copy of every finite p-subgroup of G;*

   (iii) *$P \cong Q$.*

PROOF. It is clear that (i) implies (ii) by definition of Wehrfritz p-subgroup. It is also clear that (i) follows from (iii), so it therefore suffices to prove that (ii) implies (iii).

In this case P contains an isomorphic copy of every finite p-subgroup of Q, so Lemma 3.3.2 implies that there is a monomorphism $\tau : Q \longrightarrow P$. On the other hand, Q is a Wehrfritz p-subgroup, so there is a monomorphism $\mu : P \longrightarrow Q$. Hence there is a monomorphism $\nu = \mu \circ \tau : Q \longrightarrow Q$. If the image of ν is denoted by Q_1 and if $Q_1 \neq Q$ then, since $Q_1 \cong Q$, we have $\nu(Q_1) = Q_2 \neq Q_1$. If we have constructed a subgroup Q_n of Q, isomorphic to Q, for some $n \geq 1$, then we inductively define $\nu(Q_n) = Q_{n+1}$. Note that $Q_n \neq Q_{n+1}$ and this enables us to construct a strictly descending chain

$$Q = Q_0 \gneq Q_1 \gneq Q_2 \cdots \gneq Q_n \gneq \cdots$$

of subgroups of Q. However, this is impossible since Q is Chernikov and hence satisfies the minimal condition on subgroups. This contradiction shows that $Q = Q_1$ and hence ν is an isomorphism. Consequently P and Q are isomorphic. $\qquad\qquad\qquad\qquad\qquad\qquad\qquad\qquad\qquad\qquad\square$

We have shown that the Wehrfritz p-subgroups of a locally finite group with Chernikov p-subgroups are isomorphic. However, the isomorphism need not be induced by any automorphism of the whole group, even when the Wehrfritz p-subgroups are locally cyclic (see [49, Example 2.5.6], for example) so the Wehrfritz p-subgroups need not be conjugate in general.

3.3.5. THEOREM. *Let G be a locally finite group and let p be a prime. The following are equivalent.*

(i) *G has finite section p-rank;*
(ii) *The p-subgroups of G are Chernikov;*
(iii) *The abelian p-subgroups of G have finite section p-rank.*

PROOF. We first prove that (i) and (ii) are equivalent. If G has finite section p-rank, then it follows from Corollary 3.2.21 that every p-subgroup of G is Chernikov.

Conversely, suppose that the p-subgroups of G are Chernikov. By Theorem 3.3.3, G has a Wehrfritz p-subgroup W. Let U, V be subgroups of G such that U is a normal subgroup of V and V/U is a finite elementary abelian p-group. Since G is locally finite, there is a finite subgroup F such that $V = FU$. If P is a Sylow p-subgroup of F then it is clearly the case that $V = PU$. By Proposition 3.3.4, P is isomorphic to some finite subgroup of W. Since W is Chernikov it has finite section p-rank $\mathbf{sr}_p(W) = k$ say, by Corollary 3.2.21. Then $V/U = PU/U \cong P/(P \cap U)$ is isomorphic to a section of W and hence $|V/U| \leq p^k$. Consequently, the order of every finite elementary abelian p-section of G is at most p^k and hence every elementary abelian p-section of G is finite. Furthermore, our proof shows that $\mathbf{sr}_p(G) = \mathbf{sr}_p(W) = k$ and the result follows.

Next we prove that (ii) and (iii) are equivalent. Since a Chernikov p-group has finite section p-rank it is clear that (iii) is implied by (ii). Next we assume that (iii) holds. If P is a p-subgroup of G then P is locally nilpotent and its abelian subgroups have finite section p-rank. By Proposition 3.2.3, the abelian p-subgroups of P are Chernikov. By Theorem 3.2.20, P is a Chernikov group so that (ii) follows. □

Suppose that G is a Chernikov group and suppose that D, the divisible part of G, has finite index n. Then, following Kegel and Wehrfritz [**134**, 3.9], G has a conjugacy class $\mathcal{K}(G)$ of subgroups of G such that

(i) $\mathcal{K}(G)$ is invariant under all automorphisms of G.
(ii) If $K \in \mathcal{K}(G)$ then $D \cap K = \Omega_n(D)$ and $G = KD$.

The proof of this result is not particularly difficult, but is omitted. Using this conjugacy class, Wehrfritz [**257**] proved the following result.

3.3.6. PROPOSITION. *Let G be a locally finite group. Suppose that the p-subgroups of G are Chernikov for some prime p and that P, Q are Wehrfritz p-subgroups of G with divisible parts D, E respectively. Then*

(i) $\Omega_n(D)$ *and* $\Omega_n(E)$ *are conjugate in G, for all $n \in \mathbb{N}$;*
(ii) *If $U \in \mathcal{K}(P)$ and $V \in \mathcal{K}(Q)$ then $U\Omega_n(D)$ and $V\Omega_n(E)$ are conjugate in G, for all $n \in \mathbb{N}$;*
(iii) *U and V are conjugate in G.*

We use this to deduce the following result of Wehrfritz [**257**].

3.3.7. COROLLARY. *Let G be a locally finite group. Suppose that the p-subgroups of G are Chernikov for some prime p. Then the p-subgroup P is a Wehrfritz p-subgroup of G if and only if P contains a conjugate of every finite p-subgroup of G.*

PROOF. The sufficiency of the condition is a consequence of Proposition 3.3.4(ii). To prove necessity, suppose that R is a finite p-subgroup of G and that P is a Wehrfritz p-subgroup. By Theorem 3.3.3, G has a Wehrfritz p-subgroup Q containing R. If E is the divisible part of Q and if $K \in \mathcal{K}(Q)$ then there exists a positive integer n such that $K\Omega_n(E)$ contains R. However, by Proposition 3.3.6, there exists $g \in G$ such that $(K\Omega_n(E))^g \leq P$. Hence some conjugate of R lies in P, and the result follows. □

Our next goal is to use the existence of Wehrfritz p-subgroups to obtain important structural results concerning locally finite groups with Chernikov Sylow p-subgroups for the prime p. First, however, we show that the Wehrfritz p-subgroups are well behaved with respect to forming quotients and normal subgroups. The results should be compared to the well-known analogous results for the Sylow p-subgroups of a finite group.

3.3.8. PROPOSITION. *Let G be a locally finite group. Suppose that the p-subgroups of G are Chernikov for some prime p and that H is a normal subgroup of G. Then the p-subgroups of H and G/H are Chernikov. Furthermore*

(i) *If P is a Wehrfritz p-subgroup of G, then PH/H is a Wehrfritz p-subgroup of G/H and $P \cap H$ is a Wehrfritz p-subgroup of H;*

(ii) *If Q/H is a Wehrfritz p-subgroup of G/H, then there exists a Wehrfritz p-subgroup W of G such that $Q/H = WH/H$.*

PROOF. By Theorem 3.3.5, $\mathbf{sr}_p(G)$ is finite and from Lemma 3.2.2 it follows that $\mathbf{sr}_p(G/H)$ and $\mathbf{sr}_p(H)$ are finite. Using Theorem 3.3.5 again we see that the Sylow p-subgroups of G/H and H are Chernikov.

Let A/H be a finite p-subgroup of G/H. Then there exists a finite p-subgroup B of G such that $A = BH$. By Corollary 3.3.7, $B^g \le P$ for some element $g \in G$ so that

$$PH/H \ge B^g H/H = (BH)^g/H = A^g/H = (A/H)^{gH}.$$

Again by Corollary 3.3.7 we deduce that PH/H is a Wehrfritz p-subgroup of G/H.

Conversely, let Q/H be a Wehrfritz p-subgroup of G/H. By Theorem 3.3.3 , Q has a Wehrfritz p-subgroup W. As above WH/H is a Wehrfritz p-subgroup of Q/H and hence $Q/H = WH/H$, so $Q = WH$. Let C be an arbitrary finite p-subgroup of G. Then Corollary 3.3.7 implies that there is an element $x \in G$ such that

$$Q/H \ge (CH/H)^{xH} = (CH)^x/H = C^x H/H.$$

Hence $C^x \le Q$ and, since W is a Wehrfritz p-subgroup of Q, Corollary 3.3.7 implies that there is an element $y \in Q$ such that $C^{xy} = (C^x)^y \le W$. Again by Corollary 3.3.7, we deduce that W is a Wehrfritz p-subgroup of G. Thus (ii) holds.

To prove the remainder of (i) suppose that F is a finite p-subgroup of H. There exists $z \in G$ such that $F^z \le P$, by Corollary 3.3.7. Hence $F^z \le P \cap H$, since $H \lhd G$, so $P \cap H$ contains an isomorphic copy of F. It follows from Proposition 3.3.4 that $P \cap H$ is a Wehrfritz p-subgroup of H, as required. This completes the proof. $\square$

We also need the following result which can be found in Kegel and Wehrfritz [**134**, 3.16]. We omit the proof.

3.3.9. LEMMA. *Let G be a locally finite group and suppose that $H \lhd G$.*

(i) *If H is finite and G/H is a π-group, for some set of primes π, then $G/\boldsymbol{O}_\pi(G)$ is finite;*

(ii) *If H is a Chernikov p-group and G/H is a p'-group, for some prime p, then $G/(H \times \boldsymbol{O}_{p'}(G))$ is finite.*

The proof of this result relies on the Schur–Zassenhaus Theorem. Clearly its purpose is to enable us to move certain types of factor about, which will prove rather useful in the proof of the next result. With these results in hand we can now prove Kargapolov's fundamental theorems concerning locally finite groups of finite section p-rank, following the proof outlined in the book of Kegel and Wehrfritz [**134**]. As usual, we say that the group G *involves* a group H if H is isomorphic to a section of G. First we need the following useful concept which allows us to use induction arguments.

Suppose that G is a Chernikov group with divisible part D. Then D is a direct product of finitely many Prüfer subgroups and the number $\mathbf{mmx}(G)$ of Prüfer subgroups occurring in this direct product is an invariant of D. The pair $\mathbf{s}(G) = (\mathbf{mmx}(G), |G/D|)$ is called the *size* of G. The sizes of Chernikov groups can be well-ordered lexicographically. Thus, if H is a Chernikov group with divisible part E, then we say that *the size of G is smaller than the size of H*, symbolically $\mathbf{s}(G) < \mathbf{s}(H)$, if $\mathbf{mmx}(G) < \mathbf{mmx}(H)$ or $\mathbf{mmx}(G) = \mathbf{mmx}(H)$ and $|G/D| < |H/E|$.

If G is a locally finite group with Chernikov p-subgroups for the prime p, then, by Theorem 3.3.3, G has Wehrfritz p-subgroups. Furthermore every Wehrfritz p-subgroup contains an isomorphic copy of every p-subgroup of G and the Wehrfritz p-subgroups are isomorphic. Thus if W, V are Wehrfritz p-subgroups of G, then it is clear that $\mathbf{s}(W) = \mathbf{s}(V)$. This shows that $\mathbf{s}(W)$ is an invariant of the group. We call the size of the Wehrfritz p-subgroups occurring the *p-size* of G, and denote this by $\mathbf{s}_p(G)$. Furthermore, if W is a Wehrfritz p-subgroup of G and P is an arbitrary p-subgroup, then $\mathbf{s}(P) \leq \mathbf{s}(W)$.

Let G be a group and π a set of primes. Let

$$\mathcal{M} = \{H | H \lhd G \text{ such that } G/H \text{ is a } \pi\text{-group}\}$$

and let $\mathbf{O}^\pi(G)$ denote the intersection of all members of $\mathcal{M}$. If G is a locally finite group, then $G/\mathbf{O}^\pi(G)$ is also a π-group. Hence, in this case, $\mathbf{O}^\pi(G)$ is the smallest normal subgroup whose factor group is a π-group. Thus $\mathbf{O}^\pi(G)$ is what we might term the π-residual of G. It is clearly a characteristic subgroup of G.

3.3.10. THEOREM (Kargapolov [**126**]). *Let G be a locally finite group of finite section p-rank for the prime p. Then $G/\boldsymbol{O}_{p',p}(G)$ is* *finite if and only if every simple section of G containing elements of order p is finite.*

PROOF. We first prove that if $G/\mathbf{O}_{p',p}(G)$ is finite, then the given condition holds and to do this we prove the contrapositive. Suppose that G has subgroups U, V such that V is normal in U and U/V is an infinite simple group containing an element xV of order p. Let $H = \mathbf{O}_{p'}(G)$ and note that $U \cap H$ is normal in U. Since U/V is simple either $(U \cap H)V = V$ or $(U \cap H)V = U$. However, the latter case cannot arise since $(U \cap H)V/V$ is a p'-group whereas U/V contains elements of order p. Consequently, $U \cap H \leq V$ and the section $UH/VH \cong U/V$ is involved in G/H. Thus we may replace G by G/H and assume that $\mathbf{O}_{p'}(G) = 1$. If P is a normal p-subgroup of G of finite index, then as above, either $(U \cap P)V = V$ or $(U \cap P)V = U$. In the former case $UP/VP \cong U/V$ is a section of G/P, which is impossible since P has finite index. In the latter case $U \cap P/V \cap P$ is an infinite simple p-group which is impossible, by Proposition 1.2.20. Hence $G/\mathbf{O}_{p',p}(G)$ cannot be finite.

Suppose now that there is a group G of finite section p-rank that does not involve any infinite simple group containing elements of order p, but that $G/\mathbf{O}_{p',p}(G)$ is infinite. Then there is a counter example to the theorem of minimal p-size, which we also denote by G. By Proposition 3.3.8, $G/\mathbf{O}_{p'}(G)$ has the same p-size as G and also involves no infinite simple group containing elements of order p. We may therefore assume that $\mathbf{O}_{p'}(G) = 1$. Let $\sigma = \Pi(G) \setminus \{p\}$ and let $M = \mathbf{O}^\sigma(G)$. Clearly $\mathbf{O}^\sigma(M) = M$ and since $\mathbf{O}_p(G) \leq M$ we have $\mathbf{O}_p(M) = \mathbf{O}_p(G)$. Now the Wehrfritz p-subgroups of G lie in M, so M has the same p-size as G. If $M/\mathbf{O}_p(M)$ is finite then we can apply Lemma 3.3.9(i) to deduce that $|G/\mathbf{O}_p(M) : \mathbf{O}_{p'}(G/\mathbf{O}_p(M))|$ is finite. Let $U/\mathbf{O}_p(M) = \mathbf{O}_{p'}(G/\mathbf{O}_p(M))$. Then G/U is finite, $\mathbf{O}_p(M)$ is Chernikov and $U/\mathbf{O}_p(M)$ is a p'-group so, by Lemma 3.3.9(ii), $|U : \mathbf{O}_p(M) \times \mathbf{O}_{p'}(U)|$ is finite. However, $\mathbf{O}_{p'}(U)$ is trivial, so $|G : \mathbf{O}_p(G)| = |G : U| \cdot |U : \mathbf{O}_p(M)|$ is finite, contrary to the choice of G. Hence $M/\mathbf{O}_p(M)$ is infinite, so we may replace G by M and therefore assume that $G = \mathbf{O}^\sigma(G)$ is generated by p-elements.

Let H be a proper normal subgroup of G and let W be a Wehrfritz p-subgroup of G. Clearly G/H contains elements of order p, so Proposition 3.3.8 implies that its Wehrfritz p-subgroup WH/H is nontrivial. Thus $\mathbf{s}(W \cap H) < \mathbf{s}(W)$. By Proposition 3.3.8, $W \cap H$ is a Wehrfritz p-subgroup of H and consequently every proper normal subgroup of G has smaller p-size than G. Since H inherits the hypotheses on G it follows, by the choice of G, that $H/\mathbf{O}_p(H)$ is finite.

Suppose, for a contradiction, that H is an infinite proper normal subgroup of G. Then $\mathbf{O}_p(H)$ is also infinite and hence the p-size of G/H is smaller than the p-size of G. By the choice of G it follows that

(3.3) $|G/H : \mathbf{O}_{p',p}(G/H)|$ is finite.

Let $Y/H = \mathbf{O}_{p'}(G/H)$. Since $H/\mathbf{O}_p(H)$ is finite Lemma 3.3.9(i) can be applied to $Y/\mathbf{O}_p(H)$ and we deduce that

(3.4) $|Y/\mathbf{O}_p(H) : \mathbf{O}_{p'}(Y/\mathbf{O}_p(H))|$ is finite.

Let $S/\mathbf{O}_p(H) = \mathbf{O}_{p'}(Y/\mathbf{O}_p(H))$. We know $\mathbf{O}_p(H)$ is Chernikov and $S/\mathbf{O}_p(H)$ is a p'-group so, by Lemma 3.3.9(ii), we have that

$$|S : \mathbf{O}_p(H) \times \mathbf{O}_{p'}(S)| \text{ is finite.}$$

Since $\mathbf{O}_{p'}(G) = 1$ it follows that $\mathbf{O}_{p'}(S) = 1$, so $S/\mathbf{O}_p(H)$ is finite and hence $Y/\mathbf{O}_p(H)$ is finite, by Equation (3.4). Let $L/H = \mathbf{O}_{p',p}(G/H)$. We can apply Lemma 3.3.9(i) to the group $L/\mathbf{O}_p(H)$ to deduce that

$$|L/\mathbf{O}_p(H) : \mathbf{O}_p(L/\mathbf{O}_p(H))| \text{ is finite.}$$

However, if $R/\mathbf{O}_p(H) = \mathbf{O}_p(L/\mathbf{O}_p(H))$ then R is a normal p-subgroup of G and $|G : R|$ is finite, by Equation (3.3), contradicting the choice of G.

Consequently every proper normal subgroup of G is finite. If X is a proper normal subgroup of G, then $G/C_G(X)$ is finite and if $C_G(X) \neq G$, then G is finite. Hence every proper normal subgroup of G is central and, since G is clearly non-abelian, $G/Z(G)$ is an infinite simple group which, by choice of G, is generated by its elements of p-power order. This is a final contradiction.   □

3.3.11. COROLLARY. *Let G be a periodic locally soluble group of finite section p-rank for the prime p. Then $G/\mathbf{O}_{p'}(G)$ is a Chernikov group.*

PROOF. By Theorem 3.3.10 and Corollary 1.2.19, $G/\mathbf{O}_{p',p}(G)$ is finite. Since $\mathbf{O}_{p',p}(G)/\mathbf{O}_{p'}(G)$ is a p-group it is Chernikov, by Proposition 3.3.8. Hence $G/\mathbf{O}_{p'}(G)$ is also Chernikov.   □

We obtain the following consequence of this result due to S. N. Chernikov [**40**].

3.3.12. COROLLARY. *Let G be a periodic locally soluble group. If the Sylow p-subgroups of G are finite for the prime p, then $G/\mathbf{O}_{p'}(G)$ is finite.*

PROOF. It is easy to see, using Sylow's theorem, that the Sylow p-subgroups of G are conjugate. It follows from Proposition 3.3.8 that the p-subgroups of $G/\mathbf{O}_{p'}(G)$ are also finite and hence $\mathbf{O}_{p',p}(G)/\mathbf{O}_{p'}(G)$ is finite. By Theorem 3.3.10 and Corollary 1.2.19, $G/\mathbf{O}_{p',p}(G)$ is finite, whence so is $G/\mathbf{O}_{p'}(G)$, as required. □

The following result of S. N. Chernikov is also now very easy.

3.3.13. COROLLARY (Chernikov [40]). *Let G be a periodic locally soluble group. If the Sylow p-subgroups of G are finite for all primes p, then G is residually finite.*

PROOF. By Corollary 3.3.12, $G/\mathbf{O}_{p'}(G)$ is finite for every prime $p \in \Pi(G)$. Clearly $\cap_{p \in \Pi(G)} \mathbf{O}_{p'}(G) = 1$ and, using Remak's Theorem, there is an embedding

$$G \longrightarrow \underset{p \in \Pi(G)}{\mathrm{Cr}} \, G/\mathbf{O}_{p'}(G).$$

This shows that G is residually finite. □

We shall see later how to exploit these results further.

3.4. Structure of Locally Generalized Radical Groups with Finite Section p-Rank

In the final section of this chapter, we now take the results of Section 3.3 and combine these with the results of Chapter 2 to obtain strong structural results for locally generalized radical groups of finite section p-rank. Of course, locally (soluble-by-finite) groups are locally generalized radical.

3.4.1. LEMMA. *Let G be a generalized radical group. Suppose that R is the maximal normal radical subgroup of G and L/R is the maximal normal locally finite subgroup of G/R. If G/R is infinite, then L/R is infinite.*

EXERCISE 3.12. *Prove Lemma 3.4.1.*

We now obtain a further result from [67].

3.4.2. THEOREM. *Let G be a locally generalized radical group of finite section p-rank $\mathbf{sr}_p(G) = r$, for some prime p. Then G has finite 0-rank at most $2r$. Furthermore, G has normal subgroups $T \le L \le K \le S \le G$ such that*

(i) *T is locally finite and G/T is soluble-by-finite;*
(ii) *L/T is a torsion-free nilpotent group;*

(iii) K/L is a finitely generated torsion-free abelian group;

(iv) G/K is finite and S/T is the soluble radical of G/T.

Moreover, there are functions $f_5, f_6 : \mathbb{N} \longrightarrow \mathbb{N}$ such that $|G/K| \leq f_5(r)$ and $dl(S/T) \leq f_6(r)$.

PROOF. We proceed to prove that G has finite 0-rank and to this end we let E denote an arbitrary finitely generated subgroup of G. Let $U = \mathbf{Tor}\,(E)$, let $H = E/U$ and let R denote the maximal normal radical subgroup of H. Let P denote the Hirsch–Plotkin radical of R. Clearly P is torsion-free, so it follows from Corollary 3.2.24 that P is nilpotent of class at most $\mathbf{sr}_p(G) = r$ and of 0-rank at most r. Using the arguments of the proof of Theorems 2.4.11 and 2.4.13 it follows that R/P is abelian-by-finite. Then, by Lemma 3.2.2 and Proposition 3.2.3 $r_0(R/P) \leq \mathbf{sr}_p(G) = r$. Using Lemma 3.2.2 again, we deduce that R has finite 0-rank at most $r + r = 2r$.

Let F/R be the maximal normal locally finite subgroup of H/R. Then clearly F has 0-rank at most $2r$ and it follows from Theorem 2.4.13 that F/R is finite. Lemma 3.4.1 shows that H/R is also finite and hence H has finite 0-rank at most $2r$. Consequently, E also has finite 0-rank at most $2r$ and it follows from Proposition 2.5.2 that the 0-rank of G is finite, at most $2r$. We now apply Theorem 2.4.13 to obtain the result and remark that $f_5(r) = f_4(2r)$ and $f_6(r) = f_2(2r)$. The result follows.　　　□

Theorem 3.4.2 has been proved in the paper [**67**], but here we obtain better bounds for the functions f_5 and f_6.

The following result is an immediate consequence of Theorems 3.3.5 and 3.4.2.

3.4.3. THEOREM. *Let G be a locally generalized radical group of finite section p-rank $\mathbf{sr}_p(G) = r$, for some prime p. Then G has finite 0-rank and $r_0(G) \leq 2r$. Furthermore, G has normal subgroups $T \leq L \leq K \leq S \leq G$ such that T is a locally finite group whose Sylow p-subgroups are Chernikov, L/T is torsion-free nilpotent, K/L is finitely generated torsion-free abelian, G/K is finite and S/K is soluble. Moreover, $|G/K| \leq f_5(r)$ and $dl(S/T) \leq f_6(r)$.*

We also note the following consequence.

3.4.4. COROLLARY. *Let G be a locally generalized radical group of finite section p-rank $\mathbf{sr}_p(G) = r$, for some prime p. Suppose that every simple section of G, containing elements of order p, is finite. Then G contains normal subgroups $Q \leq T \leq L \leq K \leq S \leq G$ such that Q is a locally finite p'-subgroup, T/Q is a Chernikov group whose divisible part* *is a p-group, L/T is torsion-free nilpotent, K/L is finitely generated torsion-free abelian, G/K is finite and S/K is soluble. Furthermore, G has finite 0-rank and $r_0(G) \leq 2r$. Also $|G/K| \leq f_5(r)$ and $dl(S/T) \leq f_6(r)$.*

The corollary follows immediately from Theorems 3.4.3 and 3.3.10. Finally, in this section we note:

3.4.5. COROLLARY. *Let G be a locally radical group of finite section p-rank $sr_p(G) = r$, for some prime p. Then G contains normal subgroups $Q \leq T \leq L \leq K \leq G$ such that Q is a periodic locally soluble p'-subgroup, T/Q is a soluble Chernikov group whose divisible part is a p-group, L/T is torsion-free nilpotent, K/L is finitely generated torsion-free abelian, and G/K is a finite soluble group such that $|G/K| \leq f_5(r)$ and $dl(G/T) \leq f_6(r)$. Furthermore, G has finite 0-rank and $r_0(G) \leq 2r$.*

This theorem is an immediate consequence of Theorem 3.4.3 and Corollary 3.3.11.

CHAPTER 4

Groups of Finite Section Rank

In the previous chapter, we looked at the concept of section p-rank for a prime number p and obtained information about the structure of a group having finite section p-rank. The natural next step is to consider the situation where the group has finite section p-rank for all primes p. This does not simply mean that we extract the results from Chapter 3 since the class of groups with finite section p-rank for all primes p has many interesting properties. As we have seen previously, the study of groups with finite section rank naturally splits into two cases: the consideration of the maximal normal torsion subgroup $\mathbf{Tor}\,G$ and the consideration of the factor group $G/\mathbf{Tor}\,G$. In the case when G is a locally generalized radical group the subgroup $\mathbf{Tor}\,G$ is locally finite. Consequently, we must begin our study of groups with finite section rank by considering the locally finite case, which is fundamental and gives us an opportunity for substantial progress.

4.1. Locally Finite Groups with Finite Section Rank

4.1.1. DEFINITION. Let G be a group. The group G is said to have *finite section rank* if $\mathbf{sr}_p(G)$ is finite for all primes p.

This definition can be made a little more precise. Let $\mathbb{P}$ be the set of all primes and let $\sigma : \mathbb{P} \longrightarrow \mathbb{N}_0$ be a function. Then the group G has *finite section rank* σ if $\mathbf{sr}_p(G) = \sigma(p)$ is finite, for each prime p.

In the paper [**228**], D. J. S. Robinson introduced the following class of groups:

A group G has *finite abelian sectional rank* if it has no infinite elementary abelian p-sections for any prime p.

It is clear that if a group G has finite section rank σ, for some function $\sigma : \mathbb{P} \longrightarrow \mathbb{N}_0$, then G has finite abelian sectional rank. Thus a group having finite section rank is a specific type of group having finite sectional rank. It quite often happens, however, that a group with finite abelian sectional rank has finite section rank for some function σ. These two classes of groups coincide in various classes of generalized soluble groups; for example, they coincide when the groups are abelian, as the next lemma observes.

Ranks of Groups: The Tools, Characteristics, and Restrictions
By Martyn R. Dixon, Leonid A. Kurdachenko and Igor Ya Subbotin

4.1.2. LEMMA. *Let G be an abelian group and suppose that every elementary abelian section of G is finite. Then there exists a function $\sigma : \mathbb{P} \longrightarrow \mathbb{N}_0$ such that G has finite section rank σ. Moreover, $\sigma(p) = r_p(G) + r_0(G)$, for all primes p. In particular, if G is periodic, then $\sigma(p) = r_p(G)$, for all primes p and if G is torsion-free then $\sigma(p) = r_0(G)$, for each prime p.*

EXERCISE 4.1. *Prove Lemma 4.1.2.*

In Theorem 3.3.5, we started to consider the relationship between the section p-rank of a group and the section p-rank of its abelian subgroups. The following basic result can be thought of as a continuation of the theme presented in Theorem 3.3.5.

4.1.3. THEOREM. *Let G be a locally finite group. The following are equivalent:*

(i) *G has finite section rank σ;*
(ii) *each p-subgroup of G is Chernikov, for every prime p;*
(iii) *if A is an arbitrary abelian subgroup of G, then the p-rank of A is finite for all primes p;*
(iv) *if A is an arbitrary abelian subgroup of G, then the section p-rank of A is finite for all primes p.*

Moreover, if G has finite section rank σ, then $r_p(A) \leq \sigma(p)$ for all abelian subgroups A.

PROOF. The fact that (i), (ii) and (iv) are equivalent follows from Theorem 3.3.5. Let A be an arbitrary abelian p-subgroup of G for some prime p. By Lemma 3.1.3, $r_p(A) = sr_p(A)$, which shows that (iii) and (iv) are equivalent. If G has finite section rank σ, then Lemma 3.2.2 shows that $sr_p(A) \leq sr_p(G) = \sigma(p)$. By Lemma 3.1.3, $r_p(A) = sr_p(A)$, so that $r_p(A) \leq \sigma(p)$, which finishes the proof.   □

This theorem is rather interesting since in parts (iii) and (iv) there is no assumption that the p-ranks and the section p-ranks are bounded, for the prime p. However, part (i) implies that such a bound indeed exists. In particular, for locally finite groups, the property of having finite abelian sectional rank is equivalent to the property of having finite section rank.

The following result of V. V. Belyaev [21] is fundamental to the rest of the work in this book. During its proof we refer to the Feit–Thompson theorem, a tour de force in the theory of finite groups. We recall, for the convenience of the reader, that this theorem states that a finite $2'$-group is soluble. We shall also use the classification of finite simple groups during the proof.

4.1.4. THEOREM. *Let G be a locally finite group. If G has finite section rank, then G is almost locally soluble.*

PROOF. Suppose that there is a counterexample to the theorem. The Feit–Thompson Theorem [84] implies that all counterexamples have elements of order 2, and among these counterexamples we choose a group G, with minimal 2-size. Let $\sigma = \Pi(G) \setminus \{2\}$. Then $G/\mathbf{O}^\sigma(G)$ is a $2'$-group so is locally soluble, by the Feit–Thompson Theorem. In this case, if $M = \mathbf{O}^\sigma(G)$ is almost locally soluble, then there is a characteristic subgroup N of M such that N is locally soluble and M/N is finite. Let $H = C_G(M/N)$. Then, of course, G/H is finite. In fact H is locally soluble. To see this suppose that X is a finite subgroup of H. Then $XM/M \cong X/X \cap M$ is soluble. Also $X \cap M \cap H = X \cap M$ and since $(H \cap M)' \leq H \cap N$ it follows that $(X \cap M)' \leq X \cap N$. Since $X \cap N$ is also soluble, this shows that X is soluble and hence H is locally soluble, as claimed. Consequently, G is almost locally soluble.

Hence, without loss of generality, we may assume $G = \mathbf{O}^\sigma(G)$, and hence that G is generated by its 2-elements. Let H be a proper normal subgroup of G and let W be a Wehrfritz 2-subgroup of G. Then WH/H is a Wehrfritz 2-subgroup of G/H, by Proposition 3.3.8. Since G/H contains elements of order 2, WH/H is nontrivial. Also by Proposition 3.3.8, $W \cap H$ is a Wehrfritz 2-subgroup of H. It follows that $\mathbf{s}(W \cap H) < \mathbf{s}(W)$ so that, in particular, every proper normal subgroup of G has smaller 2-size than G. From the choice of G we see that every proper normal subgroup of G is almost locally soluble.

Let R be the product of all the proper normal subgroups of G. Then every simple section of R is finite and by Kargapolov's theorem, Theorem 3.3.10, R is almost locally soluble. In particular, $G \neq R$ and hence G/R is an infinite simple group. Consequently, there is a simple counterexample to the theorem which we again denote by G.

Evidently the classification of finite simple groups implies that a simple group with Chernikov Sylow p-subgroups for even a single prime p is necessarily linear. For, Kegel [133] showed that an infinite locally finite simple group with Chernikov Sylow p-subgroups for some prime p is either linear over an infinite locally finite field of prime characteristic or there are infinitely many new sporadic finite simple groups. The classification of finite simple groups implies that this latter possibility cannot happen, so our counterexample G is linear over a field of characteristic q, say. Then the maximal q-subgroups of G must be finite and we can now apply a corollary to the Brauer–Feit theorem (see [258, Theorem 9.7]). In particular this theorem shows that G contains an abelian normal subgroup of finite index, a contradiction. $\square$

Belyaev [21] originally obtained this theorem in the early 1980's without using the classification of finite simple groups, as we have done here. Belyaev instead used the notion of a signalizer functor, a concept of some interest in finite group theory, as well as a result of Aschbacher on finite simple groups. He also used certain characterizations of $PSL_2(F)$, where F is an infinite locally finite field of odd characteristic. Before the classification of finite simple groups, such characterizations were of immense value when studying simple locally finite groups.

4.1.5. LEMMA. *Let G be a group. The subgroup generated by all the $\mathfrak{F}$-perfect subgroups is $\mathfrak{F}$-perfect. In particular, every group G contains a unique maximal, characteristic $\mathfrak{F}$-perfect subgroup.*

EXERCISE 4.2. *Prove Lemma 4.1.5.*

This unique largest $\mathfrak{F}$-perfect subgroup of a group G is called the $\mathfrak{F}$-*perfect part* of G. In Chapter 1, we already mentioned that every $\mathfrak{F}$-perfect abelian group is divisible.

The following strong structural result, due to M. I. Kargapolov [126], can now be obtained for locally soluble groups of finite section rank, since such groups do not contain infinite simple sections by Corollary 1.2.19.

4.1.6. THEOREM. *Let G be a locally finite group of finite section rank. Then G contains a normal divisible abelian subgroup R such that G/R is residually finite and the Sylow p-subgroups of G/R are finite for each prime p.*

PROOF. It follows from Theorem 4.1.4 that G is almost locally soluble. It therefore suffices to prove the result in the locally soluble case. Let p be a prime. A locally soluble group has no infinite simple section, by Corollary 1.2.19, so $G/\mathbf{O}_{p'}(G)$ is Chernikov, by Corollary 3.3.11. Then there exists a normal subgroup D_p of G such that G/D_p is finite and $D_p/\mathbf{O}_{p'}(G)$ is a divisible abelian p-group. Let $D = \bigcap\{\, D_p \mid p \in \Pi(G)\,\}$. By Remak's Theorem there is an embedding

$$G/D \longrightarrow \operatorname*{Cr}_{p\in\Pi(G)} G/D_p$$

which shows that G/D is residually finite.

Clearly $\bigcap\{\, \mathbf{O}_{p'}(G) \mid p \in \mathbb{P}\,\} = 1$. Again using Remak's Theorem we obtain an embedding

$$D \longrightarrow \operatorname*{Cr}_{p\in\Pi(G)} D_p/\mathbf{O}_{p'}(G).$$

The fact that $D_p/\mathbf{O}_{p'}(G)$ is abelian, for each $p \in \Pi(G)$, implies that D is abelian.

Suppose that R is the $\mathfrak{F}$-perfect part of G. Note that RD/D is also $\mathfrak{F}$-perfect. On the other hand, as we saw above, G/D is residually finite and so cannot contain $\mathfrak{F}$-perfect subgroups. Consequently, RD/D is trivial and hence $R \leq D$. Thus R is abelian and hence divisible (see [88, §20], for example). Let p be a prime and let W_p be a Wehrfritz p-subgroup of G, whose existence follows from 3.3.3. If V_p is the divisible part of W_p then V_p is $\mathfrak{F}$-perfect so that $V_p \leq R$. However, $R = \underset{p \in \Pi(G)}{\mathrm{Dr}}\ R_p$, where R_p is a direct product of finitely many Prüfer p-subgroups (see [88, Theorem 23.1]). Clearly R_p is G-invariant and hence every Sylow p-subgroup of G contains R_p. In particular, $R_p \leq V_p$ and hence $R_p = V_p$, for all primes $p \in \Pi(G)$. Since

$$W_p R/R \cong W_p/(W_p \cap R) = W_p/V_p$$

it follows that $W_p R/R$ is finite. By Proposition 3.3.8, $W_p R/R$ is a Wehrfritz p-subgroup of G/R and, using the definition of a Wehrfritz p-subgroup and Lemma 3.3.2, we see that every Sylow p-subgroup of G/R is finite. An application of Corollaries 3.3.12 and 3.3.13 completes the proof.   □

4.1.7. PROPOSITION. *Let G be a periodic radical group of finite section rank. Then G is countable.*

EXERCISE 4.3. *Prove Proposition 4.1.7.*

We note that the condition that G be radical in Proposition 4.1.7 is crucial, since the result is false otherwise. R. Baer [15, Folgerung 5.4] constructed an example of an uncountable periodic locally soluble group with finite Sylow p-subgroups for all primes p.

4.1.8. LEMMA. *Let G be a locally finite group with finite Sylow p-subgroups for all primes p. If $\Pi(G)$ is finite, then G is finite.*

EXERCISE 4.4. *Prove Lemma 4.1.8.*

We recall that a group is *bounded* or has *finite exponent* if there is a natural number b such that $G^b = 1$. The following pair of results are but two consequences of Lemma 4.1.8.

4.1.9. COROLLARY. *Let G be a locally finite group with finite Sylow p-subgroups for all primes p. If G is bounded, then G is finite.*

4.1.10. COROLLARY. *Let G be a locally finite group of finite section rank. If G is bounded, then G is finite.*

PROOF. It follows from Theorem 4.1.3 that the Sylow p-subgroups of G are Chernikov for all primes p. Since G is bounded, the Sylow p-subgroups of G are finite, for all primes p, and Corollary 4.1.9 now applies.   □

We recall that a finite group G is called *semisimple* if G contains no nontrivial normal abelian subgroups.

A normal subgroup H of a finite group G is called *completely reducible* if H is a direct product of simple groups. Every finite semisimple group G has a unique non-trivial maximal normal completely reducible subgroup. This subgroup is called the *completely reducible radical* of G (see [**172**, §61], for example).

The classification of finite simple groups implies that if S is a non-abelian finite simple group then the Schreier conjecture (see, for example, [**48**, p. 133]) is true, namely that the outer automorphism group of S is soluble. We shall use this fact in the proof of the next proposition.

4.1.11. PROPOSITION. *Let G be a finite group of section 2-rank d. Then G has normal subgroups $R \leq S \leq V \leq G$ such that R is the soluble radical of G, S/R is the direct product of at most d finite non-abelian simple groups, V/S is soluble and $|G/V| \leq d!$.*

PROOF. Throughout the proof we shall just consider the group G/R, so without loss of generality we may assume that R is trivial. Let S be the completely reducible radical of G, so that $S = S_1 \times S_2 \times \cdots \times S_m$, for certain non-abelian simple groups S_j, where $1 \leq j \leq m$. Since the subgroups S_j have even order, it follows that $m \leq d$. Let $X = \{S_1, \ldots, S_m\}$.

For each element $g \in G$ define the map $\sigma_g : S \longrightarrow S$ by $\sigma_g(x) = x^g$ for each $x \in S$. By a corollary of the Krull-Remak-Schmidt Theorem (see [**229**, 3.3.10], for example), $X = \{S_1, \ldots, S_m\} = \{S_1^g, \ldots, S_m^g\}$ for every element $g \in G$. Hence the map $\bar\sigma_g : X \longrightarrow X$ defined by $\bar\sigma_g(S_j) = S_j^g$ is a permutation of X. It is easy to see that, if $\mathrm{Sym}(X)$ denotes the symmetric group on X, then the mapping $\nu : G \longrightarrow \mathrm{Sym}(X)$ defined by $\nu(g) = \bar\sigma_g$ is a homomorphism. Hence $\mathbf{Im}(\nu) \leq \mathrm{Sym}(X)$ and

$$\mathbf{ker}(\nu) = \{g \in G : S_j^g = S_j, \text{ for each } j, 1 \leq j \leq m\} = \cap_{a \leq j \leq m} N_G(S_j).$$

Let $V = \mathbf{ker}(\nu)$ and let $V_j = S_j C_V(S_j) = S_j \times C_V(S_j)$. Certainly, V/V_j is embedded in $\mathbf{Aut}\,(S_j)/\mathbf{Inn}(S_j)$, a soluble group, by our remarks preceding the statement of the proposition, for each j. Hence $V/C_V(S_j)$ contains the normal subgroup $V_j/C_V(S_j) \cong S_j$ such that V/V_j is soluble, for each j with $1 \leq j \leq m$. Since G contains no nontrivial soluble normal subgroups, $\cap_{1\le j\le m}C_V(S_j) = C_V(S) = 1$. By Remak's Theorem, we obtain the embedding $f : V \longrightarrow \operatorname*{Dr}_{1\le j\le m} V/C_V(S_j)$. Since

$$f(S) = \operatorname*{Dr}_{1\le j\le m} SC_V(S_j)/C_V(S_j) = \operatorname*{Dr}_{1\le j\le m} S_jC_V(S_j)/C_V(S_j)$$
$$= \operatorname*{Dr}_{1\le j\le m} V_j/C_V(S_j),$$

V/S is isomorphic to a subgroup of $\operatorname*{Dr}_{1\le j\le m} V/V_j$, which is soluble. Finally G/V is isomorphic to a subgroup of $\operatorname{Sym}(X)$ so that $|G/V| \le m! \le d!$, as required. $\qquad\square$

We make the following observations from this.

4.1.12. COROLLARY. *Let G be a finite group of section 2-rank d. Then the number of non-abelian composition factors of G is at most $d + d!$.*

4.1.13. COROLLARY. *Let G be a group of finite section rank. Then G contains a characteristic subgroup L of finite index whose finite factor groups are soluble.*

PROOF. Let $\mathbf{sr}_2(G) = d$. If G contains no proper normal subgroups of finite index then of course we may take $L = G$. Therefore suppose that the family

$$\mathcal{R} = \{H \le G|H \text{ is a proper normal subgroup of finite index}\}$$

is non-empty. If F is an arbitrary finite group let $\mathbf{sc}(F)$ denote the number of non-abelian composition factors in a given composition series of F. Corollary 4.1.12 shows that $\mathbf{sc}(G/H) \le d + d!$ for every $H \in \mathcal{R}$. Hence there is subgroup $M \in \mathcal{R}$ such that $\mathbf{sc}(G/M)$ is maximal. Let

$$\mathcal{B} = \{H \in \mathcal{R}||G/H| \le |G/M|\} \text{ and}$$

let $L = \cap_{H\in\mathcal{B}}H$. Then L is a characteristic subgroup of G and G/L is locally finite by Proposition 2.5.4. Clearly G/L is bounded and Corollary 4.1.10 implies that G/L is finite. Since $M \in \mathcal{B}$ we have $\mathbf{sc}(G/L) = \mathbf{sc}(G/M)$. Let K be a normal subgroup of L such that L/K is finite and let $U = \mathbf{core}_G K$. It is easy to see that G/U is finite. By the choice of L we have $\mathbf{sc}(G/L) = \mathbf{sc}(G/U)$ and it follows that the composition factors of L/U are abelian. Hence L/U is soluble, as is L/K. $\qquad\square$

4.2. Structure of Locally Generalized Radical Groups with Finite Section Rank

Using the results of Chapter 3 and Section 4.1 we arrive at the following far reaching description of the structure of generalized radical groups with finite section rank.

4.2.1. THEOREM. *Let G be a locally generalized radical group of finite section rank. Then G has finite 0-rank at most $2t$, where $t = \min\{sr_p(G)|p \in \mathbb{P}\}$. Furthermore, G has normal subgroups*

$$D \leq T \leq L \leq K \leq S \leq G$$

satisfying the following conditions.

(i) *T is a periodic almost locally soluble group;*
(ii) *the Sylow p-subgroups of G are Chernikov for all primes p;*
(iii) *D is a divisible abelian subgroup;*
(iv) *the Sylow p-subgroups of T/D are finite for all primes p and T/D is residually finite;*
(v) *L/T is a torsion-free nilpotent group;*
(vi) *K/L is a finitely generated torsion-free abelian group;*
(vii) *G/K is finite and $|G/K| \leq f_5(t)$;*
(viii) *S/T is the soluble radical of G/T and $dl(S/T) \leq f_6(t)$.*

In particular, G is a generalized radical group.

This theorem is an immediate consequence of Theorems 3.4.3, 4.1.3, 4.1.4 and 4.1.6.

4.2.2. COROLLARY. *Let G be a radical group of finite section rank. Then G is countable.*

PROOF. Let $T = \mathbf{Tor}(G)$. By Theorem 4.2.1, G/T is countable. Proposition 4.1.7 implies that T is also countable and hence G is countable. $\square$

As we already mentioned in Chapter 1, the product of two normal locally soluble subgroups need not be locally soluble. However such groups cannot be locally finite, as the following result shows.

4.2.3. LEMMA. *Let G be a locally finite group. Then the subgroup generated by the normal locally soluble subgroups of G is also locally soluble. In particular, G contains a unique maximal characteristic locally soluble subgroup.*

EXERCISE 4.5. *Prove Lemma 4.2.3.*

By Lemma 4.2.3, a locally finite group G contains a greatest normal locally soluble subgroup which is called the *locally soluble radical* of G.

4.2.4. COROLLARY. *Let G be a locally generalized radical group of finite section rank. Then G is locally soluble-by-soluble-by-finite.*

PROOF. By Theorem 4.2.1, G has a series of normal subgroups

$$D \leq T \leq L \leq K \leq S \leq G.$$

By condition (i) of that theorem, T contains a normal locally soluble subgroup of finite index and so it follows that the locally soluble radical R of T has finite index in T. Since R is characteristic in T, it is normal in G. Let $C = C_G(T/R)$ and $B = S \cap C$. Since G/C and G/S are both finite it follows that G/B is also finite. Clearly B has a series of normal subgroups

$$D \leq R \leq T \cap C \leq L \cap C \leq K \cap C \leq B$$

such that R is locally soluble and B/R is soluble. This completes the proof. $\square$

Certainly locally (soluble-by-finite) groups are locally generalized radical, so the following result is easy to deduce.

4.2.5. COROLLARY. *Let G be a locally (soluble-by-finite) group of finite section rank. Then G is almost locally soluble.*

In Lemma 4.1.2 we showed that an abelian group whose elementary abelian sections are finite has finite section rank. Now we extend this statement to generalized radical groups. We will need the following generalization of Theorem 1.5.19.

4.2.6. PROPOSITION. *Let G be a group and suppose that $G/\zeta_k(G)$ is locally finite for some natural number k. Then $\gamma_{k+1}(G)$ is locally finite.*

EXERCISE 4.6. *Prove Proposition 4.2.6.*

We shall also require the following lemma.

4.2.7. LEMMA. *Let G be a group and suppose that $G/\zeta_k(G)$ is locally nilpotent for some natural number k. Then G is locally nilpotent.*

EXERCISE 4.7. *Prove Lemma 4.2.7.*

We extract part of the proof of Theorem 2.4.11 in the next result.

4.2.8. PROPOSITION. *Let G be a generalized radical group and let L be the Hirsch–Plotkin radical of G. Suppose that $\textbf{Tor}\,(G) = 1$ and that L has a finite series*

$$1 = L_0 \leq L_1 \leq L_2 \leq \cdots \leq L_n = L$$

of G-invariant subgroups such that the factors L_j/L_{j-1} are L-central, for $1 \leq j \leq n$. Then $\cap_{1 \leq j \leq n} C_G(L_j/L_{j-1}) = L$.

PROOF. Let $C = \cap_{j=1}^{n} C_G(L_j/L_{j-1})$. Since $L_j/L_{j-1} \leq \zeta(L/L_{j-1})$ we have $L \leq C_G(L_j/L_{j-1})$, for each j such that $1 \leq j \leq n$ and hence $L \leq C$.

Suppose, for a contradiction, that $C \neq L$. Since G is generalized radical either $\mathbf{Ln}(C/L) \neq 1$ or $\mathbf{Lf}(C/L) \neq 1$. If we suppose that $D/L = \mathbf{Ln}(C/L) \neq 1$, then $D \neq L$ and the choice of D shows that $L \leq \zeta_n(D)$. Then Lemma 4.2.7 shows that D is locally nilpotent, contrary to the choice of L. Hence $K/L = \mathbf{Lf}(C/L) \neq 1$. Again we have that $L \leq \zeta_n(K)$ and, by Proposition 4.2.6, $\gamma_{n+1}(K)$ is also locally finite. Since $\mathbf{Tor}\,(G) = 1$ it follows that $\gamma_{n+1}(K) = 1$. Thus K is nilpotent and hence $K \leq L$, again a contradiction. Thus $C = L$. □

4.2.9. LEMMA. *Let G be a generalized radical group and suppose that $\mathbf{Tor}\,(G) = 1$. If every elementary abelian section of G is finite, then G has finite 0-rank.*

PROOF. Let L be the Hirsch–Plotkin radical of G. Using Proposition 1.2.11 and the fact that $\mathbf{Tor}\,(G) = 1$, we deduce that L is torsion-free. Lemma 4.1.2 shows that every abelian subgroup of L has finite 0-rank. By Theorem 2.3.3, L is nilpotent of class at most $2k$ and has finite 0-rank at most $k(k+1)/2$, where $k = \mathbf{r}_0(A)$ and A is a maximal normal abelian subgroup of L. Employing the same arguments that were used in the proof of Theorem 2.4.11, we deduce that L has a finite series of G-invariant subgroups,

$$1 = A_0 \leq A_1 \leq \cdots \leq A_n = L,$$

every factor of which is torsion-free, central in L and G-rationally irreducible. In particular, $n \leq 2k$. Let $C_j = C_G(A_{j+1}/A_j)$, for $0 \leq j \leq n-1$. Then we can view $G_j = G/C_j$ as a soluble irreducible subgroup of $GL_m(\mathbb{Q})$ where $m = r_0(A_{j+1}/A_j)$. By Corollary 1.4.12, G_j is abelian-by-finite and indeed G_j contains a normal free-abelian subgroup U_j such that G_j/U_j is finite. By Proposition 4.2.8, $L = \cap_{j=0}^{n-1} C_j$, and Remak's Theorem implies that there is an embedding

$$G/L \hookrightarrow \mathrm{Dr}_{j=0}^{n-1} G/C_j.$$

Hence G/L contains a free abelian subgroup K/L such that G/K is finite. Let $t = \mathbf{r}_0(K/C)$. Then K is a polyrational subgroup of G and $\mathbf{r}_0(K) = k(k+1)/2 + t$. Since G/K is finite $\mathbf{r}_0(G) = \mathbf{r}_0(K)$, so that G has finite 0-rank. □

Uniting Theorem 4.1.3 and Lemma 4.2.9 we obtain

4.2.10. THEOREM. *Let G be a generalized radical group and suppose that every elementary abelian section of G is finite. Then there exists a function $\sigma : \mathbb{P} \longrightarrow \mathbb{N}_0$ such that G has finite section rank σ.*

We notice again that the hypotheses do not suggest the existence of a bound on the p-ranks of the elementary abelian sections, but the conclusion illustrates that such a bound exists. Finally, in this section we consider the influence of the locally soluble subgroups on the structure of certain groups of finite section rank.

4.2.11. THEOREM. *Let G be a locally (soluble-by-finite) group. If every locally soluble subgroup of G has finite section rank, then G has finite section rank. In particular, G is almost locally soluble.*

PROOF. Let F be a finitely generated subgroup of G. Then F is almost soluble and we let S denote the maximal normal soluble subgroup of F. By hypothesis, S has finite section rank and Theorem 4.2.1 shows that S has finite 0-rank. Since F/S is finite it follows that $\mathbf{r}_0(F)$ is finite also.

Suppose that the 0-ranks of the finitely generated subgroups of G are not bounded. Then there is a family $\{F_n | n \in \mathbb{N}\}$ of finitely generated subgroups such that

$$\mathbf{r}_0(F_1) < \mathbf{r}_0(F_2) < \cdots < \mathbf{r}_0(F_n) < \cdots .$$

For each $n \in \mathbb{N}$ let $K_n = \langle F_1, \ldots, F_n \rangle$. Then each subgroup K_n is finitely generated and each K_n contains a soluble normal subgroup R_n of finite index in K_n. Then R_n is also finitely generated and has finite section rank. Since $\mathbf{r}_0(R_n) = \mathbf{r}_0(K_n) \geq \mathbf{r}_0(F_n)$, for each $n \in \mathbb{N}$, we have

$$\mathbf{r}_0(R_1) < \mathbf{r}_0(R_2) < \cdots < \mathbf{r}_0(R_n) < \cdots .$$

Now $R_n \lhd K_n$, for all $n \in \mathbb{N}$, and hence $\langle R_1, R_2, \ldots, R_n \rangle = R_1 R_2 \ldots R_n$. It follows that $R_1 \ldots R_n$ has a finite series of normal subgroups whose factors are soluble and consequently $R_1 \ldots R_n$ is soluble, for all $n \in \mathbb{N}$. Hence the subgroup $E = \cup_{n \in \mathbb{N}} (R_1 \ldots R_n)$ is locally soluble and by hypothesis E has finite section rank. Theorem 4.2.1 shows that E has finite 0-rank and hence there is a natural number m such that $\mathbf{r}_0(R_m) > \mathbf{r}_0(E)$. On the other hand, R_m is a subgroup of E so Lemma 2.2.3 implies that $\mathbf{r}_0(R_m) \leq \mathbf{r}_0(E)$, which is a contradiction.

This contradiction shows that there is a natural number k such that $\mathbf{r}_0(F) \leq k$ for every finitely generated subgroup F of G. By Proposition 2.5.2, G then must have finite 0-rank and Theorem 2.4.13 shows that G contains normal subgroups $T \leq L \leq K \leq G$ such that T

is locally finite, L/T is torsion-free nilpotent, K/L is finitely generated, torsion-free abelian and G/K is finite. Since the Sylow p-subgroups of T are locally nilpotent, they have finite section p-rank for all primes p. By Corollary 3.2.21, each such Sylow p-subgroup is Chernikov and Theorem 4.1.3 implies that T has finite section rank. We have that $\mathbf{r}_0(G) = \mathbf{r}_0(K) = \mathbf{r}_0(K/T)$. By Corollary 3.2.24, the fact that $\mathbf{r}_0(L/T)$ is finite implies that L/T has finite section rank. Similarly, K/L has finite section rank. Lemma 3.2.2 implies that K/T, and therefore G/T, has finite section rank. Again using Lemma 3.2.2 we deduce that the entire group G has finite section rank. Finally, Corollary 4.2.5 proves that G is almost locally soluble. $\qquad\Box$

We end this section by giving the structure of locally nilpotent groups of finite section rank.

4.2.12. THEOREM. *Let G be a locally nilpotent group of finite section rank. Then $\mathbf{Tor}\,(G)$ is a direct product of hypercentral Chernikov p-groups, for each prime p, $G/\mathbf{Tor}\,(G)$ is torsion-free nilpotent and G is hypercentral.*

PROOF. By Corollary 3.2.21 the p-component of G is a hypercentral Chernikov group, so the claim concerning $\mathbf{Tor}\,(G)$ is immediate. Then $G/\mathbf{Tor}\,(G)$ is clearly a torsion-free locally nilpotent group and, by Theorem 3.2.26, G has finite 0-rank. Hence $G/\mathbf{Tor}\,(G)$ is nilpotent, by Theorem 2.3.3.

It remains to prove that G is hypercentral. To see this, let $x \in \mathbf{Tor}\,(G)$. Then there is a finite normal subgroup N of G such that $x \in N$ and $G/C_G(N)$ is finite. Let T be a transversal to $C_G(N)$ in G and note that T is finite. Clearly, $G = \langle T \rangle C_G(N)$. Furthermore, $H = \langle N, T \rangle$ is a finitely generated nilpotent group. Of course, $N \cap \zeta(H)$ centralizes T and it is also centralized by $C_G(N)$, so $N \cap \zeta(H) \leq \zeta(G)$. We use induction on n to prove that $N \cap \zeta_n(H) \leq \zeta_n(G)$, for all $n \in \mathbb{N}$; the argument above illustrates the case $n = 1$, so suppose that the result is true for $n - 1$. We note that

$$[N \cap \zeta_n(H), C_G(N)] = 1$$

and using the induction hypothesis we have

$$[N \cap \zeta_n(H), T] \leq N \cap \zeta_{n-1}(H) \leq \zeta_{n-1}(G).$$

Therefore, $[N \cap \zeta_n(H), G] \leq \zeta_{n-1}(G)$, from which it follows that $N \cap \zeta_n(H) \leq \zeta_n(G)$.

In particular, if c is the nilpotency class of H, then we have $N \leq \zeta_c(G)$. Hence $x \in \zeta_\omega(G)$ and since $x \in \mathbf{Tor}\,(G)$ was arbitrary we deduce that $\mathbf{Tor}\,(G) \leq \zeta_\omega(G)$. Hence $G/\zeta_\infty(G)$ is nilpotent, so that G is hypercentral, as required. $\square$

4.3. Connections Between the Order of a Finite Group and Its Section Rank

In Corollary 4.1.10 we showed that a bounded locally finite group having finite section rank is finite. In this section, we strengthen this result considerably by finding a bound for the order of such a group.

Since the order of a finite group is the product of the orders of its Sylow subgroups, the main work here is concerned with the determination of a bound for the order of a finite p-group P with the property that $\mathbf{sr}_p(P) = r$ and $P^b = 1$, where $b = p^d$, for some $d \in \mathbb{N}$. In P we choose a maximal normal abelian subgroup A. Since P is nilpotent, $A = C_P(A)$ and hence $P/A = P/C_P(A)$ is isomorphic to some p-subgroup of $\mathbf{Aut}\,(A)$. Now $A = \underset{1 \leq j \leq n}{\mathrm{Dr}} \langle a_j \rangle$, for certain elements a_j, where $n \leq r$. Hence $\mathbf{Aut}\,(A)$ can be embedded in $GL_n(\mathbb{Z}/p^d\mathbb{Z})$. Consequently, in order to proceed, we need some additional facts concerning the Sylow p-subgroups of linear groups over the ring $\mathbb{Z}/p^d\mathbb{Z}$.

4.3.1. LEMMA. *Let p be a prime and let V be a vector space over the prime field $\mathbb{F}_p$. Let*

$$|\mathcal{L}_k| = \{S | S \text{ is a linearly independent subset of } V \text{ such that } |S| = k\}.$$

Then $\mathcal{L}_k = (p^n - 1)(p^n - p)\ldots(p^n - p^{k-1})$, where $n = \mathbf{dim}_{\mathbb{F}_p}(V)$.

EXERCISE 4.8. *Prove Lemma 4.3.1.*

4.3.2. COROLLARY. *Let p be a prime and let V be a vector space over the prime field $\mathbb{F}_p$. Then the number of distinct bases of V is*

$$|\mathcal{L}_n| = (p^n - 1)(p^n - p)\ldots(p^n - p^{n-1}),$$

where $n = \dim_{\mathbb{F}_p}(V)$.

4.3.3. COROLLARY. *Let p be a prime. Then*

$$|GL_n(\mathbb{Z}/p\mathbb{Z})| = (p^n - 1)(p^n - p)\ldots(p^n - p^{n-1})$$
$$= p^{n(n-1)/2}(p^n - 1)(p^{n-1} - 1)\ldots(p - 1).$$

EXERCISE 4.9. *Prove Corollary 4.3.3.*

4.3.4. PROPOSITION. *Let p be a prime. Then, for all natural numbers k,*

$$|GL_n(\mathbb{Z}/p^k\mathbb{Z})| = (p^{nk} - p^{nk-n})(p^{nk} - p^{nk-n+1})\ldots(p^{nk} - p^{nk-1})$$
$$= p^{n(2nk-n-1)/2}(p^n - 1)(p^{n-1} - 1)\ldots(p - 1).$$

PROOF. We use additive notation for the group operation in A. Let $A = C_1 \oplus \cdots \oplus C_n$, where $C_j = \langle c_j \rangle \cong \mathbb{Z}/p^k\mathbb{Z}$, for $1 \leq j \leq n$. Also let $A_m = \Omega_m(A)$, for $1 \leq m \leq k$ and note that $G = \mathbf{Aut}\,(A) \cong GL_n(\mathbb{Z}/p^k\mathbb{Z})$.

We use induction on k and note that, for the case $k = 1$, the result follows from Corollary 4.3.3. Suppose that $k > 1$ and that we have already proved that

$$|GL_n(\mathbb{Z}/p^{k-1}\mathbb{Z})| = p^{n(2n(k-1)-n-1)/2}(p^n - 1)(p^{n-1} - 1)\ldots(p - 1).$$

Let $D = \Omega_{k-1}(A)$ and let $L = C_G(D)$ so that, by the induction hypothesis,

$$|G/L| = p^{n(2n(k-1)-n-1)/2}(p^n - 1)(p^{n-1} - 1)\ldots(p - 1).$$

Let $g \in L$ and if $a \in A$, then let $g(a) = b$. We have $pb = pg(a) = g(pa) = pa$ and hence $p(b-a) = 0$, so that $b-a \in A_1$. Thus $g(a) = a+d$ for some element $d \in A_1$. Let ε denote the identity automorphism of A. Then our work above proves that $g - \varepsilon$ is an endomorphism of A such that $\mathbf{Im}(g - \varepsilon) = A_1$ and $\mathbf{ker}(g - \varepsilon) = A_{k-1}$. Hence $g - \varepsilon$ induces an endomorphism $\bar{g} : A/A_{k-1} \longrightarrow A_1$.

We show that the mapping $g \longmapsto \bar{g}$ is a bijection. Indeed, if $g, h \in L$ and $g \neq h$ then clearly $g - \varepsilon \neq h - \varepsilon$ and hence $\bar{g} \neq \bar{h}$. To show that this mapping is surjective, let $\phi : A/A_{k-1} \longrightarrow A_1$ be an arbitrary homomorphism. Define $f : A \longrightarrow A$ by $f(a) = a + \phi(a + A_{k-1})$. Observe that $f(a) \in a + A_1$. We have, for $a, b \in A$,

$$\begin{aligned}
f(a + b) &= (a + b) + \phi((a + b) + A_{k-1}) \\
&= a + b + \phi((a + A_{k-1}) + (b + A_{k-1})) \\
&= a + b + \phi(a + A_{k-1}) + \phi(b + A_{k-1}) \\
&= a + \phi(a + A_{k-1}) + b + \phi(b + A_{k-1}) \\
&= f(a) + f(b),
\end{aligned}$$

so that f is an endomorphism of A. Furthermore, if $a \in A_{k-1}$, then $a + A_{k-1} = A_{k-1}$ and hence $f(a) = a$. Suppose that $a \in A$ and that $f(a) = 0$. Then $a + \phi(a + A_{k-1}) = 0$ so $\phi(a + A_{k-1}) = -a \in A_1 \leq A_{k-1}$. Hence $\phi(a + A_{k-1}) = 0$ and $a = 0$. Thus f is a monomorphism.

Let c be an arbitrary element of A and let $v = \phi(c + A_{k-1})$. We have

$$f(c - v) = c - v + \phi(c - v + A_{k-1}) = c - \phi(c + A_{k-1}) + \phi(c + A_{k-1}) = c$$

and it follows that f is also an epimorphism. Hence $f \in \mathbf{Aut}\,(A)$. We saw above that $f(a) = a$ whenever $a \in A_{k-1}$ so that $f \in L$.

Finally, we see from the definition of f that $\bar{f} = \phi$, which completes the proof that the map $g \longmapsto \bar{g}$ is a bijection. From this it follows that $|L| = |\mathbf{Hom}(A/A_{k-1}, A_1)|$. We note that A/A_{k-1} and A_1 are direct sums of n copies of $\mathbb{Z}/p\mathbb{Z}$. Using [**88**, Theorem 43.1], for example, we observe that

$$\mathbf{Hom}(A/A_{k-1}, A_1) \cong \underbrace{\mathbf{Hom}(\mathbb{Z}/p\mathbb{Z}, A_1) \oplus \cdots \oplus \mathbf{Hom}(\mathbb{Z}/p\mathbb{Z}, A_1)}_{n}.$$

Applying [**88**, Example 2 of Chapter 43] gives $\mathbf{Hom}(\mathbb{Z}/p\mathbb{Z}, A_1) \cong A_1$, so that

$$|\mathbf{Hom}(A/A_{k-1}, A_1)| = |\mathbf{Hom}(\mathbb{Z}/p\mathbb{Z}, A_1)|^n = |A_1|^n = (p^n)^n.$$

Finally,

$$|GL_n(\mathbb{Z}/p^k\mathbb{Z})| = |G| = |G/L| \cdot |L| = p^r(p^n - 1)(p^{n-1} - 1)\ldots(p - 1),$$

where $r = n(2n(k-1)-n-1)/2+n^2 = n(2nk-n-1)/2$, as required. $\square$

Let $P_n(\mathbb{Z}/p^k\mathbb{Z})$ denote the set of matrices

$$\{A = [a_{jk}] \in GL_n(\mathbb{Z}/p^k\mathbb{Z}) | a_{jj} \equiv 1 \pmod{p}, a_{jk} \in p\mathbb{Z}, \text{ if } j > k\}.$$

It is easy to see that $|P_n(\mathbb{Z}/p^k\mathbb{Z})| = p^{n(2nk-n-1)/2}$ and using Proposition 4.3.4 we deduce the following.

4.3.5. COROLLARY. *Let p be a prime. Then $P_n(\mathbb{Z}/p^k\mathbb{Z})$ is a Sylow p-subgroup of $GL_n(\mathbb{Z}/p^k\mathbb{Z})$.*

In particular, if $k = 1$ we see that the subgroup $UT_n(\mathbb{F}_p)$ is a Sylow p-subgroup of $GL_n(\mathbb{F}_p)$.

Using these results, we are now able to obtain the promised bounds.

4.3.6. THEOREM. *Let p be a prime and let P be a finite p-group. Suppose that $\boldsymbol{sr}_p(P) = r$ and $P^b = 1$, where $b = p^d$, for some natural number d. Then $|P| \leq p^{d(r+1)r-r(r+1)/2} = p^{r(r+1)(2d-1)/2}$.*

PROOF. Let A be a maximal normal abelian subgroup of P. Since P is nilpotent we have $A = C_P(A)$, so that $P/A = P/C_P(A)$ is isomorphic to some p-subgroup of $\mathbf{Aut}\,(A)$. We know that $A = \mathrm{Dr}_{1 \leq j \leq n}\langle a_j \rangle$, where $n \leq r$, so $\mathbf{Aut}\,(A)$ can be embedded in $GL_n(\mathbb{Z}/p^d\mathbb{Z})$. By Proposition 4.3.4, the order of the Sylow p-subgroups of $GL_n(\mathbb{Z}/p^d\mathbb{Z})$ is $p^{n(2dn-n-1)/2}$. Since $|A| \leq p^{nd}$ we have

$$|P| = |A| \cdot |P/A| \leq p^{nd}p^{(2nd-n-1)n/2} = p^{(2nd-n-1+2d)n/2} = p^{d(n+1)n-(n+1)n/2}$$

$$\leq p^{d(r+1)r-r(r+1)/2},$$

since $n \leq r$, which is the required bound. $\square$

There are many interesting consequences and special cases of this result which are interesting in their own right and we give some of these here.

4.3.7. COROLLARY. *Let G be a finite group and let b be the least natural number such that $G^b = 1$. Let $m = \max\{sr_p(G)|p \in \Pi(G)\}$. Then*

$$|G| \leq b^{(m+1)m}/s^{(m+1)m/2}$$

where s is the product of all the primes in $\Pi(G)$. In particular, $|G| \leq b^{(m+1)m}$.

PROOF. Let $\Pi(G) = \{p_1, \ldots, p_n\}$ and let P_j be a Sylow p_j-subgroup of G, for $1 \leq j \leq n$. Then $|G| = |P_1||P_2|\ldots|P_n|$. Let $b = p_1^{d_1} \ldots p_n^{d_n}$. By Theorem 4.3.6 it follows that $|P_j| \leq p_j^{d_j(m+1)m-(m+1)m/2}$, so that

$$|G| \leq p_1^{d_1(m+1)m-(m+1)m/2} \ldots p_n^{d_n(m+1)m-(m+1)m/2}$$
$$= (p_1^{d_1} \ldots p_n^{d_n})^{(m+1)m}/(p_1 \ldots p_n)^{(m+1)m/2}$$
$$= b^{m(m+1)}/(p_1 \ldots p_n)^{(m+1)m/2} = b^{(m+1)m}/s^{(m+1)m/2}.$$

Since $s \geq 1$ we clearly obtain $|G| \leq b^{(m+1)m}$, as required. □

4.3.8. COROLLARY. *Let G be a bounded locally finite group and let b be the least natural number such that $G^b = 1$. If G has finite section rank, then G is finite and $|G| \leq b^{(m+1)m}$, where $m = \max\{sr_p(G)|p \in \Pi(G)\}$.*

This follows immediately since, by Corollary 4.1.10, G is finite and so we can apply Corollary 4.3.7.

We say that a group G is *locally graded* if every finitely generated subgroup of G has a proper subgroup of finite index. The class of locally graded groups is very broad, containing all soluble groups, all almost soluble groups, all locally soluble groups, all locally (almost soluble) groups, all locally residually finite groups, all locally generalized radical groups and so on.

4.3.9. COROLLARY. *Let G be a bounded locally graded group and let b be the least natural number such that $G^b = 1$. If G has finite section rank, then G is finite and $|G| \leq b^{(m+1)m}$, where $m = \max\{sr_p(G)|p \in \Pi(G)\}$.*

EXERCISE 4.10. *Prove Corollary 4.3.9.*

Since residually finite groups are locally graded, we note the following result.

4.3.10. COROLLARY. *Let G be a bounded residually finite group and let b be the least natural number such that $G^b = 1$. If G has finite section rank, then G is finite and $|G| \leq b^{(m+1)m}$, where $m = \max\{sr_p(G)|p \in \Pi(G)\}$.*

4.3.11. COROLLARY. *Let G be a group of finite section rank and let*

$$\mathcal{B}(d) = \{H \leq G | H \text{ has finite index and } |G : H| \leq d\}.$$

Let B denote the intersection of all subgroups from the family $\mathcal{B}(d)$. Then G/B is finite of order at most $(d!)^{(m+1)m}$, where

$$m = \max\{sr_p(G)|p \text{ is a divisor of } d!\}.$$

PROOF. Clearly B is a characteristic subgroup of G. If $H \in \mathcal{B}(d)$, then $G/\text{core}_G H$ is isomorphic to some subgroup of the symmetric group S_d, from which it follows that $|G/\text{core}_G H|$ divides $d!$. By Remak's Theorem we obtain the embedding $G/B \longrightarrow \underset{H \in \mathcal{B}(m)}{\text{Cr}} G/\text{core}_G H$, which shows that G/B is residually finite. Furthermore, $(G/B)^{d!} = 1$ and we now apply Corollary 4.3.10. □

The bounds obtained above can be improved, which was done in the work of A. Lubotzky and A. Mann [179]. In this paper the important concept of a *powerful p-group* was introduced. Additionally, it was shown how powerful p-subgroups influence the structure of finite p-groups. Here if p is an odd prime, then a finite p-group G is called *powerful* if $G' \leq G^p$ and a finite 2-group is called powerful if $G' \leq G^4$.

Here we exhibit some of the main results of the article [179] and give a bound for the order of a p-group P such that $\mathbf{sr}_p(P) = r$ and $P^b = 1$, where $b = p^d$, for some natural number d. We use the notation $\iota(\alpha)$ to denote the least integer that is at least the real number α.

4.3.12. THEOREM (Lubotzky and Mann [179]). *Let p be a prime and let G be a finite p-group. Suppose that $\mathbf{sr}_p(G) = s$. Then G contains a powerful characteristic subgroup H such that $|G/H| \leq p^m$, where*

$$m = s \cdot \iota(\log_2 s), \text{ when } p \neq 2 \text{ and } m = s(\iota(\log_2 s) + 1) \text{ when } p = 2.$$

As usual, if G is a finitely generated group then let $\mathbf{d}(G)$ denote the minimal number of generators of G.

4.3.13. THEOREM (Lubotzky and Mann [179]). *Let p be a prime and let G be a finite p-group. Suppose that $\mathbf{d}(G) = d$. If G is powerful, then G is a product of d cyclic subgroups.*

4.3.14. THEOREM (Lubotzky and Mann [**179**]). *Let p be a prime and let G be a finite p-group. Suppose that $\mathbf{d}(G) = d$. If G is powerful and H is a subgroup of G, then $\mathbf{d}(H) \leq \mathbf{d}(G)$.*

These very important results can now be used to obtain a relationship between the order of a finite p-group, its section p-rank and exponent as we show in the next corollary.

4.3.15. COROLLARY. *Let p be a prime and let P be a finite p-group. Suppose that $\boldsymbol{sr}_p(P) = r$ and $P^b = 1$, where $b = p^d$, for some natural number d. Then $|P| \leq p^{r(d+s)}$, where $s = \iota(\log_2 r) + 1$.*

PROOF. By Theorem 4.3.12, P contains a powerful subgroup V such that $|P/V| \leq p^m$, where $m = r(\iota(\log_2 r)+1)$. Let $s = \iota(\log_2 r)+1$. The section $V/\mathbf{Frat}(V)$ is elementary abelian so that $|V/\mathbf{Frat}(V)| = p^t \leq p^r$. However, the number of generators of V coincides with the number of generators of $V/\mathbf{Frat}(V)$ and it follows that $\mathbf{d}(V) = t \leq r$. By Theorem 4.3.13, V is a product of d cyclic subgroups. Each cyclic subgroup occurring has order at most p^d, so that $|V| \leq p^{dt}$. Hence

$$|P| = |V| \cdot |P/V| \leq p^{dt}p^m \leq p^{dr}p^{rs} \leq p^{rd+rs} = p^{r(d+s)},$$

as required. $\qquad\qquad\square$

4.4. Groups of Finite Bounded Section Rank

Let G be a group of finite section rank σ. In this final section of Chapter 4 we consider an important special case, the case in which σ is bounded.

4.4.1. DEFINITION. A group G has *bounded section rank* $\mathbf{bs}(G) = b$, for some natural number b, if $\mathbf{sr}_p(G) \leq b$, for all primes p and there exists a prime q such that $\mathbf{sr}_q(G) = b$. Thus

$$\mathbf{bs}(G) = \max\{\mathbf{sr}_p(G) | p \in \mathbb{P}\},$$

assuming that such a maximum exists. If there is no such natural number b we say that the group does not have bounded section rank.

For groups of bounded section rank it is possible to obtain more detail concerning their structure. But first we note some elementary properties of bounded section rank.

4.4.2. LEMMA. *Let G be a group and let H be a subgroup of G. Let L be a normal subgroup of G.*

(i) *If G has bounded section rank, then H has bounded section rank and $\boldsymbol{bs}(H) \leq \boldsymbol{bs}(G)$;*

(ii) *If G has bounded section rank, then G/L has bounded section rank and $bs(G/L) \leq bs(G)$;*

(iii) *If L and G/L have bounded section rank, then G has bounded section rank and $bs(G) \leq bs(L) + bs(G/L)$.*

EXERCISE 4.11. *Prove Lemma 4.4.2.*

4.4.3. THEOREM. *Let G be a locally radical group of finite bounded section rank b. Then $Ln(G)$ is hypercentral and $G/Ln(G)$ contains a normal abelian subgroup $A/Ln(G)$ such that G/A is finite. Furthermore, there is an integer valued function f_7 such that $|G/A| \leq f_7(r)$. In particular, G is hyperabelian.*

PROOF. We have that $sr_p(G) \leq b$ for all primes p and we let $T = \mathbf{Tor}\,(G)$. Since T is periodic and G is locally radical, T is locally finite and therefore locally soluble. Let R be the divisible part of T and note that R is abelian, by Theorem 4.1.6. Hence $R = \underset{p \in \Pi(G)}{\mathrm{Dr}}\, R_p$, where R_p is a direct product of finitely many Prüfer p-groups. Thus R has an ascending series of G-invariant subgroups whose factors are G-irreducible. By Theorem 4.1.6, T/R is residually finite and its Sylow p-subgroups are finite, for all primes p. It follows that T/R has a family of G-invariant subgroups of finite index whose intersection is trivial and hence T/R has a descending series of G-invariant subgroups whose factors are G-irreducible. By Theorem 4.2.1, G has a series of normal subgroups $T \leq L \leq K \leq G$ such that L/T is torsion-free nilpotent, K/L is finitely generated torsion-free abelian and G/K is finite soluble. Using the arguments of Lemma 2.4.10, we construct a finite series

$$T = A_0 \leq A_1 \leq \cdots \leq A_n = K$$

of G-invariant subgroups whose factors are torsion-free abelian and rationally G-irreducible, for $0 \leq j \leq n-1$. Finally G/K is soluble and therefore it has a finite series of G-invariant subgroups whose factors are G-irreducible. We unite all these four series into one series $\mathcal{S}$ of G and let V/U be one factor of this series.

Suppose first that $V \leq T$. Since T is locally soluble, V/U is an elementary abelian p-group for some prime p. Then $sr_p(V/U) \leq b$ and we may think of $G/C_G(V/U)$ as a subgroup of $GL_b(\mathbb{F}_p)$. Since V/U is finite, $G/C_G(V/U)$ is a finite soluble group. By Theorem 1.4.6, G contains a normal subgroup $H \geq C_G(V/U)$ such that $H/C_G(V/U)$ is abelian and G/H is finite of order at most $\mu(b)$ where $\mu(n)$ is the Maltsev function defined in Section 1.4.

Next, let V/U be a factor of $\mathcal{S}$ such that $T \leq U \leq V \leq K$. Since G/T is soluble, $G/C_G(V/U)$ is also soluble. By Lemma 3.2.4,

$r_0(V/U) \leq sr_p(V/U)$ for every prime p and hence $r_0(V/U) \leq sr_p(V/U) \leq b$. Therefore we can think of $G/C_G(V/U)$ as an irreducible soluble subgroup of $GL_b(\mathbb{Q})$. Again by Theorem 1.4.6, we see that G contains a normal subgroup $H \geq C_G(V/U)$ such that $H/C_G(V/U)$ is abelian and G/H is finite of order at most $\mu(b)$.

Finally, if $K \leq U \leq V \leq G$ then V/U is finite and using the arguments above we see that G again contains a normal subgroup $H \geq C_G(V/U)$ such that $H/C_G(V/U)$ is abelian and G/H is finite of order at most $\mu(b)$.

Let L be the intersection of all the subgroups $C_G(V/U)$ for all factors of the series $\mathcal{S}$ and let $\mathcal{L} = \{W \cap L | W \in \mathcal{S}\}$. If V/U is a factor of $\mathcal{S}$ then $C_G(V/U) \geq L$ and, in particular, $L = C_L(V \cap L/U \cap L)$. It follows that L has a central series. Since $L \cap T$ is locally finite it is locally nilpotent and hence $L \cap T = \underset{p \in \pi(L \cap T)}{Dr} S_p$, where S_p is a Sylow p-subgroup of $L \cap T$, for $p \in \Pi(L \cap T)$. Since $sr_p(S_p) \leq b$, S_p is a Chernikov group by Corollary 3.2.21. The subgroup $L \cap R$ has an ascending series of G-invariant subgroups whose factors are L-central. Since it is abelian and its Sylow p-subgroups are Chernikov, $L \cap R \leq \zeta_\omega(L)$.

The factor group $(L \cap T)/(L \cap R)$ has finite Sylow p-subgroups for all primes p. Let $P/(L \cap R)$ be the Sylow p-subgroup of $(L \cap T)/(L \cap R)$. As we saw above, T/R has a family of G-invariant subgroups of finite index whose intersection is trivial. It follows that T contains a normal subgroup $X \geq R$ such that $PR/R \cap X/R$ is trivial. The subgroup L acts trivially on the factors of the finite group T/X so there is a positive integer k such that $[T,_k L] \leq X$ and, in particular, $[P,_k L] \leq X$. Since $L \cap T$ is locally nilpotent, P is normal. Hence $[P,_k L] \leq X \cap P \leq R$ and since $P \leq L \cap T$ we have $[P,_k L] \leq R \cap (L \cap T) \leq L \cap R$. This shows that every Sylow p-subgroup of $(L \cap T)/(L \cap R)$ lies in some term $\zeta_k(L/L \cap R)$ of the upper central series of $L/(L \cap R)$, for some natural number k. Hence $L \cap T \leq \zeta_{2\omega}(L)$. By construction, $L/(L \cap T)$ has a finite series whose factors are L-central and hence $L/(L \cap T)$ is nilpotent. Consequently L is hypercentral.

By Remak's Theorem, we obtain an embedding

$$G/L \longrightarrow \mathrm{Cr}_{V/U \in \mathcal{S}} G/C_G(V/U).$$

As we saw above, for each factor $V/U \in \mathcal{S}$, G contains a normal subgroup $H = H_{V/U} \geq C_G(V/U)$ such that $H_{V/U}/C_G(V/U)$ is abelian and $G/H_{V/U}$ is finite of order at most $\mu(b)$. Let D be the intersection of all such $H_{V/U}$ for all factors V/U of $\mathcal{S}$. Then

$$D/L \longrightarrow \mathrm{Cr}_{V/U \in \mathcal{S}} H_{V/U}/C_G(V/U)$$

and
$$G/D \longrightarrow \mathrm{Cr}_{V/U \in \mathcal{S}} G/H_{V/U}.$$
This shows that D/L is abelian and G/D is a subgroup of a Cartesian product of finite groups whose order is at most $\mu(b)$. By Proposition 2.5.4, G/D is locally finite and, as it is a bounded group of finite section rank, G/D is finite by Corollary 4.1.10. By Corollary 4.3.11, G/D is finite and $|G/D| \leq (\mu(b)!)^{b(b+1)} = f_7(b)$.

Let $Z = \mathbf{Ln}(G)$. Clearly $L \leq Z$ so G/Z is abelian-by-finite. Furthermore, $Z \cap T = \underset{p \in \Pi(D)}{\mathrm{Dr}} Z_p$, where Z_p is the Sylow p-subgroup of $Z \cap T$, for $p \in \Pi(Z \cap T)$. Since $\mathbf{sr}_p(Z_p) \leq b$, Corollary 3.2.21 implies that Z_p is Chernikov. Then $Z \cap T$ has an ascending Z-chief series of length at most 2ω. Since Z is locally nilpotent, Proposition 1.2.20 shows that each factor of this series is Z-central. This means that $Z \cap T \leq \zeta_{2\omega}(Z)$. The factor group $Z/(Z \cap T)$ is torsion-free, locally nilpotent, by Proposition 1.2.11. By Theorem 4.2.1, G has finite 0-rank and Lemma 2.2.3 shows that $\mathbf{r}_0(Z/(Z \cap T))$ is finite. By Corollary 2.3.4, $Z/(Z \cap T)$ is nilpotent. This shows that Z is hypercentral. □

4.4.4. COROLLARY. *Let G be a locally generalized radical group of finite bounded section rank b. Then G contains a normal hypercentral subgroup L such that G/L is abelian-by-finite. In particular, G is hyperabelian-by-finite.*

PROOF. Consider the series of normal subgroups
$$D \leq T \leq L \leq K \leq S \leq G$$
guaranteed by Theorem 4.2.1. By part (i) of that theorem, T contains a normal locally soluble subgroup of finite index. Let R be the locally soluble radical of T, a characteristic subgroup of finite index in T, so R is normal in G. Let $C = C_G(T/R)$ and $B = S \cap C$. Since G/C and G/S are finite, G/B is also finite and there is a series of normal subgroups
$$D \leq R \leq T \cap C \leq L \cap C \leq K \cap C \leq B$$
such that B/R is soluble, R/D is locally soluble with finite Sylow p-subgroups for all primes p and D is abelian.

Let F be an arbitrary finitely generated subgroup of B. Since F is generalized radical it has an ascending series of normal subgroups
$$1 = F_0 \leq F_1 \leq \ldots F_\alpha \leq F_{\alpha+1} \leq \ldots F_\gamma = F$$
whose factors are locally nilpotent or locally finite. Let $F_{\alpha+1}/F_\alpha$ be a locally finite factor of this series and consider the refinement
$$F_\alpha \leq (F_{\alpha+1} \cap R)F_\alpha \leq (F_{\alpha+1} \cap L)F_\alpha \leq (F_{\alpha+1} \cap K)F_\alpha \leq F_{\alpha+1}.$$

By Theorem 4.4.3, each factor of this refinement is hyperabelian and hence $F_{\alpha+1}/F_\alpha$ is radical. It follows that F is radical and therefore B is locally radical. We may now apply Theorem 4.4.3 to deduce the result. $\qquad\square$

4.4.5. COROLLARY. *Let G be a locally generalized radical group of finite bounded section rank. Then G is countable.*

EXERCISE 4.12. *Prove Corollary 4.4.5.*

To conclude this section, we see the impact of the locally radical subgroups on the structure of locally generalized radical groups of finite bounded section rank.

4.4.6. THEOREM. *Let G be a locally generalized radical group. If every locally radical subgroup of G has finite bounded section rank, then G has finite bounded section rank.*

PROOF. Let F be a finitely generated subgroup of G, so F is a generalized radical group. Let R be the maximal normal radical subgroup of F and let L/R denote the locally finite radical of F/R. Let p be a prime and let P/R be a Sylow p-subgroup of L/R. If U is an arbitrary finitely generated subgroup of P then U is an extension of the radical subgroup $U \cap R$ by the finite p-group $U/(U \cap R) \cong UR/R \leq P/R$. In particular, U is radical and hence P is a locally radical subgroup of G. Consequently, P has finite bounded section rank and in particular $\mathrm{sr}_p(P/R)$ is finite. By Corollary 3.2.21, P/R is Chernikov. Theorem 4.1.4 shows that L/R contains a normal locally soluble subgroup S/R of finite index. If V is an arbitrary finitely generated subgroup of S then V is an extension of the radical subgroup $V \cap R$ by the finite soluble group $V/(V \cap R) \cong VR/R \leq S/R$. Thus V is radical and hence S is a locally radical subgroup of G, so has finite bounded section rank. Theorem 4.4.3 implies that S is hyperabelian and hence $S \leq R$. In turn it follows that L/R is finite. By Lemma 3.4.1, F/R is finite. Our hypotheses imply that R has finite bounded section rank and Theorem 4.2.1 implies that R has finite 0-rank. Consequently, $\mathrm{r}_0(F)$ is finite.

Suppose that the 0-ranks of the finitely generated subgroups of G are unbounded. Then there is a family $\{F_n | n \in \mathbb{N}\}$ of finitely generated subgroups of G such that

$$\mathrm{r}_0(F_1) < \mathrm{r}_0(F_2) < \cdots < \mathrm{r}_0(F_n) < \ldots.$$

Let $K_1 = F_1, K_2 = \langle F_1, F_2 \rangle, \ldots, K_n = \langle F_1, \ldots, F_n \rangle$, for each $n \in \mathbb{N}$. As above, since K_n is finitely generated it contains a normal radical subgroup R_n of finite index. Then R_n is also finitely generated and it has finite bounded section rank. Since $\mathbf{r}_0(R_n) = \mathbf{r}_0(K_n) \geq \mathbf{r}_0(F_n)$ for $n \in \mathbb{N}$ we have

$$\mathbf{r}_0(R_1) < \mathbf{r}_0(R_2) < \cdots < \mathbf{r}_0(R_n) < \ldots.$$

The subgroup R_n is normal in K_n for all $n \in \mathbb{N}$ and hence $\langle R_1, \ldots, R_n \rangle = R_1 R_2 \ldots R_n$. It follows that $R_1 \ldots R_n$ has a finite series of normal subgroups whose factors are radical and it follows that $R_1 \ldots R_n$ is radical for all $n \in \mathbb{N}$. Hence $E = \bigcup_{n \in \mathbb{N}}(R_1 \ldots R_n)$ is locally radical, so E has finite bounded section rank. Theorem 4.1.4 implies that E has finite 0-rank. Consequently, there is an integer $m \in \mathbb{N}$ such that $\mathbf{r}_0(R_m) > \mathbf{r}_0(E)$. On the other hand, R_m is a subgroup of E and Lemma 2.2.3 implies that $\mathbf{r}_0(R_m) \leq \mathbf{r}_0(E)$, a contradiction which shows that there is a positive integer k such that $\mathbf{r}_0(F) \leq k$ for all finitely generated subgroups F of G. By Proposition 2.5.2 it follows that G has finite 0-rank.

Theorem 2.4.13 shows that G has normal subgroups $T \leq L \leq K \leq G$ such that T is locally finite, L/T is torsion-free nilpotent, K/L is finitely generated torsion-free abelian and G/K is finite. As above, we see that for each prime p the Sylow p-subgroups of T are Chernikov. Theorem 4.1.4 implies that T contains a normal locally soluble subgroup C of finite index. It follows that C has finite bounded section rank and, by Theorem 4.4.3, C is hyperabelian. Using the arguments of Corollary 4.4.4, we see that G is almost hyperabelian and hence G has finite bounded section rank, as required.   □

CHAPTER 5

Zaitsev Rank

In this chapter, we shall consider a different numerical invariant which makes sense only for infinite groups. The concept we discuss comes from a seemingly unrelated part of infinite group theory and there appears to be no relationship with the concept of dimension. However, this concept also gives us some "measure of largeness" and, as we shall see later, this rank is connected very closely with the 0-rank and the special rank. The rank which we discuss now is connected with both the minimal and maximal condition and unites them in some sense. It was introduced by D. I. Zaitsev and first appeared in the study of groups with the weak minimal condition in the paper [**263**]. There Zaitsev used the term "index of minimality", but since this term was not really precise and not totally suitable, Zaitsev [**276**] later proposed another term, the "minimax rank". However, here we shall use the term "Zaitsev rank".

5.1. The Zaitsev Rank of a Group

Let G be a group and let

$$1 = H_0 \leq H_1 \leq \cdots \leq H_{n-1} \leq H_n = G$$

be a finite chain of subgroups of G. Let $\mathcal{C} = \{H_j | 0 \leq j \leq n\}$ and let $\mathbf{il}\,(\mathcal{C})$ denote the number of links $H_j \leq H_{j+1}$ in this chain such that the index $|H_{j+1} : H_j|$ is infinite.

5.1.1. DEFINITION. A group G is said to have *finite Zaitsev rank* $\mathbf{r}_Z(G) = m$ if $\mathbf{il}\,(\mathcal{C}) \leq m$ for every finite chain of subgroups $\mathcal{C}$ and provided there exists a chain $\mathcal{D}$ for which this number is exactly m. Otherwise we shall say that G has infinite Zaitsev rank. Of course, if G is a finite group, then $\mathbf{r}_Z(G) = 0$.

5.1.2. DEFINITION. Let H, K be subgroups of the group G and suppose that $H \leq K$. We say that the link $H \leq K$ is *infinite* if the index $|K : H|$ is infinite and say that the link $H \leq K$ is *minimal infinite* if it is infinite and for every subgroup L such that $H \leq L \leq K$ one of the indices $|L : H|$ or $|K : L|$ is finite.

Ranks of Groups: The Tools, Characteristics, and Restrictions
By Martyn R. Dixon, Leonid A. Kurdachenko and Igor Ya Subbotin

Suppose that the group G has finite Zaitsev rank and let $\mathcal{D}$ be a finite chain of subgroups such that $\mathbf{il}\,(\mathcal{D}) = \mathbf{r}_Z(G)$. Let $H \leq K$ be a link in this chain such that the index $|K : H|$ is infinite and let L be a subgroup of G such that $H \lneq L \lneq K$. If both indices $|K : L|$ and $|L : H|$ are infinite, then the chain $\mathcal{D} \cup \{L\}$ is finite and $\mathbf{il}\,(\mathcal{D} \cup \{L\}) = \mathbf{il}\,(\mathcal{D}) + 1$, contradicting the choice of $\mathcal{D}$. This shows that every link $H \leq K$ of $\mathcal{D}$ with $|K : H|$ infinite is necessarily minimal infinite.

EXERCISE 5.1. *Find the Zaitsev rank of the following groups G:*

(i) G *is an infinite cyclic group;*

(ii) G *is a Prüfer p-group;*

(iii) $G = \mathbb{Q}_p = \{m/p^n | m, n \in \mathbb{Z}\}$ *is the additive group of rational numbers with p-power denominators, for some prime p.*

We need some information about cosets. We note that if $L \lhd G$ and $H \leq K \leq G$ and if T is a left transversal to LH in LK, then the elements of T can be chosen to lie in K. For if xLH is an arbitrary coset of LH in LK, where $x \in T$, then $x = va$ for certain elements $v \in L$ and $a \in K$. Since $LK = KL$ we have $xLH = aLH$. In the next lemma we assume that $T \subseteq K$.

5.1.3. LEMMA (Zaitsev [**263**]). *Let G be a group. Suppose that L is a normal subgroup of G and that H, K are subgroups of G such that $H \leq K$. If T is a left transversal to LH in LK and U is a left transversal to $L \cap H$ in $L \cap K$, then $TU = \{xu | x \in T, u \in U\}$ is a left transversal to H in K. Furthermore, $|T| = |K : H(L \cap K)|$ and $|U| = |H(L \cap K) : H|$.*

PROOF. Let zH be an arbitrary coset to H in K, where $z \in K$. There is an element $x \in T$ such that $zLH = xLH$ and it follows that $z = xw$ for some element $w \in LH$. Then $w = vh$ for elements $v \in L$ and $h \in H$, so $z = xw = xvh$ and hence $v = x^{-1}zh^{-1}$. Since $x, z, h \in K$ it follows that $v \in K$, so $v \in L \cap K$. There exists an element $u \in U$ such that $v(L \cap H) = u(L \cap H)$, so $v = uy$ for some element $y \in L \cap H$. Consequently, $z = xvh = xuyh$ and $zH = xuyhH = xuH$. Hence TU certainly contains a transversal to H in K.

Assume now that $x_1 u_1 H = x_2 u_2 H$ where $x_1, x_2 \in T, u_1, u_2 \in U$. Then $x_1 u_1 HL = x_2 u_2 HL$. Since $U \subseteq L$ and $LH = HL$ we have $x_1 LH = x_2 LH$ and since T is a left transversal to LH in LK we have $x_1 = x_2$. It follows that $u_1 H = u_2 H$, so $u_2^{-1} u_1 \in H$. On the other hand, $u_2^{-1} u_1 \in L$, so $u_2^{-1} u_1 \in L \cap H$ and hence $u_1(L \cap H) = u_2(L \cap H)$. Since U is a transversal to $L \cap H$ in $L \cap K$ we have $u_1 = u_2$ and it follows that TU is a transversal to H in K.

Now let $a \in K$. If b is an element of K such that $a(L \cap K)H = b(L \cap K)H$, then $b = av_1 h$ where $v_1 \in L \cap K$ and $h \in H$. Hence $bLH = av_1 hLH = aLH$. Let ϕ be the mapping $a(L \cap K)H \longmapsto aLH$ from the set of all cosets of $(L \cap K)H$ in K to the set of cosets of LH in KH. From the remark preceding the lemma, if xLH is a coset of LH in KH, then there exists $a \in K$ such that $xLH = aLH$. Hence ϕ is a surjection. Furthermore, if $a, b \in K$ and $aLH = bLH$, then $b = avh$ for certain elements $v \in L$ and $h \in H$. It follows that $v = a^{-1}bh^{-1} \in K$ so that $v \in L \cap K$. This implies that $a(L \cap K)H = b(L \cap K)H$ and the mapping ϕ is injective. Thus ϕ is bijective and it follows that $|LK : LH| = |K : H(L \cap K)|$. Similar arguments enable us to show that $|H(L \cap K) : H| = |L \cap K : L \cap H|$ and the result follows. $\square$

5.1.4. LEMMA (Zaitsev [**263**]). *Let G be a group, H a subgroup of G and L a normal subgroup of G.*

(i) *If G has finite Zaitsev rank and H has finite index in G, then $r_Z(G) = r_Z(H)$.*

(ii) *If G has finite Zaitsev rank, then H has finite Zaitsev rank and $r_Z(H) \le r_Z(G)$.*

(iii) *If G has finite Zaitsev rank and L is finite, then $r_Z(G) = r_Z(G/L)$.*

(iv) *If G has finite Zaitsev rank, then G/L has finite Zaitsev rank and $r_Z(G) = r_Z(L) + r_Z(G/L)$.*

(v) *If L and G/L have finite Zaitsev rank, then G has finite Zaitsev rank.*

PROOF. The assertions (i) and (iii) are clear. To prove (ii) let

$$1 = H_0 \le H_1 \le \cdots \le H_k = H$$

be an arbitrary finite chain of subgroups of H. From this chain we obtain the following chain of subgroups of G:

$$1 = H_0 \le H_1 \le \cdots \le H_k = H \le H_{k+1} = G.$$

Let $\mathcal{C} = \{H_j | 0 \le j \le k+1\}$ and $\mathcal{D} = \{H_j | 0 \le j \le k\}$. Since $r_Z(G)$ is finite, $il(\mathcal{C}) \le r_Z(G)$ and hence $il(\mathcal{D}) \le r_Z(G)$. It follows that $r_Z(H)$ is finite and $r_Z(H) \le r_Z(G)$. Hence (ii) holds.

To prove (iv) let

$$1 = U_0/L \le U_1/L \le \cdots \le U_{k-1}/L \le U_k/L = G/L$$

be an arbitrary finite chain of subgroups of G/L. From this chain we obtain the chain of subgroups

$$1 = V_0 \le V_1 \le \cdots \le V_{k-1} \le V_k \le V_{k+1} = G$$

where $V_1 = L, V_2 = U_1, \ldots, V_k = U_{k-1}$. Let $\mathcal{V} = \{V_j | 0 \leq j \leq k+1\}$ and $\mathcal{U} = \{U_j/L | 0 \leq j \leq k\}$. Since $\mathbf{r}_Z(G)$ is finite, $\mathbf{il}\,(\mathcal{V}) \leq \mathbf{r}_Z(G)$ and hence $\mathbf{il}\,(\mathcal{U}) \leq \mathbf{r}_Z(G)$. It follows that $\mathbf{r}_Z(G/L)$ is finite and $\mathbf{r}_Z(G/L) \leq \mathbf{r}_Z(G)$.

By (ii) $\mathbf{r}_Z(L)$ is finite. Let $\mathbf{r}_Z(L) = t$ and $\mathbf{r}_Z(G/L) = d$. Choose a finite chain $\mathcal{L}$ in L and a finite chain $\mathcal{G}$ in G/L such that $\mathbf{il}(\mathcal{L}) = t$ and $\mathbf{il}(\mathcal{G}) = d$. For every term $U \in \mathcal{G}$ there is a subgroup W_U such that $U = W_U/L$. Let $\mathcal{W} = \{W_U | U \in \mathcal{G}\}$ and $\mathcal{D} = \mathcal{L} \cup \mathcal{W}$. Then $\mathcal{D}$ is a finite chain of subgroups of G and clearly it follows that $\mathbf{r}_Z(L) + \mathbf{r}_Z(G/L) \leq \mathbf{r}_Z(G)$.

Suppose that $\mathbf{r}_Z(G) > \mathbf{r}_Z(G/L) + \mathbf{r}_Z(L)$ and consider the finite chain $\mathcal{B}$ of subgroups

$$1 = B_0 \leq B_1 \leq \cdots \leq B_r = G$$

in which $\mathbf{il}(\mathcal{B}) = \mathbf{r}_Z(G)$. Next, consider the chain

$$1 = B_0 \cap L \leq B_1 \cap L \leq \cdots \leq B_r \cap L = L$$
$$= LB_0 \leq LB_1 \leq \cdots \leq LB_r = G.$$

Denote this chain by $\mathcal{A}$ and let $B_{j-1} \leq B_j$ be an infinite link of the chain $\mathcal{B}$. As we saw above, this link is minimal infinite. Then exactly one of the indices $|B_j : (B_j \cap L)B_{j-1}|$ and $|(B_j \cap L)B_{j-1} : B_{j-1}|$ is infinite and the other is finite. If $|B_j : (B_j \cap L)B_{j-1}|$ is infinite, then Lemma 5.1.3 shows that $|LB_j : LB_{j-1}|$ is infinite. In this case $|(B_j \cap L)B_{j-1} : B_{j-1}|$ is finite and Lemma 5.1.3 implies that $|B_j \cap L : B_{j-1} \cap L|$ is also finite. On the other hand, if $|(B_j \cap L)B_{j-1} : B_{j-1}|$ is infinite, then, by Lemma 5.1.3, $|B_j \cap L : B_{j-1} \cap L|$ is infinite and the index $|LB_j : LB_{j-1}|$ is finite. This shows that $\mathbf{il}(\mathcal{B}) = \mathbf{il}(\mathcal{A}) > t + d$. However, in this case either $\mathbf{r}_Z(L) > t$, or $\mathbf{r}_Z(G/L) > d$, and we obtain a contradiction which proves (iv).

Finally we prove (v). To this end, let $\mathbf{r}_Z(L) = t$ and $\mathbf{r}_Z(G/L) = d$. Let $\mathcal{B}$ be the arbitrary finite chain

$$1 = B_0 \leq B_1 \leq \cdots \leq B_r = G$$

of subgroups of G. From this chain we obtain the chain

$$1 = B_0 \cap L \leq B_1 \cap L \leq \cdots \leq B_r \cap L = L = LB_0 \leq LB_1 \leq \cdots \leq LB_r = G,$$

which we denote by $\mathcal{A}$. If both $|B_j \cap L : B_{j-1} \cap L|$ and $|LB_j : LB_{j-1}|$ are finite then $|B_j : B_{j-1}|$ must also be finite. It follows that

$$\mathbf{il}(\mathcal{B}) \leq \mathbf{il}(\mathcal{A}) \leq \mathbf{r}_Z(L) + \mathbf{r}_Z(G/L).$$

and hence $\mathbf{r}_Z(G)$ is finite. The result follows. $\qquad\square$

EXERCISE 5.2. *Find the Zaitsev rank for the group of Exercise 3.10.*

EXERCISE 5.3. *Find the Zaitsev rank for the group of Exercise 3.11.*

Our next result gives us some information about minimal infinite links.

5.1.5. LEMMA. *Let G be a group and suppose that H, K are subgroups of G such that $H \leq K$ and $K' \leq H$. If the link $H \leq K$ is minimal infinite, then either K/H contains a Prüfer subgroup of finite index, or K/H is finite-by-(infinite cyclic).*

EXERCISE 5.4. *Prove Lemma 5.1.5.*

5.1.6. COROLLARY. *Let G be a group and suppose that H, K are subgroups of G such that $H \leq K$ and $K' \leq H$. If G has finite Zaitsev rank, then K/H has a finite series of subgroups whose factors either are finite, or infinite cyclic, or are Prüfer groups.*

PROOF. By Lemma 5.1.4, $\mathbf{r}_Z(K/H)$ is finite. As we saw above, K/H has a finite chain of subgroups whose infinite links are minimal infinite. By Lemma 5.1.5, every infinite link of this chain defines a section which either is finite-by-(infinite cyclic) or contains a Prüfer subgroup of finite index. It follows that K/H has a finite series of subgroups whose factors either are finite, or infinite cyclic, or are Prüfer groups. □

Since every Prüfer group satisfies the minimal condition for all subgroups and every infinite cyclic group satisfies the maximal condition for all subgroups (the condition max), we obtain the following result.

5.1.7. COROLLARY. *Let G be a group and suppose that H, K are subgroups of G such that $H \leq K$ and $K' \leq H$. If G has finite Zaitsev rank, then K/H has a finite series of subgroups whose factors satisfy either min or max.*

Groups with a finite series of factors satisfying max or min now play a very important role in infinite group theory.

5.1.8. DEFINITION. A group G is called *minimax* if G has a finite subnormal series whose factors either satisfy min or max.

Minimax groups were first studied in a paper of R. Baer [9]. However, the first fundamental study of soluble minimax groups was begun by D. J. S. Robinson [222], where the term "minimax group" was introduced. In the paper [14], Baer used the term "polyminimax group" but "minimax group" has become the standard terminology in use. It is easy to see that subgroups and homomorphic images of minimax groups are also minimax. It is also very easy to see that an extension of a minimax group by another minimax group is minimax.

The theory of soluble-by-finite minimax groups is now an established part of infinite group theory, and more details of the theory of soluble minimax groups is available in the book of J. C. Lennox and D. J. S. Robinson [**174**]. These groups have been studied by many authors from different points of view. Minimax groups occur naturally in the study of many different finiteness conditions.

It is very easy to see, using Corollary 5.1.7, that for abelian groups the property of having finite Zaitsev rank is equivalent to being minimax. We record this as the next result.

5.1.9. COROLLARY. *Let G be an abelian group. Then G has finite Zaitsev rank if and only if G is minimax.*

With a little more effort, we can establish the following result.

5.1.10. COROLLARY. *Let G be a soluble-by-finite group. Then G has finite Zaitsev rank if and only if it has a finite subnormal series of subgroups whose factors either are finite or infinite cyclic or are Prüfer groups.*

PROOF. Suppose that G has a series of subgroups

$$1 = D_0 \lhd D_1 \lhd \ldots \lhd D_{r-1} \lhd D_r \lhd G$$

where the factors D_j/D_{j-1} are abelian, for $1 \leq j \leq r$ and G/D_r is finite. If G has finite Zaitsev rank, then by Lemma 5.1.4, each factor D_j/D_{j-1} has finite Zaitsev rank for $1 \leq j \leq r$. We may then apply Corollary 5.1.7 to deduce the result. To prove the converse we can apply Corollary 5.1.9 and Lemma 5.1.4. $\qquad\square$

The following result summarizes much of the previous work.

5.1.11. PROPOSITION. *Let G be a soluble-by-finite group. Then G has finite Zaitsev rank if and only if G is minimax.*

A natural question arises as to whether the Zaitsev rank of every minimax group is finite. However, Obraztsov [**207**] has constructed a periodic uncountable group G satisfying the minimal condition. This group has an ascending chain

$$1 = D_0 \leq D_1 \leq \ldots D_\alpha \leq D_{\alpha+1} \leq \ldots D_\gamma = G$$

where $\gamma = \omega_1$ is the first uncountable ordinal. In particular, for each $n \in \mathbb{N}$, the group G has a finite chain $\mathcal{C}$ such that $\mathbf{il}(\mathcal{C}) = n$. This shows that G has infinite Zaitsev rank.

5.2. Zaitsev Rank and 0-Rank

We now consider some other corollaries of Proposition 5.1.11 which show the connection between the Zaitsev rank and certain other ranks that we have considered. Clearly, when G is periodic, then $\mathbf{r}_0(G) = 0$ and so it is easy to construct examples where $\mathbf{r}_0(G) \neq \mathbf{r}_Z(G)$. It is certainly possible for $\mathbf{r}_0(G)$ to be finite, with $\mathbf{r}_Z(G)$ infinite, and even if $\mathbf{r}_Z(G)$ is finite we need not have $\mathbf{r}_0(G) = \mathbf{r}_Z(G)$.

However, we note the following result.

5.2.1. COROLLARY. *Let G be a soluble-by-finite group. If G has finite Zaitsev rank, then G has finite 0-rank.*

This corollary, together with Corollary 5.1.6 and Lemma 5.1.4, enables us to calculate the Zaitsev rank of a soluble-by-finite minimax group by using the formula

$$\mathbf{r}_Z(G) = \mathbf{r}_0(G) + \mathbf{qf}(G),$$

where $\mathbf{qf}(G)$ is the number of Prüfer factors in a subnormal series of a minimax group. It is not hard to see that $\mathbf{qf}(G)$ is an invariant of a soluble-by-finite minimax group. In particular, for a periodic soluble-by-finite group we have $\mathbf{r}_Z(G) = \mathbf{qf}(G)$ and we have:

5.2.2. COROLLARY. *Let G be a periodic abelian group. If G has finite Zaitsev rank, then G is a Chernikov group.*

Consequently, the following more general result holds.

5.2.3. COROLLARY. *Let G be a periodic soluble-by-finite group. If G has finite Zaitsev rank, then G is a Chernikov group.*

Our next result shows that abelian groups with finite Zaitsev rank are max-by-min.

5.2.4. COROLLARY. *Let G be an abelian group. If G has finite Zaitsev rank, then G contains a finitely generated subgroup A such that G/A is a Chernikov group.*

PROOF. By Corollary 5.2.1, G has finite 0-rank and hence G contains a finite maximal $\mathbb{Z}$-independent subset M. Let $A = \langle M \rangle$, a finitely generated group. By the maximal choice of M the factor group G/A is periodic. By Lemma 5.1.4 and Corollary 5.2.3, G/A is Chernikov. $\square$

Let G be an abelian group of finite 0-rank and choose a maximal $\mathbb{Z}$-independent subset M of G. Let $A = \langle M \rangle$. Then G/A is periodic. We let $\mathbf{Sp}(G)$ denote the set of all primes p such that the Sylow p-subgroup of G/A is infinite. If B is another free abelian subgroup of G

such that G/B is periodic, then the factors $A/(A\cap B)$ and $B/(A\cap B)$ are both finite. This shows that the set $\mathbf{Sp}(G)$ is independent of the choice of the subgroup A and hence $\mathbf{Sp}(G)$ is an invariant of G. The set $\mathbf{Sp}(G)$ is called the *spectrum* of the group G. If H is a subgroup of G then it is not hard to see that $\mathbf{Sp}(G) = \mathbf{Sp}(H) \cup \mathbf{Sp}(G/H)$.

Now let G be a soluble group of finite 0-rank. We define $\mathbf{Sp}(G)$ to be the union of the spectrums of the factors of the derived series of G. It is not hard to see that $\mathbf{Sp}(G)$ can be defined as the union of the spectrums of the factors of an arbitrary series of normal subgroups of G with abelian factors. Corollary 5.2.4 shows that a soluble-by-finite minimax group has finite spectrum.

To finish this section we determine the structure of finitely generated generalized radical groups of finite 0-rank, our purpose being to show that for such a group G the factor group $G/\mathbf{Tor}\,(G)$ is minimax. But first we establish the following preliminary results. We shall require some standard facts concerning tensor products which can be found in Fuchs [**88**].

5.2.5. LEMMA. *If A, B are abelian minimax groups then their tensor product $A \otimes B$ is also minimax.*

PROOF. By Corollary 5.1.9, A, B have finite Zaitsev rank and hence $r = \max\{\mathbf{r}_0(A), \mathbf{r}_0(B)\}$ is finite. Also, let $s_1 = \mathbf{r}_0(A) + \max\{\mathbf{sr}_p(A)|p \in \Pi(A)\}$, $s_2 = \mathbf{r}_0(B) + \max\{\mathbf{sr}_p(B)|p \in \Pi(B)\}$ and $s = \max\{s_1, s_2\}$. We note that the sets $\mathbf{Sp}(A)$ and $\mathbf{Sp}(B)$ are finite. Since A, B are minimax, Corollary 5.2.4 implies that they contain finitely generated subgroups H, K respectively such that $A/H = \underset{p\in\mathbf{Sp}(A)}{\mathrm{Dr}}\ A_p$ and $B/K = \underset{p\in\mathbf{Sp}(B)}{\mathrm{Dr}}\ B_p$ where A_p, B_p are direct products of finitely many Prüfer p-groups. For the subgroup H we have

$$H = \underset{1\leq j\leq k}{\mathrm{Dr}}\ Z_j \times \underset{p\in\Pi(H)}{\mathrm{Dr}}\ C_p,$$

where Z_j is an infinite cyclic group, for $1 \leq j \leq k \leq r$ and C_p is a direct product of at most s cyclic p-groups.

Now the sequence

$$1 \longrightarrow H \longrightarrow A \longrightarrow A/H \longrightarrow 1$$

is exact so, by a theorem of Dieudonné, (see [**88**, Theorem 60.6], for example) the sequence

$$1 \longrightarrow H \otimes (B/\mathbf{Tor}\,(B)) \longrightarrow A \otimes (B/\mathbf{Tor}\,(B))$$
$$\longrightarrow A/H \otimes (B/\mathbf{Tor}\,(B)) \longrightarrow 1$$

is also exact. It follows from the properties of tensor products of abelian groups (see [**88**, p.255]) that

$$H \otimes (B/\mathbf{Tor}\,(B)) \cong \operatorname*{Dr}_{1 \le j \le k} Y_j \times \operatorname*{Dr}_{p \in \mathbf{Sp}(A)} D_p,$$

where $Y_j = B/\mathbf{Tor}\,(B)$, for $1 \le j \le k$, and $D_p = B/B^{t_p}\mathbf{Tor}\,(B)$, for certain $t_p = p^{m_p}$, whenever $p \in \mathbf{Sp}(A)$.

This implies, that $H \otimes (B/\mathbf{Tor}\,(B))$ is also minimax. Furthermore,

$$A/H \otimes (B/\mathbf{Tor}\,(B)) \cong (\operatorname*{Dr}_{p \in \mathbf{Sp}(A)} A_p) \otimes (B/\mathbf{Tor}\,(B)).$$

Let $E = \operatorname*{Dr}_{1 \le j \le u} X_j$ be a p-basic subgroup of $B/\mathbf{Tor}\,(B)$, where X_j is an infinite cyclic group, for $1 \le j \le u$. Then, by [**88**, Theorem 61.1],

$$C_{p^\infty} \otimes (B/\mathbf{Tor}\,(B)) \cong C_{p^\infty} \otimes E \cong \operatorname*{Dr}_{1 \le j \le u} W_j,$$

where $W_j \cong C_{p^\infty}$, for $1 \le j \le u$. From this it follows that $A/H \otimes (B/\mathbf{Tor}\,(B))$ is minimax with spectrum $\mathbf{Sp}(A)$. Consequently, $A \otimes (B/\mathbf{Tor}\,(B))$ is also minimax. Now also

$$1 \longrightarrow \mathbf{Tor}\,(A) \longrightarrow A \longrightarrow A/\mathbf{Tor}\,(A) \longrightarrow 1$$

is an exact sequence so, again by [**88**, Theorem 60.6] we have

$$1 \longrightarrow \mathbf{Tor}\,(A) \otimes (B/\mathbf{Tor}\,(B)) \longrightarrow A \otimes (B/\mathbf{Tor}\,(B))$$
$$\longrightarrow (A/\mathbf{Tor}\,(A)) \otimes (B/\mathbf{Tor}\,(B)) \longrightarrow 1$$

and since a homomorphic image of an abelian minimax group with spectrum π is again minimax with spectrum $\rho \subseteq \pi$, it follows that $A/\mathbf{Tor}\,(A) \otimes B/\mathbf{Tor}\,(B)$ is minimax with spectrum a subset of $\mathbf{Sp}(A)$.

Next note that, by [**89**, Theorem 61.5], we have

$$\mathbf{Tor}\,(A \otimes B) \cong (\mathbf{Tor}\,(A) \otimes \mathbf{Tor}\,(B)) \times \mathbf{Tor}\,(A) \otimes (B/\mathbf{Tor}\,(B))$$
$$\times (A/\mathbf{Tor}\,(A)) \otimes \mathbf{Tor}\,(B)$$

and

$$A \otimes B/\mathbf{Tor}\,(A \otimes B) \cong A/\mathbf{Tor}\,(A) \otimes B/\mathbf{Tor}\,(B).$$

We have already seen that $\mathbf{Tor}\,(A) \otimes (B/\mathbf{Tor}\,(B))$ is minimax since it is a subgroup of $A \otimes (B/\mathbf{Tor}\,(B))$ and likewise $(A/\mathbf{Tor}\,(A)) \otimes \mathbf{Tor}\,(B)$ is minimax. Further, $(A/\mathbf{Tor}\,(A)) \otimes (B/\mathbf{Tor}\,(B))$ is minimax from our work above. So it remains to prove that $\mathbf{Tor}\,(A) \otimes \mathbf{Tor}\,(B)$ is minimax. However $\mathbf{Tor}\,(A)$ and $\mathbf{Tor}\,(B)$ both have the minimal condition so are direct products of Prüfer p-groups and finite cyclic groups. Therefore to compute $\mathbf{Tor}\,(A) \otimes \mathbf{Tor}\,(B)$ it is necessary to compute expressions of the form $C_{p^\infty} \otimes C_{q^\infty}$, $C_{p^\infty} \otimes C_{q^k}$ and $C_{p^m} \otimes C_{q^k}$, where k, m are positive integers and p, q are (not necessary distinct) primes. However it is well known, from the properties of tensor products (see, [**88**, p.255], for

example), that $C_{p^\infty} \otimes C_{p^\infty} = 1, C_{p^\infty} \otimes C_{q^\infty} = C_{p^\infty} \otimes C_{q^k} = C_{p^m} \otimes C_{q^k} = 1,$
whenever $p \neq q$ and that $C_{p^m} \otimes C_{p^k} = C_{p^t}$, where t is the minimum
of k, m. From this it follows that $\mathbf{Tor}\,(A) \otimes \mathbf{Tor}\,(B)$ is finite and the
result now follows.   □

To accomplish our goal for the remainder of this section, we require
some further useful facts. For an arbitrary group G there is a well-
known epimorphism

$$\vartheta : G/G' \otimes \gamma_j(G)/\gamma_{j+1}(G) \longrightarrow \gamma_{j+1}(G)/\gamma_{j+2}(G)$$

defined as follows. If $a \in G$ and $b \in \gamma_j(G)$ then $[a, b] \in \gamma_{j+1}(G)$ and
the map ϑ is given by

$$\vartheta(aG' \otimes b\gamma_{j+1}(G)) = [a, b]\gamma_{j+2}(G).$$

That this mapping is a homomorphism is established using the universal property of tensor products. The homomorphism itself is particularly useful for studying nilpotent groups. As evidence for this we now
prove the following result of D. I. Zaitsev [**268**].

5.2.6. PROPOSITION. *Let G be a nilpotent group and suppose that
G/G' is a minimax group. Then G is also minimax.*

PROOF. Since a homomorphic image of a minimax group is min-
imax, Lemma 5.2.5 and induction imply that $\gamma_j(G)/\gamma_{j+1}(G)$ is also
minimax for all $j \geq 1$. Since an extension of a minimax group by
another minimax group is also minimax it follows that the nilpotent
group G is minimax. This completes the proof.   □

Corollary 2.3.4 shows that a torsion-free locally nilpotent group
with finite 0-rank is nilpotent, so Proposition 5.2.6 implies the following
result.

5.2.7. COROLLARY (Zaitsev [**268**]). *Let G be a torsion-free locally
nilpotent group of finite 0-rank. If G/G' is minimax and $\mathbf{Sp}(G/G') =
\pi$, then G is minimax and $\mathbf{Sp}(G) = \pi$.*

Our final result in this section extends Theorem 2.4.13 just a little further, but requires some module theoretic ideas. The reader is
referred to [**152**] for further details concerning such matters.

5.2.8. THEOREM. *Let G be a locally generalized radical group of
finite 0-rank. If G is finitely generated, then $G/\mathbf{Tor}\,(G)$ is soluble-by-
finite and minimax.*

PROOF. By Theorem 2.4.13, G has normal subgroups $T \leq L \leq K \leq G$ such that T is locally finite, L/T is torsion-free nilpotent, K/L is finitely generated torsion-free abelian, and G/K is finite. Since G is finitely generated, K is finitely generated, by Proposition 1.2.13. Let $D/T = (L/T)'$. By [152, Corollary 1.8], L/D contains a free abelian subgroup C/D such that L/C is a periodic group and $\Pi(L/C)$ is finite. By Theorem 3.2.26, L/T has finite section p-rank for each prime p. Lemma 3.2.2 and Theorem 3.2.26 show that the Sylow p-subgroup of L/C is Chernikov for each prime p. Since $\Pi(L/C)$ is finite, L/C is also Chernikov. Since L/D has finite 0-rank, C/D is finitely generated so that L/D is minimax. Using Proposition 5.2.6 we see that L/T is minimax. It follows that G/T is minimax, because G/L is finitely generated and abelian-by-finite.   □

5.3. Weak Minimal and Weak Maximal Conditions

In this short section, we indicate how the study of minimax groups (and hence groups of finite Zaitsev rank) is related to certain finiteness conditions, in much the same way that Chernikov groups are related to the study of groups with the minimal condition and polycyclic-by-finite groups are related to the study of groups with the maximal condition. In the case of minimax groups the finiteness conditions known as the weak minimal condition and the weak maximal condition are the ones of interest.

5.3.1. DEFINITION. Let G be a group and $\mathcal{M}$ a family of subgroups of G. We say that $\mathcal{M}$ satisfies the *weak maximal (respectively minimal) condition* or G satisfies the *weak maximal (respectively minimal) condition for $\mathcal{M}$-subgroups* if, for every ascending (respectively descending) chain, $\{H_n | n \in \mathbb{N}\}$, of subgroups of the family $\mathcal{M}$ there exists a natural number m such that the indices $|H_{n+1} : H_n|$ (respectively $|H_n : H_{n+1}|$) are finite for all $n \geq m$.

EXERCISE 5.5. *Let G be a group and let H be a normal subgroup of G. Prove that if H and G/H satisfy the weak minimal (respectively maximal) condition on subgroups, then G satisfies the weak minimal (respectively maximal) condition on subgroups.*

Almost soluble groups of finite Zaitsev rank can be very neatly characterized in terms of the weak minimal or weak maximal conditions. These conditions were introduced by R. Baer [14] and D. I. Zaitsev [263, 264]. As with the minimal and maximal conditions, groups satisfying the weak minimal and weak maximal conditions on certain important families of subgroups have been studied by many authors. Many of the results obtained have been connected with groups of finite rank, but not directly so and we here just outline some of these results. The first results about groups with the weak minimal and weak maximal conditions on all subgroups were obtained by D. I. Zaitsev in the papers [**263**] and [**269**]. The main result he obtained is the following theorem.

5.3.2. THEOREM. *Let G be a locally (soluble-by-finite) group. Then G satisfies the weak minimal (respectively weak maximal) condition on subgroups if and only if G is a soluble-by-finite minimax group.*

Groups with the weak minimal and weak maximal conditions on abelian subgroups were studied by R. Baer [**14**] and D. I. Zaitsev [**265**]. The dual situation of groups satisfying the weak minimal condition for non-abelian subgroups was discussed by D. I. Zaitsev [**267**] and groups with the weak maximal condition for non-abelian subgroups were studied by L. S. Kazarin, L. A. Kurdachenko and I. Ya. Subbotin [**132**]. The generalization of these results to groups with the weak minimal condition on non-nilpotent subgroups has been studied by H. Smith [**249**] and M. R. Dixon, M. J. Evans and H. Smith [**60**]. Groups with the weak maximal condition on non-nilpotent subgroups have been studied by L. A. Kurdachenko, P. Shumyatsky and I. Ya. Subbotin [**162**], and L. A. Kurdachenko and N. N. Semko [**159**].

Groups with the weak minimal or weak maximal condition on normal subgroups have been considered in the papers [**138, 141, 142, 143, 144, 146, 169, 277**]. The first class to consider here is the class of locally nilpotent groups. Unlike with the maximal and minimal conditions on normal subgroups, locally nilpotent groups with the weak minimal (respectively weak maximal) condition on normal subgroups need not satisfy the weak minimal (respectively maximal) condition on all subgroups. A locally nilpotent group with the weak minimal condition on normal subgroups has the following structure.

5.3.3. THEOREM (Kurdachenko [**141**]). *Let G be a locally nilpotent group with torsion subgroup $T = \mathbf{Tor}\,(G)$. Then G satisfies the weak minimal condition for normal subgroups if and only if the following conditions hold.*

(i) *G/T is a nilpotent minimax group;*
(ii) *T satisfies the minimal condition on G-invariant subgroups;*
(iii) *G is hypercentral;*
(iv) *G is soluble;*

(v) *The maximal $\mathfrak{F}$-perfect subgroup of G is a periodic divisible abelian group.*

The following consequences of this result hold, although these had appeared in the literature earlier.

5.3.4. COROLLARY (Kurdachenko [**138**]). *A periodic locally nilpotent group satisfies the weak minimal condition on normal subgroups if and only if it is Chernikov.*

5.3.5. COROLLARY (Kurdachenko [**138**]). *A torsion-free locally nilpotent group satisfies the weak minimal condition on normal subgroups if and only if it is a nilpotent minimax group.*

The dual situation of groups with the weak minimal conditions for non-normal subgroups was studied by L. A. Kurdachenko and V. E. Goretsky [**147**]. Groups with the weak minimal or weak maximal condition for subnormal subgroups were studied by L. A. Kurdachenko [**140**] and the dual situation of groups with the weak minimal or weak maximal condition for non-subnormal subgroups has been studied by L. A. Kurdachenko and H. Smith [**163, 164, 166**]. More detailed information can be found in the survey articles [**69, 131, 276**].

EXERCISE 5.6. *Let G be a nilpotent group of nilpotency class c and suppose that there is a positive integer m such that $(\boldsymbol{Tor}(\zeta(G))^m = 1$. Prove that $(\boldsymbol{Tor}(G))^k = 1$, for some divisor k of m^c.*

EXERCISE 5.7. *Let G be a locally nilpotent group and let H be a normal subgroup of G such that G/H is periodic. Prove that if $\Pi(H) \cap \Pi(G/H) = \emptyset$, then every subgroup of $\boldsymbol{Tor}(H)$ which is normal in H is normal in G.*

EXERCISE 5.8. *Let G be a hypercentral group and let H be a normal subgroup of G such that G/H is $\mathfrak{F}$-perfect. Prove that every subgroup of $\boldsymbol{Tor}(H)$ which is normal in H is normal in G.*

EXERCISE 5.9. *Let G be a group.*

(a) *If L, K are normal subgroups of G such that $K \leq L$ and if K and L/K satisfy the minimal condition on G-invariant subgroups, then prove that L satisfies the minimal condition on G-invariant subgroups.*

(b) *If G has a finite series of normal subgroups*

$$1 = H_0 \leq H_1 \leq \cdots \leq H_n = L,$$

each factor of which satisfies the minimal condition on G-invariant subgroups, for $1 \leq j \leq n$, then prove that L satisfies the minimal condition on G-invariant subgroups.

EXERCISE 5.10. Let G be a group and let $K \lhd G$. Suppose that K satisfies the weak minimal condition on G-invariant subgroups and that G/K satisfies the weak minimal condition on normal subgroups. Prove that G satisfies the weak minimal condition on normal subgroups.

EXERCISE 5.11. Let p be a prime and let $A_j = \underset{k \in \mathbb{N}}{Dr} \langle a_{jk} \rangle$ be an elementary abelian p-group, for $1 \leq j \leq n$. Let $A = A_1 \times A_2 \times \cdots \times A_n$ and define the automorphism κ of A by

$$\kappa(a_{jk}) = \begin{cases} a_{jk} & \text{if } k = 1, \\ a_{jk}a_{j,k-1} & \text{if } k \geq 2. \end{cases}$$

Prove that $A \rtimes \langle \kappa \rangle$ satisfies the weak minimal condition on normal subgroups.

EXERCISE 5.12. Let p be a prime and let $A_j = \underset{k \in \mathbb{N}}{Dr} \langle a_{jk} \rangle$ be an elementary abelian p-group, for $j \in \mathbb{N}$. Let $A = \underset{j \in \mathbb{N}}{Dr} A_j$ and define automorphisms κ, λ of A by

$$\kappa(a_{jk}) = \begin{cases} a_{jk} & \text{if } k = 1, \\ a_{jk}a_{j,k-1} & \text{if } k \geq 2, \end{cases} \quad \lambda(a_{jk}) = \begin{cases} a_{jk} & \text{if } j = 1, \\ a_{jk}a_{j-1,k} & \text{if } j \geq 2. \end{cases}$$

Prove that $G = A \rtimes \langle \kappa, \lambda \rangle$ satisfies the weak minimal condition on normal subgroups.

CHAPTER 6

Special Rank

If A is a vector space of finite dimension k over a field F and B is a subspace of A, then it is well known that B is finite dimensional, of dimension at most k. Similarly, it is a well-known consequence of the structure theorem for finitely generated abelian groups that if G is an abelian group with k generators and B is a subgroup of G then B is finitely generated and has at most k generators. Thus a subgroup H of a finitely generated abelian group G is also finitely generated and the minimal number of generators of H is at most the minimal number of generators of G. However, for non-abelian groups, such inherited properties are well known to be false. For example, the standard restricted wreath product $\mathbb{Z} \wr \mathbb{Z}$ of two infinite cyclic groups is a 2-generator group and yet its base group is infinitely generated. Thus a subgroup of a finitely generated group need not even be finitely generated.

As we have seen, and as is well known, a subgroup of a finitely generated nilpotent group is finitely generated. However, even in this case, it is possible to have a subgroup with more generators than the original group has, as can be seen by considering the group $\mathbb{Z}_p \wr \mathbb{Z}_p$, where p is an odd prime. This group is 2-generator but its base group is p-generator of course. In this chapter, we investigate the important concept of the *special* rank (sometimes called the *Prüfer rank*) of a group which might be thought of as reconciling this situation.

6.1. Elementary Properties of Special Rank

We immediately define the concept we are interested in within this chapter.

6.1.1. DEFINITION. The group G has *finite special rank* $\boldsymbol{r}(G) = r$ if every finitely generated subgroup of G can be generated by r elements and r is the least positive integer with this property. If there is no such integer r then G is said to be of infinite special rank.

The general concept of special rank (and also the term) was introduced by A. I. Maltsev in [186]. Later, in the paper [13], R. Baer

Ranks of Groups: The Tools, Characteristics, and Restrictions
By Martyn R. Dixon, Leonid A. Kurdachenko and Igor Ya Subbotin
Copyright © 2017 John Wiley & Sons, Inc.

called the "special rank" the *"Prüfer rank"*. Hence the special rank of
a group G is sometimes called the *Prüfer rank* of G and is often just
called the *rank* of G. We shall endeavour to use the full terminology to
avoid possible confusion. H. Prüfer first defined groups of special rank
1 in his famous 1924 paper [**220**]. However, it is an exaggeration to
say that this article contained the roots of the concept of special rank.
It should also be noted that in the theory of abelian groups, the term
special rank is rarely, if at all, used. There the basic terms in use are
the p-rank and the 0-rank.

We shall denote the class of groups of finite special rank by $\mathfrak{R}$ and,
for each natural number n, we shall let $\mathfrak{R}_n$ denote the class of groups
which are of special rank at most n. It is clear that $\mathfrak{R} = \bigcup_{n \in \mathbb{N}} \mathfrak{R}_n$.

We first make some elementary observations concerning groups of
finite special rank.

6.1.2. LEMMA. *Let G be a group, let H be a subgroup of G and let
L be a normal subgroup of G.*

   (i) *If G has finite special rank, then H has finite special rank and
   $r(H) \leq r(G)$;*

   (ii) *If G has finite special rank, then G/L has finite special rank
   and $r(G/L) \leq r(G)$;*

   (iii) *If L and G/L have finite special rank, then G has finite special
   rank and $r(G) \leq r(L) + r(G/L)$.*

EXERCISE 6.1. *Prove Lemma 6.1.2.*

Unlike with the 0-rank and Zaitsev rank, the special rank has no
additivity property; that is, in general it is not true that $\mathbf{r}(G) = \mathbf{r}(L) +
\mathbf{r}(G/L)$. There are many easy examples that show this; for example,
if p is a prime and $G = \langle g \rangle$ is a cyclic group of order p^2 then G is of
special rank 1. However, if $L = \langle g^p \rangle$, then $G/L \cong C_p \cong L$, the cyclic
group of order p, so L and G/L also are of special rank 1.

Lemma 6.1.2 shows that the property of having finite special rank
at most r is closed under taking subgroups and factor groups. Furthermore, the property of having finite special rank is closed under taking
extensions. We also have the following result, the first part of which
shows that the class of groups of special rank at most r is locally closed.

6.1.3. LEMMA. *Let r be a natural number and let G be a group.*

   (i) *If every finitely generated subgroup of G has finite special rank
   at most r, then G has special rank at most r.*

   (ii) *The class of groups of finite special rank is countably recognizable.*

EXERCISE 6.2. *Prove Lemma 6.1.3.*

The class of groups of finite special rank is not locally closed, however, since the direct product of countably many copies of a cyclic group of prime order p is locally of finite special rank, but is not itself of finite special rank. The following property however is very useful.

6.1.4. LEMMA. *Let G be a periodic group and suppose that $G = \underset{n\in\mathbb{N}}{Dr}\,G_n$. If there is a positive integer r such that $r(G_n) \leq r$, for each $n \in \mathbb{N}$, and $\Pi(G_k) \cap \Pi(G_l) = \emptyset$ whenever $k \neq l$, then G has finite special rank at most r. Also $r(G) = \max\{r(G_n)|n \in \mathbb{N}\}$.*

EXERCISE 6.3. *Prove Lemma 6.1.4.*

The following corollary is immediate.

6.1.5. COROLLARY. *Let $\{p_n|n \in \mathbb{N}\}$ be a set of distinct primes. For each $n \in \mathbb{N}$ let G_n be a p_n-group and suppose that $G = \underset{n\in\mathbb{N}}{Dr}\,G_n$. If there is a positive integer r such that $r(G_n) \leq r$, for each $n \in \mathbb{N}$, then G has finite special rank at most r. Also $r(G) = \max\{r(G_n)|n \in \mathbb{N}\}$.*

If G is a group of special rank 1, then every finitely generated subgroup of G is cyclic. Thus G is a locally cyclic group; such groups were first considered in [**220**]. Every locally cyclic group either is periodic or torsion-free and it is not hard to prove that every torsion-free locally cyclic group is isomorphic to some subgroup of the additive group $\mathbb{Q}$ of rational numbers. Indeed, the torsion-free locally cyclic groups have been described completely (see [**172**, §30], for example). On the other hand, if G is a locally cyclic p-group then either G is a cyclic p-group or G is the union of an ascending chain of finite cyclic p-groups and in this latter case G is a Prüfer p-group. Consequently, if G is a periodic locally cyclic group then $G = \underset{p\in\Pi(G)}{Dr}\,G_p$, where G_p is either a cyclic p-group or a Prüfer p-group, for each $p \in \Pi(G)$. Thus a periodic group G is locally cyclic if and only if G can be embedded in $L = \underset{p\in\mathbb{P}}{Dr}\,G_p$, where G_p is a Prüfer p-group for each $p \in \mathbb{P}$, the set of all primes. Clearly L is isomorphic to $\mathbb{Q}/\mathbb{Z}$ and hence a group G is locally cyclic if and only if it is isomorphic to a section of $\mathbb{Q}$. These characterizations can be extended to abelian groups of finite special rank r as we now demonstrate.

EXERCISE 6.4. *Prove that every torsion-free locally cyclic group is isomorphic to some subgroup of the additive group $\mathbb{Q}$ of rational numbers.*

EXERCISE 6.5. *Prove the following assertions. Suppose that G is an abelian group. If G is finite, then $r(G) = r(G/\mathbf{Frat}(G))$. In particular, if G is a finite p-group, for some prime p, then $r(G) = r(G/G^p)$. If G is a finitely generated torsion-free group, then $r(G) = r_0(G)$ and hence $r(G) = r(G/G^p)$ for all primes p.*

We next explore the connection between the special rank and some of the other ranks that we have so far encountered. In particular we shall consider in detail the relationship between the special rank, the section rank and the 0-rank. The following result is very easy.

6.1.6. PROPOSITION. *Let G be a group of finite special rank r. Then the section p-rank of G is at most r for each prime p. In particular G has finite bounded section rank at most r.*

6.1.7. LEMMA. *Let p be a prime and let G be an abelian group whose p'-component is trivial. Then $sr_p(G)$ is finite if and only if $r(G)$ is finite and in this case $sr_p(G) = r(G)$.*

EXERCISE 6.6. *Prove Lemma 6.1.7.*

6.1.8. LEMMA. *Let p be a prime and let G be a finite p-group. Then $sr_p(G) = r(G)$.*

PROOF. Clearly G has finite special rank and Proposition 6.1.6 shows that $\mathbf{sr}_p(G) \leq r(G)$. Let $\mathbf{sr}_p(G) = s$ and let K be an arbitrary subgroup of G. Then $K/\mathbf{Frat}(K)$ is elementary abelian so that $|K/\mathbf{Frat}(K)| \leq p^s$. The Burnside basis theorem (see [**96**, Theorem 5.1.1], for example) implies that the minimal number of generators of K coincides with the minimal number of generators of $K/\mathbf{Frat}(K)$ and hence K is at most s-generator. Hence $r(G) \leq s = \mathbf{sr}_p(G)$ and we deduce that $\mathbf{sr}_p(G) = r(G)$, as required. □

6.1.9. COROLLARY. *Let p be a prime and let G be a locally finite p-group. Then $sr_p(G)$ is finite if and only if $r(G)$ is finite and in this case $sr_p(G) = r(G)$.*

PROOF. If G has finite special rank, then Proposition 6.1.6 shows that $\mathbf{sr}_p(G)$ is finite and $\mathbf{sr}_p(G) \leq r(G)$. Suppose that $\mathbf{sr}_p(G) = s$ is finite and let K be an arbitrary finite subgroup of G. Then $\mathbf{sr}_p(K) \leq s$ and Lemma 6.1.8 implies that $r(K) = \mathbf{sr}_p(K) \leq s$. Since this is true for every finite subgroup, we have $r(G) \leq \mathbf{sr}_p(G)$. In any case, we have $\mathbf{sr}_p(G) = r(G)$, as required. □

6.1.10. THEOREM. *Let P be a locally finite p-group for some prime p. The following are equivalent.*

(i) P has finite special rank;

(ii) P has finite section p-rank;

(iii) P is Chernikov.

Furthermore, in this case $sr_p(P) = r(P)$.

PROOF. Property (iii) is equivalent to property (ii) using Corollary 3.2.21. Corollary 6.1.9 implies that property (ii) is equivalent to property (i). □

We note that the implication that (i) implies (iii) was proved by N. N. Myagkova [200].

By contrast, for each suitably large prime p, A. Yu. Ol'shanskii [208] and E. Rips have constructed infinite 2-generator groups all of whose proper subgroups are cyclic of order p. These "Tarski monsters" (so named after A. Tarski, who first proposed their possible existence) clearly are p-groups of rank 2 and section p-rank 1 and certainly are not Chernikov groups.

We observe the following easy result.

6.1.11. LEMMA. Let G be an abelian group with torsion subgroup $Tor(G)$. Then $r(G) = r(G/Tor(G)) + r(Tor(G))$.

EXERCISE 6.7. Prove Lemma 6.1.11.

6.1.12. LEMMA. Let G be an abelian group of finite special rank. Then $r(G) = r_0(G) + \max\{sr_p(Tor(G))|p \in \Pi(G)\}$.

PROOF. Indeed, by Lemma 6.1.11,

$$r(G) = r(G/Tor(G)) + r(Tor(G)).$$

Since $G/Tor(G)$ is torsion-free we have

$$r(G/Tor(G)) = r_0(G/Tor(G)) = r_0(G).$$

Let G_p be the Sylow p-subgroup of G. Then $Tor(G) = \underset{p \in \Pi(G)}{\mathrm{Dr}} G_p$. By Corollary 6.1.5, $r(Tor(G)) = \max\{r(G_p)|p \in \Pi(G)\}$. Finally, Theorem 6.1.10 shows that $r(G_p) = sr_p(G_p) = sr_p(Tor(G))$ and the result follows. □

6.1.13. LEMMA. Let G be an abelian group of rank r. Then G can be embedded in the direct product of k copies of $\mathbb{Q}/\mathbb{Z}$ and t copies of $\mathbb{Q}$, where $r = k + t$. Furthermore, $k = r(Tor(G))$ and $t = r_0(G/Tor(G)) = r(G/Tor(G))$.

PROOF. If p is a prime we denote the Sylow p-subgroup of a group H by H_p. Let U be a maximal torsion-free subgroup of our group G, so that G/U is periodic. Since G has finite rank it follows that $\Omega_1(G_p)$

and $\Omega_1((G/U)_p)$ are both finite and it is easy to see that $\Omega_1(G_p)U/U = \Omega_1((G/U)_p)$. It follows that $\mathbf{sr}_p(G/U) = \mathbf{sr}_p(\mathbf{Tor}\,(G))$. Using Corollary 6.1.5 and Lemma 6.1.7 we see that $\mathbf{r}(G/U) = \mathbf{r}(\mathbf{Tor}\,(G))$. Since $(G/U)_p$ is a direct product of Prüfer p-groups and cyclic p-groups, by Lemma 6.1.7 and Proposition 3.2.3, and since every cyclic p-subgroup may be embedded in a Prüfer p-group we obtain the embedding

$$G/U \longrightarrow \underbrace{\mathbb{Q}/\mathbb{Z} \times \mathbb{Q}/\mathbb{Z} \times \cdots \times \mathbb{Q}/\mathbb{Z}}_{k}$$

where $k = \mathbf{r}(G/U) = \mathbf{r}(\mathbf{Tor}\,(G))$.

On the other hand, $G/\mathbf{Tor}\,(G)$ is torsion-free, so $\mathbf{r}(G/\mathbf{Tor}\,(G)) = \mathbf{r}_0(G/\mathbf{Tor}\,(G))$. We may embed $G/\mathbf{Tor}\,(G)$ in its divisible envelope

$$G/\mathbf{Tor}\,(G) \longrightarrow \underbrace{\mathbb{Q} \times \mathbb{Q} \times \cdots \times \mathbb{Q}}_{t}$$

where $t = \mathbf{r}_0(G/\mathbf{Tor}\,(G)) = \mathbf{r}(G/\mathbf{Tor}\,(G))$. Finally we note that $\mathbf{Tor}\,(G) \cap U = 1$ and Remak's Theorem implies that there is an embedding of G into $G/\mathbf{Tor}\,(G) \times G/U$, so we obtain the embedding

$$G \longrightarrow \underbrace{\mathbb{Q}/\mathbb{Z} \times \mathbb{Q}/\mathbb{Z} \times \cdots \times \mathbb{Q}/\mathbb{Z}}_{k} \times \underbrace{\mathbb{Q} \times \mathbb{Q} \times \cdots \times \mathbb{Q}}_{t}$$

where $k + t = \mathbf{r}(\mathbf{Tor}\,(G)) + \mathbf{r}(G/\mathbf{Tor}\,(G)) = \mathbf{r}(G)$, by Lemma 6.1.11. $\qquad\square$

Note that when G is a torsion-free abelian group of special rank r then there is an r-generator subgroup $H = \langle a_1, \ldots, a_r \rangle$ that is not $(r-1)$-generator. Then $H \cong \underbrace{\mathbb{Z} \times \cdots \times \mathbb{Z}}_{r}$ by the fundamental theorem of finitely generated abelian groups. Consequently, the $\mathbb{Z}$-rank of G is r and it is easy to see that $\mathbf{r}_0(G) = r = \mathbf{r}(G)$ in this case. We shall now examine the relationship between $\mathbf{r}(G)$ and $\mathbf{r}_0(G)$ in general. As we shall see, the two invariants are equal for a wider class of groups. We recall that if G is a group and A is a normal subgroup of G, then G splits over A if there is a subgroup H of G such that $G = AH$ and $A \cap H = 1$. The subgroup H is then called a complement to A in G.

6.1.14. LEMMA. Let G be a torsion-free abelian group of finite 0-rank and let p be a prime such that $p \notin \mathbf{Sp}(G)$. Then $\mathbf{r}(G/G^p) = \mathbf{r}_0(G) = \mathbf{r}(G)$.

PROOF. As we saw above, G has finite special rank and $\mathbf{r}_0(G) = \mathbf{r}(G)$. Let $\mathbf{Sp}(G) = \sigma$ and let K be a finitely generated subgroup of G such that $\mathbf{r}_0(G) = \mathbf{r}_0(K)$, so G/K is periodic. Let q be a prime and let S/K be the Sylow q-subgroup of G/K. By Lemma 6.1.2, S/K has finite special rank and Theorem 6.1.10 shows that S/K is Chernikov. Let D/K be the divisible part of G/K, so $\Pi(D/K) \subseteq \sigma$. Since D/K is divisible it is complemented in G/K, say $G/K = D/K \times L/K$ for some subgroup L/K (see [88, Theorem 21.2], for example). Let P/K be the Sylow p-subgroup of L/K (and if $p \notin \Pi(L/K)$, then set $P = K$). The finiteness of P/K shows that P is finitely generated. Then $P \neq P^p$ and, moreover, $\mathbf{r}(P/P^p) = \mathbf{r}_0(P) = \mathbf{r}_0(G)$. It is clear that G/P is a p'-group, so P/P^p is the Sylow p-subgroup of G/P^p. It follows that $G/P^p = P/P^p \times Q/P^p$ where Q/P^p is the Sylow p'-subgroup of G/P^p. Hence $G/Q \cong P/P^p$ is an elementary abelian p-group such that $\mathbf{r}(G/Q) = \mathbf{r}(P/P^p) = \mathbf{r}_0(G)$. On the other hand, G/G^p is an elementary abelian p-group and Lemma 6.1.2 shows that $\mathbf{r}(G/G^p) \leq \mathbf{r}(G) = \mathbf{r}_0(G)$. Since $G^p \leq Q$ it follows that $Q = G^p$ and therefore $\mathbf{r}(G/G^p) = \mathbf{r}_0(G) = \mathbf{r}(G)$. $\qquad\square$

6.2. The Structure of Groups Having Finite Special Rank

In this section our goal is to investigate the relationship between the special rank and the 0-rank further.

6.2.1. LEMMA. *Let G be a group and suppose that G has a series of normal subgroups*

$$1 = A_0 \leq A_1 \leq \cdots \leq A_n \leq A_{n+1} = G$$

such that

(i) *A_{j+1}/A_j is torsion-free abelian of finite 0-rank, for $0 \leq j \leq n$;*
(ii) *$\boldsymbol{Sp}(A_{j+1}/A_j)$ is finite for $0 \leq j \leq n$.*

Then there is a prime p and normal subgroups $K \leq L$ of G such that G/L is finite and L/K is an elementary abelian p-group whose section p-rank is $\boldsymbol{r}_0(A_1) + \boldsymbol{r}_0(A_2/A_1) + \cdots + \boldsymbol{r}_0(A_{n+1}/A_n)$.

PROOF. Let p be a prime such that $p \notin \mathbf{Sp}(G)$. It follows that $p \notin \mathbf{Sp}(A_{j+1}/A_j)$ for each j such that $0 \leq j \leq n$. Let $B_1 = A_1^p$. By Lemma 6.1.14, $\mathbf{r}(A_1/B_1) = \mathbf{r}(A_1) = \mathbf{r}_0(A_1)$. Let $G_1 = C_G(A_1/B_1)$ and $C_2 = C_{A_2}(A_1/B_1) = A_2 \cap G_1$. Clearly, $A_1 \leq C_2$ and A_1/B_1 is a central factor of C_2/B_1, so C_2/B_1 is nilpotent of class at most 2. Also, $C_2' \leq A_1$. Since A_1/B_1 is finite, G/G_1 is also finite. Hence if $x, y \in C_2$ we have $[y^p B_1, x B_1] = [y B_1, x B_1]^p = 1$, since A_1/B_1 has exponent p. Therefore, $y^p B_1 \in \zeta(C_2/B_1)$, for all $y \in C_2$ and it follows that $Z_2/B_1 = (C_2/B_1)^p \leq \zeta(C_2/B_1)$. Clearly, Z_2 is G-invariant. We note that C_2/B_1 has finite special rank. By Lemma 6.1.2 its elementary abelian factor group $(C_2/B_1)/(Z_2/B_1)$ has finite special rank and therefore is finite. Since $|A_2 : Z_2| = |A_2 : C_2| \cdot |C_2 : Z_2|$ is finite we have

$$\mathbf{r}_0(A_2/A_1) = \mathbf{r}_0(A_2/B_1) = \mathbf{r}_0(Z_2/B_1).$$

Now Z_2/B_1 is abelian and has finite torsion subgroup A_1/B_1 so $Z_2/B_1 = A_1/B_1 \times U/B_1$ where $U/B_1 \cong Z_2/A_1$ is a torsion-free subgroup (see [88, Theorem 27.5], for example). Let $B_2/B_1 = (Z_2/B_1)^p = (U/B_1)^p$, a G-invariant subgroup. Since $U/B_1 \cong Z_2/A_1$ and $p \notin \mathbf{Sp}(Z_2/A_1)$ it follows from Lemma 6.1.14 that

$$\mathbf{r}((U/B_1)/(U/B_1)^p) = \mathbf{r}_0(U/B_1) = \mathbf{r}_0(Z_2/A_1) = \mathbf{r}_0(A_2/A_1).$$

Furthermore, since $Z_2/B_1 = A_1/B_1 \times U/B_1$ we have that

$$(Z_2/B_1)/(Z_2/B_1)^p \cong (A_1/B_1) \times (U/B_1)/(U/B_1)^p.$$

In turn, this implies that

$$\mathbf{r}(Z_2/B_2) = \mathbf{r}(A_1/B_1) + \mathbf{r}((U/B_1)/(U/B_1)^p) = \mathbf{r}_0(A_1) + \mathbf{r}_0(A_2/A_1).$$

Using similar arguments, after finitely many steps we construct normal subgroups K, L of G such that $K \leq L$, the index $|G : L|$ is finite, L/K is an elementary abelian p-group and

$$\mathbf{r}(L/K) = \mathbf{r}_0(A_1) + \mathbf{r}_0(A_2/A_1) + \cdots + \mathbf{r}_0(A_{n+1}/A_n),$$

as required.   □

We deduce the following important result due to Zaitsev [268].

6.2.2. THEOREM. *Let G be a polyrational group. Then $\mathbf{r}_0(G) = \mathbf{bs}(G) = \mathbf{r}(G)$.*

PROOF. Let

$$1 = A_0 \lhd A_1 \lhd \ldots \lhd A_n \lhd A_{n+1} = G$$

be a finite series such that A_{j+1}/A_j is a torsion-free abelian group of finite 0-rank, for each j such that $0 \leq j \leq n$. Let $\mathbf{r}_0(A_{j+1}/A_j) = k_{j+1}$ and choose the subset $M_{j+1} = \{a_{j+1,m} | 1 \leq m \leq k_{j+1}\}$ such that $A_{j+1}/\langle M_{j+1}\rangle A_j$ is periodic. Let

$$H = \langle a_{j+1,m} | 0 \leq j \leq n, 1 \leq m \leq k_{j+1}\rangle$$

and let $B_j = H \cap A_j$, for $0 \leq j \leq n$. The definition of H implies that $\mathbf{r}_0(A_{j+1}/A_j) = \mathbf{r}_0(B_{j+1}/B_j)$ and hence $\mathbf{r}_0(G) = \mathbf{r}_0(H)$. Since H is finitely generated, H is minimax by Theorem 5.2.8. By Theorem 2.4.13, H has a series of normal subgroups $U \leq V \leq H$ such that U is torsion-free nilpotent, V/U is finitely generated torsion-free abelian and H/V is finite. Then $\mathbf{r}_0(H) = \mathbf{r}_0(V)$ and clearly V has a series of normal subgroups whose factors are minimax torsion-free abelian groups. By

Lemma 6.2.1, V contains normal subgroups K, L such that $K \leq L$, L/K is an elementary abelian p-group for some $p \notin \mathbf{Sp}(V) = \mathbf{Sp}(H)$ and $\mathbf{r}(L/K) = \mathbf{r}_0(V) = \mathbf{r}_0(H) = \mathbf{r}_0(G)$. Since $\mathbf{r}(G) \geq \mathbf{r}(L/K)$, by Lemma 6.1.2, we have $\mathbf{r}(G) \geq \mathbf{r}_0(G)$. On the other hand, Lemma 6.1.2 implies that

$$\mathbf{r}(G) \leq \mathbf{r}(A_1) + \cdots + \mathbf{r}(A_{n+1}/A_n)$$
$$= \mathbf{r}_0(A_1) + \mathbf{r}_0(A_2/A_1) + \cdots + \mathbf{r}_0(A_{n+1}/A_n) = \mathbf{r}_0(G).$$

Hence $\mathbf{r}(G) = \mathbf{r}_0(G)$.

On the other hand, Proposition 6.1.6 implies that $\mathbf{bs}(G) \leq \mathbf{r}(G)$. However, $\mathbf{bs}(G) \geq \mathbf{sr}_p(G) = \mathbf{r}_0(G) = \mathbf{r}(G)$ so that $\mathbf{bs}(G) = \mathbf{r}(G)$. □

For the next result, we note that the function f_4 was first defined in Theorem 2.4.13. The following result appears in [**67**].

6.2.3. COROLLARY. *Let G be a locally generalized radical group of finite 0-rank r. Then $G/\mathbf{Tor}\,(G)$ has finite special rank. Moreover, $r_0(G) \leq r(G/\mathbf{Tor}\,(G)) \leq r(G)$ and $r(G/\mathbf{Tor}\,(G)) \leq r_0(G) + f_4(r)$.*

PROOF. Let $T = \mathbf{Tor}\,(G)$. By Theorem 2.4.13, G has a series of normal subgroups $T \leq L \leq K \leq S \leq G$ such that L/T is torsion-free nilpotent, K/L is finitely generated torsion-free abelian, G/K is finite and $|G/K| \leq f_4(r)$. The subgroup K/T is polyrational so we deduce, from Theorem 6.2.2, that $\mathbf{r}(K/T) = \mathbf{r}_0(K/T) = \mathbf{r}_0(K)$. It follows, by Lemma 2.2.3, that $\mathbf{r}_0(G) \leq \mathbf{r}(G/\mathbf{Tor}\,(G))$. Since G/K is finite, $\mathbf{r}_0(G) = \mathbf{r}_0(K)$ and Lemma 6.1.2 now shows that

$$\mathbf{r}(G/T) \leq \mathbf{r}(K/T) + \mathbf{r}(G/K) \leq \mathbf{r}_0(G) + |G/K| \leq \mathbf{r}_0(G) + f_4(r).$$

□

We consider next the structure of locally radical groups of finite special rank. First, we obtain the following result concerning locally nilpotent groups due to V. M. Glushkov [**93**].

6.2.4. PROPOSITION. *Let G be a torsion-free locally nilpotent group. If G has finite special rank r, then G has finite 0-rank. Moreover, G is a polyrational nilpotent group and $\mathbf{r}(G) = \mathbf{r}_0(G)$.*

PROOF. Let L be an arbitrary finitely generated subgroup of G. Then L is torsion-free nilpotent and hence its upper central series has torsion-free abelian factors. By Lemmas 6.1.2 and 6.1.12, every such factor has finite 0-rank at most r and hence L is a polyrational group. By Theorem 6.2.2, $\mathbf{r}_0(L) = \mathbf{r}(L) \leq r$. Corollary 2.3.4 implies that L is nilpotent. Since the factors of the upper central series of L are torsion-free, $\mathbf{ncl}\,(L) \leq r - 1$. This is true for every finitely generated subgroup of G so G is nilpotent of class at most $r - 1$. By Lemma 2.5.1, G has 0-rank at most r. As above, we see that G is polyrational and $\mathbf{r}_0(G) = \mathbf{r}(G) = r$. $\qquad\qquad\qquad\qquad\qquad\qquad\qquad\qquad\qquad\qquad\qquad\Box$

As we have seen, the case of locally nilpotent groups turns out to be rather interesting, perhaps even peculiar. In the case of locally finite p-groups, the special rank and the section p-rank coincide and in the case of torsion-free locally nilpotent groups we see that the special rank and the 0-rank coincide. Furthermore, torsion-free locally nilpotent groups of finite special rank are nilpotent. Naturally the question arises whether the same is true for periodic locally nilpotent groups. By Theorem 6.1.10, a periodic locally nilpotent group of finite special rank is a direct product of Chernikov p-groups. Certainly there are Chernikov p-groups which are not nilpotent, so the answer to the above question is in the negative. However, every such group is soluble. A further natural question arises as to whether the derived length of a Chernikov p-group of finite rank r is bounded by a function of r. In fact no such bound exists even for finite p-groups as the following examples of Yu I. Merzlyakov [**194**] show. In particular, we shall observe that a periodic hypercentral group of finite special rank need not be soluble.

6.2.5. EXAMPLE. Let p be a prime and let m, n be natural numbers such that $m \geq 2$. For $1 \leq k \leq m - 1$ let

$$P_k = \{E + p^k A | A \in M_n(\mathbb{Z}/p^m\mathbb{Z})\},$$

where, as usual, E denotes the (in this case) $n \times n$ identity matrix. It is clear that $P_{k+1} \leq P_k$, for all natural numbers k and that P_k is a normal subgroup of P_1.

Let $\phi_k : P_k \longrightarrow P_{m-1}$ be the mapping defined by $\phi_k(E + p^k A) = E + p^{m-1} A$, for $A \in M_n(\mathbb{Z}/p^m\mathbb{Z})$. This map is an epimorphism. Indeed, we have, for $A, B \in M_n(\mathbb{Z}/p^m\mathbb{Z})$,

$$(E + p^k A)(E + p^k B) = E + p^k A + p^k B + p^{2k} AB = E + p^k(A + B + p^k AB),$$

so that

$$\phi_k((E + p^k A)(E + p^k B)) = E + p^{m-1}(A + B + p^k AB)$$
$$= E + p^{m-1} A + p^{m-1} B + p^{k+m-1} AB$$
$$= E + p^{m-1} A + p^{m-1} B,$$

since $k + m - 1 \geq m$ and hence $p^{k+m-1} AB = 0$. On the other hand,

$$\phi_k(E + p^k A)\phi_k(E + p^k B) = (E + p^{m-1}A)(E + p^{m-1}B)$$
$$= E + p^{m-1}A + p^{m-1}B + p^{2m-2}AB$$
$$= E + p^{m-1}A + p^{m-1}B,$$

since $2m-2 \geq m$ and hence $p^{2m-2}AB = 0$. Thus ϕ_k is a homomorphism which is easily seen to be an onto map. We note also that $\mathbf{ker}(\phi_k) = P_{k+1}$ and hence $P_k/P_{k+1} \cong P_{m-1}$.

Now consider the mapping $\eta : M_n(p^{m-1}\mathbb{Z}/p^m\mathbb{Z}) \longrightarrow P_{m-1}$, defined by $\eta(p^{m-1}A) = E + p^{m-1}A$, for $A \in M_n(\mathbb{Z}/p^m\mathbb{Z})$. Again we have

$$\eta(p^{m-1}A)\eta(p^{m-1}B) = (E + p^{m-1}A)(E + p^{m-1}B)$$
$$= E + p^{m-1}A + p^{m-1}B + p^{2m-2}AB$$
$$= E + p^{m-1}A + p^{m-1}B$$
$$= E + p^{m-1}(A + B)$$
$$= \eta(p^{m-1}A + p^{m-1}B).$$

It follows that η is a homomorphism from the additive group of $M_n(p^{m-1}\mathbb{Z}/p^m\mathbb{Z})$ into P_{m-1}. It is easy to see that η is a bijection and hence η is an isomorphism. Consequently, P_{m-1} is an elementary abelian p-group and $\mathbf{r}(P_{m-1}) = \mathbf{sr}_p(P_{m-1}) = n^2$. Furthermore, this means that for each natural number k, P_k/P_{k+1} is an elementary abelian p-group.

We next show that P_1 also has rank n^2. Let H be an arbitrary subgroup of P_1. We claim that $\phi_k(H \cap P_k) \leq \phi_{k+1}(H \cap P_{k+1})$. Indeed if $Y \in \phi_k(H \cap P_k)$ then $Y = \phi_k(X)$ where $X \in H \cap P_k$. Thus $X = E + p^k A$ for some A and $Y = E + p^{m-1}A$. Now

$$X^p = E + p^{k+1}A + \binom{p}{2}p^{2k}A^2 + \cdots + p^{kp}A^p$$
$$= E + p^{k+1}\left(A + \binom{p}{2}p^{k-1}A + \cdots + p^{kp-k-1}A^p\right),$$

which is an element of $H \cap P_{k+1}$. Hence

$$\phi_{k+1}(X^p) = E + p^{m-1}\left(A + \binom{p}{2}p^{k-1}A + \cdots + p^{kp-k-1}A^p\right)$$
$$= E + p^{m-1}A = Y,$$

which proves the claim. From this it follows that

$$\phi_1(H) \le \phi_2(H \cap P_2) \le \cdots \le \phi_{m-1}(H \cap P_{m-1}) = H \cap P_{m-1}$$

and of course we also have

$$H \cap P_{m-1} \le H \cap P_{m-2} \le \cdots \le H \cap P_2 \le H.$$

We let $0 \le t_1 \le \cdots \le t_{m-2} \le t_{m-1} = t \le n^2$ be integers which are the dimensions of the subspaces $\phi_1(H), \phi_2(H \cap P_2), \ldots, \phi_{m-2}(H \cap P_{m-2})$ respectively, and we let $\{A_j | 1 \le j \le t_k\}$ be a corresponding basis for $\phi_k(H \cap P_k)$ obtained by extending the basis $\{A_j | 1 \le j \le t_{k-1}\}$ for $\phi_{k-1}(H \cap P_{k-1})$ in the usual manner, for $2 \le k \le m - 1$. Thus every element of $H \cap P_{m-1}$ can be written in the form $A_1^{s_1} \ldots A_t^{s_t}$, for certain integers s_j. We write $A_j = E + p^{m-1}D_j$, for some matrix D_j, for $1 \le j \le t$.

Now there are integers $m(j)$ such that $A_j = B_j^{b_j}$, where $b_j = p^{m(j)}$, $B_j = E + p^{m-1-m(j)}D_j \in H \cap P_{m-1-m(j)}$ and $m(j) \le m - 2$ is chosen largest with this property. Thus $A_j = B_j^{b_j}$ and B_j is not a pth power of some element of H. We claim that $H = \langle B_j | 1 \le j \le t \rangle$. Note that, by construction, $B_j \in H$. Note also that

$$H \cap P_{m-1} \le \langle A_j | 1 \le j \le t \rangle \le \langle B_j | 1 \le j \le t \rangle.$$

We suppose inductively that $H \cap P_{l+1} \le \langle B_j | 1 \le j \le t \rangle$ and now prove that $H \cap P_l \le \langle B_j | 1 \le j \le t \rangle$ also. To this end, let $X \in (H \cap P_l) \setminus (H \cap P_{l+1})$. Then $X = E + p^l A$ for some matrix A and $\phi_l(X) = E + p^{m-1}A = A_1^{s_1} \ldots A_t^{s_t}$. We note that in general

$$(E + p^k B)^p \equiv E + p^{k+1}B \pmod{P_{k+2}}.$$

We note also that if $l < m - 1 - m(j)$ then $l + 1 \le m - 1 - m(j)$ and hence $B_j = E + p^{m-1-m(j)}D_j \in P_{m-1-m(j)} \le P_{l+1}$ in this case. Consequently, in the computations which follow, we may assume that $l \ge m - 1 - m(j) \ge 1$. Working modulo P_{l+1} it is now easy to see that

$$(E + p^{m-1-m(1)}D_1)^{s_1 p^{m(1)+l+1-m}} \ldots (E + p^{m-1-m(t)}D_t)^{s_t p^{m(t)+l+1-m}}$$
$$= (E + s_1 p^l D_1) \ldots (E + s_t p^l D_t)$$
$$= E + p^l(s_1 D_1 + \cdots + s_t D_t)$$

On the other hand, $X = E + p^l A$ so

$$\phi_l(X) = E + p^{m-1}A = (E + p^{m-1}D_1)^{s_1} \ldots (E + p^{m-1}D_t)^{s_t}$$
$$= E + p^{m-1}(s_1 D_1 + \ldots s_t D_t).$$

Thus $A = s_1 D_1 + \cdots + s_t D_t + pB$ for some matrix B and hence

$$X = E + p^l A \equiv E + p^l(s_1 D_1 + \ldots s_t D_t) \pmod{P_{l+1}}.$$

It follows that we have

$$X = B_1^{s_1 p^{m(1)+l+1-m}} \ldots B_t^{s_t p^{m(t)+l+1-m}} Y,$$

where $Y \in P_{l+1}$. Also $X, B_j \in H$, for $1 \le j \le t$ and we deduce that $Y \in H$ also. Thus $Y \in H \cap P_{l+1}$. However, by the induction hypothesis, $H \cap P_{l+1} \le \langle B_j | 1 \le j \le t \rangle$ and hence $X \in \langle B_j | 1 \le j \le t \rangle$. Consequently, $H \cap P_l \le \langle B_j | 1 \le j \le t \rangle$ and it follows by induction that $H = \langle B_j | 1 \le j \le t \rangle$. Hence H is t-generator so P_1 has rank at most n^2 and since P_{m-1} has rank n^2 it follows that the rank of P_1 is precisely n^2.

Let

$$P = \{A = (a_{j,k}) \in GL_n(\mathbb{Z}/p^m \mathbb{Z}) | a_{j,j} \equiv 1 \pmod{p},$$
$$a_{j,k} \in p\mathbb{Z} \text{ whenever } j > k\}.$$

Then, according to Corollary 4.3.5, P is a Sylow p-subgroup of $GL_n(\mathbb{Z}/p^m \mathbb{Z})$. It is easy to see that P_1 is normal in P and that $P/P_1 \cong UT_n(\mathbb{F}_p)$. As we saw in Chapter 1, the upper central series of $UT_n(\mathbb{F}_p)$ is the chain of subgroups

$$E = UT_n^n(\mathbb{F}_p) \le UT_n^{n-1}(\mathbb{F}_p) \le \cdots \le UT_n^2(\mathbb{F}_p) \le UT_n^1(\mathbb{F}_p) = UT_n(\mathbb{F}_p)$$

and, if F_+ denotes the additive group of $\mathbb{F}_p$, then

$$UT_n^m(\mathbb{F}_p)/UT_n^{m+1}(\mathbb{F}_p) \cong \underbrace{F_+ \oplus \cdots \oplus F_+}_{n-m}.$$

In particular, it follows that $\mathbf{r}(UT_n(\mathbb{Z}/p\mathbb{Z})) \le n(n-1)/2$. Since P_1 has rank precisely n^2 it follows, by Lemma 6.1.2, that

$$\mathbf{r}(P) \le n(n-1)/2 + n^2 = n(3n-1)/2.$$

Now let $n \ge 3$ be a natural number and let π denote an infinite set of odd primes. Let

$$\sigma = \{t(p) | p \in \pi\}$$

be a set of natural numbers with the property that $t(p) < t(q)$ whenever $p < q$ are primes in the set π.

For each prime $p \in \pi$ let $G_p = \{E + pA | A \in M_n(\mathbb{Z}/p^{t(p)}\mathbb{Z})\}$. As we saw above, the finite p-group G_p has finite special rank at most n^2. If we let $T_{j,k}(s)$ denote the matrix $E + sE_{j,k}$, where $E_{j,k}$ is the matrix whose only non-zero entry is 1 in the (j, k) place, then $[T_{j,k}(s), T_{k,l}(w)] = T_{j,l}(sw)$ for distinct j, k, l (see [129, 3.2.1], for example). If we set $s = w = p^u$, where $u = 2^{t(p)}$, then it is easy to see that $\mathbf{dl}(G_p) < \mathbf{dl}(G_q)$

whenever $p < q$. It follows that the group $G = \underset{p \in \pi}{\mathrm{Dr}} \, G_p$ is not soluble. Also G is hypercentral and, by Corollary 6.1.5, G has finite special rank at most n^2.

6.2.6. LEMMA. *Let G be a finitely generated nilpotent group and suppose that $T = \boldsymbol{Tor}\,(G)$ is a p-group for some prime p. Then*

$$r(G) = r(T) + r_0(G/T) = sr_p(G) + r_0(G/T).$$

PROOF. It follows from Lemma 6.1.7 that $\mathbf{r}(T) = \mathbf{sr}_p(T)$. Proposition 6.2.4 shows that $\mathbf{r}(G/T) = \mathbf{r}_0(G/T)$. Therefore

$$\mathbf{r}(G) \le \mathbf{r}(T) + \mathbf{r}(G/T) = \mathbf{sr}_p(T) + \mathbf{r}_0(G/T).$$

According to Lemma 3.2.22, G has normal subgroups U and V such that $U \le V$ and V/U is an elementary abelian p-group of order p^k, where $k = \mathbf{sr}_p(T) + \mathbf{r}_0(G/T)$. Consequently, $\mathbf{sr}_p(G) \ge \mathbf{sr}_p(T) + \mathbf{r}_0(G/T)$. On the other hand, $\mathbf{sr}_p(G) \le \mathbf{r}(G)$ so we have $\mathbf{r}(G) = \mathbf{r}(T) + \mathbf{r}_0(G/T)$, as required.   □

6.2.7. COROLLARY. *Let G be a finitely generated nilpotent group and suppose that $T = \boldsymbol{Tor}\,(G)$ is a p-group for some prime p. Then $r(G) = sr_p(G)$.*

PROOF. By Lemma 6.1.7, $\mathbf{r}(T) = \mathbf{sr}_p(T)$ and it suffices to apply Corollary 3.2.25 and Lemma 6.2.6 to deduce that $\mathbf{sr}_p(G) = \mathbf{r}(G)$.   □

6.2.8. COROLLARY. *Let G be a finitely generated nilpotent group. Then*

$$r(G) = r_0(G) + \max\{sr_p(\boldsymbol{Tor}\,(G)) | p \in \Pi(G)\}.$$

EXERCISE 6.8. *Prove Corollary 6.2.8.*

6.2.9. COROLLARY. *Let G be a finitely generated nilpotent group. Then*

$$r(G) = \max\{sr_p(G) | p \in \Pi(G)\}.$$

PROOF. Let q be a prime such that $\max\{\mathbf{sr}_p(G) | p \in \Pi(G)\} = \mathbf{sr}_q(G)$. By Proposition 6.1.6, $\mathbf{sr}_p(G) \le \mathbf{r}(G)$ so we have $\mathbf{sr}_q(G) \le \mathbf{r}(G)$.

We let G_p be the Sylow p-subgroup of G, for each prime p and let $T = \boldsymbol{Tor}\,(G) = \underset{p \in \Pi(G)}{\mathrm{Dr}} \, G_p$. Let $Q_p = \underset{q \in \Pi(G),\, q \neq p}{\mathrm{Dr}} \, G_q$. Corollary 3.2.25 shows that $\mathbf{sr}_p(G/Q_p) = \mathbf{sr}_p(T/Q_p) + \mathbf{r}_0(G/T)$. Using Lemma 3.2.2 we deduce that $\mathbf{sr}_p(T/Q_p) = \mathbf{sr}_p(T)$ and $\mathbf{sr}_p(G/Q_p) = \mathbf{sr}_p(G)$. Hence $\mathbf{sr}_p(G) = \mathbf{sr}_p(T) + \mathbf{r}_0(G/T)$. Since this is true for each $p \in \Pi(G)$ we deduce that $\max\{\mathbf{sr}_p(T) | p \in \Pi(G)\} = \mathbf{sr}_q(T)$.

By Lemma 6.1.2, we have $\mathbf{r}(G) \le \mathbf{r}(T) + \mathbf{r}(G/T)$ and Proposition 6.2.4 shows that $\mathbf{r}(G/T) = \mathbf{r}_0(G/T)$. By Corollary 6.1.5, $\mathbf{r}(T) =$

$\max\{\mathbf{r}(G_p)|p \in \Pi(G)\}$ and Lemma 6.1.8 implies that $\mathbf{sr}_p(G_p) = \mathbf{r}(G_p)$ for each prime p. It follows that $\mathbf{r}(T) = \mathbf{sr}_q(G_q) = \mathbf{sr}_q(T)$. Furthermore,

$$\mathbf{r}(T) + \mathbf{r}(G/T) = \mathbf{sr}_q(T) + \mathbf{r}_0(G/T) = \mathbf{sr}_q(T/Q_q) + \mathbf{r}_0((G/Q_q)/(T/Q_q)).$$

Corollary 3.2.25 shows that

$$\mathbf{sr}_q(G/Q_q) = \mathbf{sr}_q(T/Q_q) + \mathbf{r}_0((G/Q_q)/(T/Q_q)).$$

On the other hand, Lemma 3.2.2 shows that $\mathbf{sr}_q(G/Q_q) = \mathbf{sr}_q(G)$. From this it follows that $\mathbf{r}(T) + \mathbf{r}(G/T) = \mathbf{sr}_q(G)$ and hence $\mathbf{r}(G) \leq \mathbf{sr}_q(G)$, which proves the result. $\qquad\square$

6.2.10. COROLLARY. *Let G be a locally nilpotent group and suppose that $T = \mathbf{Tor}\,(G)$ is a p-group for some prime p. Then G has finite special rank if and only if G has finite section p-rank. In this case $\mathbf{r}(G) = \mathbf{sr}_p(G)$.*

PROOF. Suppose first that $\mathbf{sr}_p(G)$ is finite. Let K be an arbitrary finitely generated subgroup of G. By Corollary 6.2.7, $\mathbf{r}(K) = \mathbf{sr}_p(K)$ and, in particular, $\mathbf{r}(K) \leq \mathbf{sr}_p(G)$. It follows that $\mathbf{r}(G)$ is finite and $\mathbf{r}(G) \leq \mathbf{sr}_p(G)$.

Conversely if $\mathbf{r}(G)$ is finite, then Proposition 6.1.6 shows that $\mathbf{sr}_p(G)$ is also finite and $\mathbf{sr}_p(G) \leq \mathbf{r}(G)$. It follows that $\mathbf{sr}_p(G) = \mathbf{r}(G)$. $\qquad\square$

6.2.11. THEOREM. *Let G be a locally nilpotent group. Then G has finite special rank if and only if G has finite bounded section rank. In this case $\mathbf{r}(G) = \mathbf{bs}(G)$.*

PROOF. If $\mathbf{r}(G)$ is finite, then Proposition 6.1.6 shows that $\mathbf{sr}_p(G)$ is finite and $\mathbf{sr}_p(G) \leq \mathbf{r}(G)$, for each prime $p \in \Pi(G)$. Hence $\mathbf{bs}(G) \leq \mathbf{r}(G)$.

Now suppose that $\mathbf{bs}(G)$ is finite. For each prime p, let G_p denote the Sylow p-subgroup of G and let $T = \mathbf{Tor}\,(G) = \underset{p \in \Pi(G)}{\mathrm{Dr}}\ G_p$. Of course $\mathbf{sr}_p(T) \leq \mathbf{sr}_p(G) \leq \mathbf{bs}(G)$, for each $p \in \Pi(G)$. Lemma 6.1.8 implies that $\mathbf{sr}_p(G_p) = \mathbf{r}(G_p)$, for each prime p and Corollary 6.1.5 allows us to deduce that

$$\mathbf{r}(T) = \max\{\mathbf{r}(G_p)|p \in \Pi(G)\} = \max\{\mathbf{sr}_p(G_p)|p \in \Pi(G)\} \leq \mathbf{bs}(G).$$

By Corollary 3.2.24, $\mathbf{sr}_p(G/T) = \mathbf{r}_0(G/T)$, for each prime $p \in \Pi(G)$. On the other hand, $\mathbf{r}_0(G/T) = \mathbf{r}(G/T)$, by Proposition 6.2.4 and this implies that $\mathbf{r}(G/T) \leq \mathbf{bs}(G)$. By Lemma 6.1.2, we deduce that $\mathbf{r}(G)$ is finite.

Now let K be a finitely generated subgroup of G such that $\mathbf{r}(K) = \mathbf{r}(G)$. By Corollary 6.2.9 we have

$$\mathbf{r}(K) = \max\{\mathbf{sr}_p(K)|p \in \Pi(G)\} \leq \max\{\mathbf{sr}_p(G)|p \in \Pi(G)\} \leq \mathbf{bs}(G).$$

It follows that $\mathbf{r}(G) \leq \mathbf{bs}(G)$, which proves the result. □

For soluble groups and their generalizations, it is in general no longer possible to deduce the equality of the ranks as we have in the previous results. In Exercise 3.1 we considered the group $G = A \rtimes \langle x \rangle$, where A is the additive group of rational numbers whose denominators are of the form p^k, for the fixed prime p, $k \in \mathbb{Z}$ and $a^x = a^p$, for every $a \in A$. This group is polyrational and $\mathbf{r}_0(G) = 2$, so that $\mathbf{r}(G) = 2$. However $\mathbf{sr}_p(G) = 1$.

The following theorem describes the structure of locally generalized radical groups of finite special rank. Such groups are in fact generalized radical, but more can be said. The theorem is an immediate consequence of Proposition 6.1.6, Corollary 4.4.4 and Theorems 6.1.10 and 6.2.2.

6.2.12. THEOREM. *Let G be a locally generalized radical group of finite special rank r. Then the Hirsch–Plotkin radical L of G is hypercentral and there is a normal abelian subgroup K/L of G/L such that G/K is finite. In particular, G is almost hyperabelian, $\mathbf{Tor}(L)$ is a direct product of Chernikov p-subgroups, $L/\mathbf{Tor}(L)$ is nilpotent and $K/\mathbf{Tor}(L)$ has finite 0-rank at most r. Also G has finite 0-rank and $\mathbf{r}_0(G) \leq r$.*

From this we can deduce a theorem of N. S. Chernikov [36].

6.2.13. COROLLARY. *Let G be a locally (soluble-by-finite) group of finite special rank. Then G is almost hyperabelian.*

A further consequence of this result is a theorem of B. I. Plotkin [217, 16.3.1] (see also [218]).

6.2.14. COROLLARY. *Let G be a locally radical group. If G has finite special rank, then G is a radical group.*

The following consequence is due to V. S. Charin [32].

6.2.15. COROLLARY. *Let G be a locally soluble group of finite special rank. Then G has a normal hypercentral subgroup D such that G/D is soluble. In particular G is hyperabelian.*

We may also deduce a theorem of M. I. Kargapolov [125].

6.2.16. COROLLARY. *Let G be a periodic locally soluble group of finite special rank and let L be the Hirsch–Plotkin radical of G. Then L is hypercentral and G/L contains a normal abelian subgroup K/L with finite Sylow p-subgroups for all primes p such that G/K is finite.*

We also observe the following immediate consequence of Proposition 6.1.6 and Corollary 4.4.5.

6.2.17. COROLLARY. *Let G be a locally generalized radical group of finite special rank. Then G is countable.*

Next we prove the following converse of Proposition 6.1.6, which at least holds in the locally generalized radical case.

6.2.18. THEOREM. *Let G be a locally generalized radical group. Then G has bounded section rank if and only if G has finite special rank.*

PROOF. We have already noted, in Proposition 6.1.6, that a group of finite special rank has bounded section rank.

We therefore suppose that G has bounded section rank b. By Corollary 4.4.4, there are normal subgroups $L \leq K \leq G$ such that L is hypercentral, K/L is abelian and G/K is finite. Theorem 6.2.11 shows that L and K/L have special rank at most b and Lemma 6.1.2 enables us to see that K has special rank at most $2b$. Since G/K is finite, it has finite special rank and again using Lemma 6.1.2, we deduce that G also has finite special rank.   □

Later, we shall consider more precise connections between the special rank and the bounded section rank, but for now we note some consequences, the first of which follows immediately from Theorem 6.2.18 and Theorem 4.4.6.

6.2.19. COROLLARY. *Let G be a locally generalized radical group. If every locally radical subgroup of G has finite special rank, then G has finite special rank.*

One consequence of this result is the main theorem of [51] which we prove next.

6.2.20. COROLLARY. *Let G be a locally (soluble-by-finite) group. If every locally soluble subgroup of G has finite rank then G has finite rank.*

PROOF. Let H be an arbitrary locally radical subgroup of G and let F be a finitely generated subgroup of H. On the one hand, F is a radical subgroup. On the other hand, F is almost soluble and hence is soluble. It follows that H is locally soluble. Hence every locally radical subgroup of G has finite special rank. By Corollary 6.2.19, G has finite special rank. $\qquad\square$

6.3. The Relationship Between the Special Rank and the Bounded Section Rank

In this section we obtain the numerical relationship between the special rank and the bounded section rank. The results we obtain are based on the results of the last section and certain theorems about finite groups which we now present. We start with the following useful technical results.

If G is a finite group and V is a finitely generated $\mathbb{F}_p G$-module then we can form the natural semidirect product $VG = V \rtimes G$. Let

$$U(G, V) = \{\sigma \in \mathbf{Aut}\,(VG) | \sigma(v) = v,\ \text{for all } v \in V \text{ and}$$
$$\sigma(Vg) = Vg,\ \text{for all } g \in G\}.$$

Thus, $U(G, V)$ is the group of automorphisms acting trivially on V and VG/V, so is the stability group of the extension $1 \leq V \leq VG$. In particular, by Theorem 1.2.22 and the fact that the additive group of V is elementary abelian, we obtain the following straightforward result which appears in a paper of M. Aschbacher and R. Guralnick [4]. This and a number of the following observations come from this paper.

6.3.1. LEMMA. *Let G be a finite group and let V be a finitely generated $\mathbb{F}_p G$-module. Then $U(G, V)$ is an elementary abelian p-group.*

As usual, let $H^1(G, V)$ denote the first cohomology group of G on V. Then $H^1(G, V)$ can be interpreted as the group $U(G, V)/\mathbf{Aut}\,_V(VG)$, where $\mathbf{Aut}\,_V(VG)$ is the group of automorphisms of VG induced by conjugation by some element of V (see [99, Section 3.5], for example).

6.3.2. LEMMA ([4]). *Let G be a finite group and let V be a finitely generated $\mathbb{F}_p G$-module. Then $U(G, V)$ acts regularly on the set of complements to V in VG.*

PROOF. Let H be a complement to V in VG. Since $VG = V \rtimes H$, each element $g \in G$ can be written uniquely in the form $g = vh$, where $v \in V, h \in H$ and there is an isomorphism $\sigma : G \longrightarrow H$ defined by $\sigma(g) = h$. Now σ extends to an element of $U = U(G, V)$ by defining $\sigma(v) = v$, for all $v \in V$ and from this it follows that U acts transitively on the set of complements of V. However, $N_U(G) = C_U(G) = 1$, so the result follows. $\qquad\square$

This allows us to deduce the following pair of results from [4] and we give a proof of the second one.

6.3.3. LEMMA. *Let G be a finite group and let V be a finitely generated $\mathbb{F}_pG$-module. Then $|H^1(G,V)|$ is the number of conjugacy classes of complements of V in VG.*

6.3.4. LEMMA. *Let G be a finite group and let V be a finitely generated $\mathbb{F}_pG$-module.*

(a) *If there exist subgroups X_i of G, for $1 \leq i \leq t$, such that $G = \langle X_1, X_2, \ldots, X_t \rangle$, then $U(G,V) = \mathop{Dr}\limits_{1 \leq j \leq t} U(X_j, V)$;*

(b) *If $G = \langle g \rangle$, then $|U(G,V)| = |\{v \in V | (vg)^m = 1\}|$, where m is the order of g;*

(c) *If G can be generated by d elements, then $|U(G,V)| \leq |V|^d$.*

PROOF. If $\sigma \in U(G,V)$, we let σ_j denote the restriction of σ to X_j. Define the mapping $\tau : U(G,V) \longrightarrow \mathop{Dr}\limits_{1 \leq j \leq t} U(X_j, V)$ by $\tau(\sigma) = (\sigma_j)_{1 \leq j \leq t}$. Since $\sigma = 1$ if and only if $\sigma_j = 1$, for $1 \leq j \leq t$ it follows that τ is an embedding and this proves assertion (a).

Since $\langle vg \rangle$ is a complement for V in $\langle V, g \rangle$ if and only if vg has order m, (b) follows. Finally, (c) follows from (a) and (b).   □

The next result establishes the connection between $H^1(G,V)$ and the generators of VG.

6.3.5. LEMMA (Aschbacher and Guralnick [4]). *Let $G = \langle x_1, \ldots, x_d \rangle$ be a finite group and let V be a simple $\mathbb{F}_pG$-module. Then VG can be generated by d elements if and only if $|U(G,V)| < |V|^d$. In particular, if $C_G(V) \neq G$, then VG can be generated by d elements if and only if $|H^1(G,V)| < |V|^{d-1}$.*

PROOF. Let $\alpha = (v_1, \ldots, v_d) \in \underbrace{V \oplus \cdots \oplus V}_{d}$, where $v_j \in V$, for $1 \leq j \leq d$ and let

$$G_\alpha = \langle v_1 x_1, \ldots, v_d x_d \rangle.$$

Now $G \leq VG_\alpha$, so $VG_\alpha = VG$. Clearly $V \cap G_\alpha \triangleleft VG$ so, as V is simple, we have $V \cap G_\alpha = 1$ or $V \cap G_\alpha = V$. Hence either G_α is a complement to V in VG or $G_\alpha = VG$. Furthermore, if $\alpha \neq \beta = (w_1, \ldots, w_d)$, but $G_\alpha = G_\beta$, then there exists i such that $v_i \neq w_i$ and hence $V \cap G_\alpha = V$ so $G_\alpha = VG$. Also every complement of V in VG has the form G_α, for some α. Thus $VG = G_\alpha$, for some α if and only if $|U(G,V)| < |V|^d$, by Lemmas 6.3.2 and 6.3.4. Since $\{x_1, \ldots, x_d\}$ is an arbitrary generating set for G, the first result follows. Finally, when V is also nontrivial and simple, then $|\mathbf{Aut}_V(VG)| = |G|$ and the second statement follows.   □

In the next result we obtain a bound for $|H^1(G, V)|$ in terms of a composition series for V.

6.3.6. LEMMA (Aschbacher and Guralnick [4]). *Let G be a finite group and let V be a finitely generated $\mathbb{F}_p G$-module. If W is an $\mathbb{F}_p G$-submodule of V, then*

$$|H^1(G, V)| \leq |H^1(G, W)| \cdot |H^1(G, V/W)|,$$

with equality if W is a direct summand of V.

PROOF. Define the mapping $\Phi : U(G, V) \longrightarrow U(G, V/W)$ by

$$(\Phi(\sigma))(x) = \sigma(x)W, \text{ whenever } x \in VG, \sigma \in U(G, V).$$

It is easy to see that Φ is a homomorphism. Clearly, $\ker(\Phi) = \{\sigma \in U(G, V) | [\sigma, G] \leq W\}$ which can be identified with a subgroup of $U(G, W)$. This makes it easy to see that

$$|U(G, V)| \leq |U(G, W)| \cdot |U(G, V/W)|$$

with equality holding if W is a direct summand of V. The result now follows because

$$|C_V(G)| \leq |C_W(G)| \cdot |C_{V/W}(G)|$$

and

$$|\mathbf{Aut}_V(VG)| \geq |\mathbf{Aut}_W(WG)| \cdot |\mathbf{Aut}_{V/W}((V/W)G)|.$$

Clearly equality holds if W is a direct summand. $\qquad\square$

We recall that $\mathbf{d}(G)$ denotes the minimum number of generators of a finitely generated group G. We now incorporate some of these results in the following proposition, which is due to A. Lucchini [183].

6.3.7. PROPOSITION. *Let G be a finite group and let L be a normal 2-subgroup of G. Suppose that the Sylow 2-subgroups of G can be generated by r elements. If G/L can be generated by d elements, where $d \geq r$, then G can be generated by d elements.*

PROOF. We assume that L is a minimal normal 2-subgroup of G and note that the general case follows by induction on the length of a G-composition series for L. Then, since L is an elementary abelian 2-group we can think of it as a simple $\mathbb{F}_2 G$-module.

Suppose, for a contradiction, that G is not d-generator. Let $G/L = \langle g_1 L, \ldots, g_d L \rangle$ and let $H = \langle g_1, \ldots, g_d \rangle$. By our supposition, $H \neq G$, so $G = HL$. Also $L \cap H$ is normal in H and L is abelian, so $L \cap H$ is normal in G. By the minimality of L we have $L \cap H = L$ or $L \cap H = 1$. The former case would imply that $L \leq H$, so that $H = G$, a contradiction. Consequently, $L \cap H = 1$, so H is a complement for L in G.

Let $|L| = q$. Then there are q^d sets $\{g_1 u_1, \ldots, g_d u_d\}$ of elements of G with $u_1, \ldots, u_d \in L$ and any two of these sets generate distinct complements of L in G, as in the proof of Lemma 6.3.5. By Lemma 6.3.2, $U(H, L)$ acts regularly on the set of complements of L in HL. We note that $H^1(H, L) = U(H, L)/\mathbf{Aut}_L(HL)$, where $\mathbf{Aut}_L(HL)$ is the group of automorphisms of HL induced by conjugation by some element of L. It follows that

$$(6.1) \qquad q^d = |U(H, L)| = |H^1(H, L)| \cdot |\mathbf{Aut}_L(HL)|.$$

There are two cases to consider. If $L \leq \zeta(G)$, then $|L| = 2$. If Q is a Sylow 2-subgroup of H, then $L \times Q$ is a Sylow 2-subgroup of G, so $\mathbf{d}(Q) < r \leq d$. On the other hand, $|\mathbf{Aut}_L(HL)| = 1$ and $|H^1(H, L)| = |H/H'H^2| = 2^{\mathbf{d}(Q)}$, which contradicts Equation (6.1).

Hence we may assume that L is not central in G. In this case $\mathbf{Aut}_L(HL) \cong L$ so Equation (6.1) implies that $q^{d-1} \leq |H^1(H, L)|$. However, if Q is again a Sylow 2-subgroup of H, then $|H^1(H, L)| \leq |H^1(Q, L)|$. By Lemma 6.3.4 we obtain

$$q^{d-1} \leq |H^1(H, L)| \leq |H^1(Q, L)| \leq q^{\mathbf{d}(Q)}.$$

We have $\mathbf{d}(Q) = r - a$, where $2^a = |L/[L, Q]|$. Since $r - 1 \leq d - 1 \leq \mathbf{d}(Q) = r - a$ we deduce that $a \leq 1$. If $a = 0$, then $L = [L, Q]$, so that $L = 1$. Hence $a = 1$, $d = r$ and $|H^1(Q, L)| = q^{\mathbf{d}(Q)}$. If L is a trivial Q-module, then $[L, Q] = 1$ and hence $|L| = 2$. In this case L would also be a trivial H-module, a contradiction. Thus L is a nontrivial Q-module and so $|\mathbf{Aut}_L(LQ)| \neq 1$. Then, using Lemma 6.3.4, we obtain,

$$q^{\mathbf{d}(Q)} \leq |H^1(Q, L)| < |U(Q, L)| \leq q^{\mathbf{d}(Q)},$$

which is the final contradiction. $\qquad\qquad\qquad\qquad\qquad\qquad \square$

We shall also require the following result of L. G. Kovács [**136**].

6.3.8. LEMMA. *Let p be a prime and let G be a finite group. Let P be a Sylow p-subgroup of G which can be generated by d elements. Suppose that G contains a nontrivial normal p-subgroup L such that $G = L \rtimes D$, for some subgroup D. Then P/L can be generated by $d - 1$ elements.*

EXERCISE 6.9. *Prove Lemma 6.3.8*

We also need the following results of the paper [**4**] whose proof requires a specific analysis of finite simple groups (but not the classification of finite simple groups), so we present these results without proof.

6.3.9. THEOREM (Aschbacher and Guralnick [4]). *Let G be a finite group, let p be a prime and let V be a simple $\mathbb{F}_p G$-module. Let L be the natural semidirect product of V by G. Let K be a normal subgroup of G minimal subject to the conditions*

(a) $K \leq C = C_G(V)$ *and* $C/K \cong \underbrace{V \oplus \cdots \oplus V}_{r}$ *(as an $\mathbb{F}_p G$-module), for some natural number r;*

(b) C/K *possesses a complement in G/K.*

Let $q = |\operatorname{Hom}_G(V,V)|$ and suppose that G can be generated by d elements.

(i) *If V is a trivial module, then L can be generated by d elements if and only if $r < d$;*

(ii) *If V is nontrivial, then L can be generated by d elements if and only if $hq^r < |V|^{d-1}$, where $h = |H^1(G/C, V)|$.*

Using the same notation, we have also

6.3.10. COROLLARY (Aschbacher and Guralnick [4]). *Let $W = \underbrace{V \oplus \cdots \oplus V}_{t}$ and let H be the natural semidirect product of W and G. If V is nontrivial, then L can be generated by d elements if and only if $hq^{r+t-1} < |V|^{d-1}$. In particular, this holds if $t \leq s(d-2) + 1 - r$, where $q^s = |V|$.*

6.3.11. PROPOSITION (Longobardi and Maj [175]). *Let G be a finite group, let p be a prime and let L be a normal p-subgroup of G. Suppose that the Sylow p-subgroups of G can be generated by r elements. If G/L can be generated by d elements, where $d \geq r + 1$, then G can be generated by d elements.*

PROOF. We assume that L is a minimal normal p-subgroup of G, so that L is elementary abelian and note that the general case follows by induction on the length of a G-composition series for L. If $L \leq \mathbf{Frat}(G)$ then the result is clear, by the Burnside Basis Theorem, so we suppose that $L \nleq \mathbf{Frat}(G)$. Then there is a maximal subgroup H such that $L \nleq H$ and hence $G = HL$. Clearly $L \cap H$ is normal in H and since L is elementary abelian it follows that $L \cap H$ is normal in G. By the minimality of L we have $L \cap H = L$ or $L \cap H = 1$. The former case would imply that $L \leq H$ so that $H = G$, a contradiction. Consequently, $L \cap H = 1$ and $G/L \cong H$ can be generated by d elements.

Let P be an arbitrary Sylow p-subgroup of G. Since $L \lhd G$ we have $L \leq P$ and hence $P = L(P \cap H)$. Also $L \cap H = 1$, so there is a maximal subgroup M of P such that $P \cap H \leq M$, but $L \nleq$

M. Hence $L \not\leq \mathbf{Frat}(P)$. Let $F/L = \mathbf{Frat}(P/L)$. Since $P/F \cong$ $(P/L)/(F/L)$ is elementary abelian, $\mathbf{Frat}(P) \leq F$. Since $L \leq F$, but $L \not\leq \mathbf{Frat}(P)$, it follows that $F \neq \mathbf{Frat}(P)$, so $|P/F| < |P/\mathbf{Frat}(P)| \leq$ p^r, by hypothesis. The Burnside Basis Theorem implies that P/L is at most $(r - 1)$-generator. Clearly P/L is a Sylow p-subgroup of G/L and since $G/L \cong H$, the Sylow p-subgroups of H are also at most $(r - 1)$-generator.

Note that L is a simple $\mathbb{F}_p H$-module and suppose, for a contradiction, that G is not d-generator. Then Theorem 6.3.9 and the hypotheses imply that we may assume that L is not a trivial module. Let $C = C_H(L)$. By Theorem 6.3.9, there is a normal subgroup K of H such that H/K splits over C/K and hence there is a subgroup D/K such that $H/K = C/K \rtimes D/K$. Furthermore, $C/K = C_1/K \times \cdots \times C_t/K$, where C_j/K is a minimal normal p-subgroup of H/K, for $1 \leq j \leq t$ and our assumption implies that $(d - 2) + (1 - t) < 1$. Thus $d - 1 < t + 1$, so that $r \leq d - 1 < t + 1$, and hence $r - 1 < t$. We now apply Lemma 6.3.8 $(r - 1)$ times to deduce that each Sylow p-subgroup of $(C_r/K \times \cdots \times C_t/K) \rtimes D/K$ is trivial, a final contradiction.   □

We next prove a result of W. Gaschütz [90], which actually holds in greater generality, but we give just the finite case.

6.3.12. PROPOSITION. *Let G be a finite n-generator group and let L be a normal subgroup of G. For each subset $\{e_1, \ldots, e_n\}$ of G such that $G/L = \langle e_1 L, \ldots, e_n L \rangle$ there is a subset $\{g_1, \ldots, g_n\}$ such that $G = \langle g_1, \ldots, g_n \rangle$ and $g_j \in e_j L$, for $1 \leq j \leq n$.*

PROOF. Let $\{e_1, \ldots, e_n\}$ be a fixed subset of G such that $G/L = \langle e_1 L, \ldots, e_n L \rangle$. If $L \leq \mathbf{Frat}(G)$, then $G = \langle e_1, \ldots, e_n \rangle$ and the result follows in this case, so we suppose that $L \not\leq \mathbf{Frat}(G)$.

Since G is finite, there are only finitely many subgroups B with the property that $G = BL$. It is easy to see that each such subgroup B is uniquely defined by $B \cap e_1 L, \ldots, B \cap e_n L$. Indeed, in this case, $B = \langle B \cap e_1 L, \ldots, B \cap e_n L \rangle$.

We consider sets of elements of the form

(6.2) $\{g_1, \ldots, g_n\}$ such that $g_j \in e_j L$, for $1 \leq j \leq n$

and we wish to show that for the given set $\{e_1, \ldots, e_n\}$ there is at least one subset, as in (6.2), with the property that $\langle g_1, \ldots, g_n \rangle = G$. We shall call each such subset a *correct* subset. For each subgroup U of G we let

$$\varepsilon(U) = \begin{cases} 0, & \text{if } UL \neq G, \\ 1, & \text{if } UL = G, \end{cases}$$

and note that there are $|L \cap U|^n \varepsilon(U)$ subsets in U with the property specified in (6.2). If $M_1, \ldots, M_r$ are the maximal subgroups of G not containing L, then it is evident that there are

$$\Phi_L = |L|^n +$$

$$\sum_{1 \leq k \leq r} \left(\sum_{j(1) < \cdots < j(k)} (-1)^k |L \cap M_{j(1)} \cap \cdots \cap M_{j(k)}|^n \varepsilon(M_{j(1)} \cap \cdots \cap M_{j(k)}) \right)$$

correct subsets of G, a number which depends on L only (and of course G). Since G is n-generator there are elements $x_1, \ldots, x_n$ such that $G = \langle x_1, \ldots, x_n \rangle$ and this choice implies that $\Phi_L \neq 0$. Hence the set $\{e_1, \ldots, e_n\}$ such that $G/L = \langle e_1 L, \ldots, e_n L \rangle$ determines a correct subset. This completes the proof. $\square$

The proof of the following result, due to R. Guralnick [102], requires a detailed analysis of the finite simple groups, so we also present this result without proof.

6.3.13. THEOREM. *Let G be a finite simple group. Then G can be generated by an involution and a Sylow 2-subgroup.*

From this we deduce yet another result due to A. Lucchini [183].

6.3.14. THEOREM. *Let G be a finite group and suppose that the Sylow subgroups of G are each at most d-generator. Then there are elements $g_1, \ldots, g_{d+1}$ such that $G = \langle g_1, \ldots, g_{d+1} \rangle$ and such that $\langle g_1, \ldots, g_d \rangle$ contains a Sylow 2-subgroup of G.*

PROOF. We use induction on the order of G. If G is simple, then Theorem 6.3.13 is easily seen to imply the result. Hence we may assume that G is not simple and choose a minimal normal subgroup, N, of G. Since G/N inherits the hypotheses, our induction hypothesis asserts that there exist elements $g_1, \ldots, g_{d+1}$ of G such that $G/N = \langle g_1 N, \ldots g_{d+1} N \rangle$ and $\langle g_1 N, \ldots, g_d N \rangle$ contains a Sylow 2-subgroup of G/N. There are several cases to consider.

If N is a 2-group, we let $H = \langle g_1, \ldots, g_d, N \rangle$ and note that H contains a Sylow 2-subgroup of G, by the induction hypothesis. By hypothesis, the Sylow 2-subgroups of H are at most d-generator and Proposition 6.3.7 asserts that H contains elements $h_1, \ldots, h_d$ such that $H = \langle h_1, \ldots, h_d \rangle$. Then $G = \langle h_1, \ldots, h_d, g_{d+1} \rangle$ and the result follows in this case.

Next, we suppose that N is a p-group for some prime $p \neq 2$. In this case we let d_p be the minimal number of generators of a Sylow p-subgroup of G. Certainly $d + 1 \geq d_p + 1$, so Proposition 6.3.11

implies that G can be generated by $d + 1$ elements. Proposition 6.3.12 asserts that there are elements $u_1, \ldots, u_{d+1}$ of N with the property that $G = \langle g_1 u_1, \ldots, g_{d+1} u_{d+1} \rangle$. We let $K = \langle g_1 u_1, \ldots, g_d u_d \rangle$. Then it follows from our induction hypothesis that KN contains a Sylow 2-subgroup of G. Let the order of such a Sylow 2-subgroup be q, so that q divides $|KN|$. Since $|KN| = |K||N|/|K \cap N|$ and since $|N|$ is odd, it follows that q divides the order of K. Hence K contains a Sylow 2-subgroup of G. In this case, the elements $g_1 u_1, \ldots, g_{d+1} u_{d+1}$ satisfy the desired conclusion for G and the result follows in this case.

Finally we suppose that N is not a soluble group. In this case N is the direct product of isomorphic non-abelian simple groups. By Theorem 6.3.13, there is an involution $x \in N$ and a Sylow 2-subgroup P of N such that $N = \langle x, P \rangle$. Of course, by Sylow's Theorem, there is an element $y \in N$ such that $x \in P^y$ and hence $N = \langle P, P^y \rangle$. If $g \in G$, then $P^g \in N$ and the conjugacy of the Sylow 2-subgroups of N implies that $G = NN_G(P)$. This shows that, without loss of generality, $g_j \in N_G(P)$, for $1 \leq j \leq d + 1$. Let $R = \langle g_1, \ldots, g_d, P \rangle$. By our inductive assumption RN contains a Sylow 2-subgroup of G. Let the order of such a Sylow 2-subgroup be q, so that q divides $|RN|$. Note that $R \cap N$ contains a Sylow 2-subgroup of N, so $|N/R \cap N|$ is odd. Since $|RN| = |R||N|/|R \cap N|$, it follows that q divides $|R|$ and hence R contains a Sylow 2-subgroup of G. By hypothesis, such a Sylow 2-subgroup is at most d-generator and, since $P \lhd R$, we may apply Proposition 6.3.7 to deduce that R contains elements $v_1, \ldots, v_d$ such that $R = \langle v_1, \ldots, v_d \rangle$. Now $\langle v_1, \ldots, v_d, g_{d+1} y \rangle$ contains P and also $P^{g_{d+1} y} = P^y$, so it also contains N. However, $G = \langle R, g_{d+1}, N \rangle$, so that $G = \langle v_1, \ldots, v_d, g_{d+1} y \rangle$, since $y \in N$. Hence in this case $v_1, \ldots, v_d, g_{d+1} y$ are the desired $d + 1$ elements of G.

This completes the proof. $\qquad \square$

This result has the following very important and interesting corollary, which has been proved by Guralnick [103] and Lucchini [183].

6.3.15. COROLLARY. *Let G be a finite group. If each Sylow subgroup of G is at most d-generator, then G is at most $(d + 1)$-generator.*

We also note the following result which is important for our purposes.

6.3.16. COROLLARY. *Let G be a finite group. Suppose that there is a natural number b such that $\mathbf{sr}_p(G) \leq b$. Then G has special rank at most $b + 1$.*

PROOF. Let H be an arbitrary subgroup of G. If p is a prime, then Lemma 6.1.8 shows that every Sylow p-subgroup of H has special rank at most b and hence every Sylow p-subgroup of H is at most b-generator. Since this is true for each prime p, Corollary 6.3.15 allows us to deduce that H is at most $(b+1)$-generator. This means that G has special rank at most $b + 1$, as required. $\square$

These results have far-reaching consequences for locally generalized radical groups. For example, we are now in a position to obtain the following quantitative version of Theorem 6.2.18.

6.3.17. THEOREM. *Let G be a locally generalized radical group.*

(i) *If G has finite special rank r, then G has finite bounded section rank at most r;*

(ii) *If G has bounded section rank b, then G has finite special rank at most $3b + 1$.*

PROOF. Our assertion (i) follows from Proposition 6.1.6, so we prove assertion (ii) and, to this end, suppose that G has finite bounded section rank b. By Corollary 4.4.4, G contains normal subgroups $L \leq K$ such that L is locally nilpotent, K/L is abelian and G/K is finite. Theorem 6.2.11 shows that L and K/L both have finite special rank at most b. Using Lemma 6.1.2 we deduce that K has special rank at most $2b$. By Lemma 3.2.2 and Corollary 6.3.16 we see that G/K has finite special rank at most $b + 1$. Using Lemma 6.1.2 again, we finally see that G has finite special rank at most $3b + 1$. $\square$

6.4. A Taste of the Exotic

As we have seen, locally generalized radical groups of finite special rank are quite well behaved. In general, however, it appears that groups of finite special rank can be rather complicated. The best known examples of bad behaviour are the Tarski monsters, so named after A. Tarski who first suggested their possible existence. Examples of Tarski monsters were constructed by A.Yu. Ol'shanskii [208]. These groups, which can be constructed for each sufficiently large prime p (greater than 10^{75}), are 2-generator infinite simple groups, and each proper subgroup is generated by a single element, so such a group has special rank 2. In [209, Theorem 28.1] the groups have exponent p; Theorem 28.2 of [209] provides a further example of a periodic group of rank 2, and [209, Theorem 28.3] gives an example of a torsion-free group whose proper subgroups are infinite cyclic and again this group has special rank 2.

The following result of Ol'shanskii [**209**, Theorem 35.1] allows us to construct further exotic examples of groups of finite special rank.

6.4.1. THEOREM. *Let* $\{G_\lambda | \lambda \in \Lambda\}$ *be a finite or countable set of non-trivial finite or countably infinite groups without involutions. Suppose* $|\Lambda| \geq 2$ *and that* n *is a sufficiently large odd number (at least* 10^{75} *). Suppose that, whenever* $\lambda, \mu \in \Lambda$ *and* $\lambda \neq \mu$ *, then* $G_\lambda \cap G_\mu = 1$ *. Then there is a countable simple group* $G = OG(G_\lambda | \lambda \in \Lambda)$ *, containing a copy of* G_λ *, for all* $\lambda \in \Lambda$ *, with the following properties:*

(i) *If* $x, y \in G$ *and* $x \in G_\lambda \setminus \{1\}, y \notin G_\lambda$ *, for some* $\lambda \in \Lambda$ *, then* $G = \langle x, y \rangle$ *;*

(ii) *Every proper subgroup of* G *is either a cyclic group of order dividing* n *or is contained in some subgroup conjugate to some* G_λ *.*

Hence, these groups of Ol'šanskii's are 2-generator and have subgroups which are restricted by the choice of the constituent groups G_λ. An application of this theorem allows us to construct the following examples. To illustrate how chaotic the situation can be in general we see that even when all proper subgroups of a group G have finite special rank, G itself need not have finite special rank.

6.4.2. PROPOSITION. *There is a 2-generator group* G *of infinite special rank, all of whose proper subgroups have finite special rank.*

PROOF. For each $n \in \mathbb{N}$ let G_n be a countable group of rank r_n, without involutions, and suppose that $r_{n+1} > r_n$ for all $n \geq 1$. Then the group $G = OG(G_n | n \in \mathbb{N})$, constructed as above, is 2-generator, but clearly does not have finite special rank. However, every proper subgroup of G does have finite special rank. $\square$

In a similar manner, we can obtain the following result by choosing the G_n to be periodic abelian groups of special rank n without involutions.

6.4.3. PROPOSITION. *There is a periodic 2-generator group of infinite special rank, all of whose proper subgroups are abelian of finite special rank.*

The groups occurring here need not be finitely generated as we now show.

6.4.4. PROPOSITION. *There is an infinitely generated group of infinite special rank, all of whose proper subgroups have finite special rank.*

PROOF. For each $n \in \mathbb{N}$ let G_n be a non-trivial countable group without involutions which has special rank n. Let $H_1 = OG(G_1, G_2)$ and, for $n \geq 2$, let $H_n = OG(H_{n-1}, G_{n+1})$. Note that

$$H_1 \lneqq H_2 \lneqq \cdots \lneqq H_n \lneqq \cdots.$$

Let $H = \bigcup_{n \in \mathbb{N}} H_n$. It is clear that each H_n has special rank n so that H has infinite special rank. We claim that in fact H is not finitely generated, in contrast to Proposition 6.4.2, that every proper subgroup of H has finite special rank, and that each proper subgroup of H is in fact contained in one of the groups H_n. To see this latter fact and hence the rest, let K be a subgroup of H and suppose that $K \nleq H_n$ for each n. We show that $K = H$ by proving that $H_n \leq K$, for all $n \in \mathbb{N}$. Let n be fixed. Since $K \nleq H_n$ there exists $x \in K$ such that $x \notin H_n$. Since $H = \bigcup_{n \in \mathbb{N}} H_n$ there exists $j > n$ such that $x \in H_j$. Since $K \nleq H_j$ there exists $y \in K$ such that $y \notin H_j$. Hence there exists $m > j$ such that $y \in H_m \setminus H_{m-1}$. However, by construction of H_m, we have $\langle x, y \rangle = H_m \leq K$. Since $H_n \leq H_m$ we have $H_n \leq K$ and the result follows. $\qquad \square$

Using Ol'shanskii's construction, V. N. Obraztsov [207] constructed the first example of an uncountable group satisfying the minimal condition. We prove next that this group has finite special rank.

6.4.5. THEOREM. *Let p be a prime such that $p > 10^{75}$. Then there exists an uncountable p-group G of finite special rank.*

PROOF. Let C be a group of order p and let $G_1 = OG(C, C)$. Let H be a subgroup of G_1. If $H = G_1$, then H has two generators. If H is a proper subgroup of G_1, then H is cyclic. It follows that G_1 has special rank 2. Put $G_2 = OG(G_1, C)$. More generally, using transfinite induction, we construct certain groups G_α, for all ordinals $\alpha < \Omega$, where Ω is the first uncountable ordinal, as follows: if $\alpha - 1$ exists, then set $G_\alpha = OG(G_{\alpha-1}, C)$. By the construction G_α contains a subgroup isomorphic to $G_{\alpha-1}$, so that there is an embedding of $G_{\alpha-1}$ in G_α. If α is a limit ordinal, then let $G_\alpha = \bigcup_{\beta < \alpha} G_\beta$. Let $G = \bigcup_{\alpha < \Omega} G_\alpha$. By construction $G_\alpha \neq G_{\alpha+1}$ for each $\alpha < \Omega$ and hence G is not countable.

Using transfinite induction we prove that G has a special rank 2 for each $\alpha < \Omega$. We have already proved this for $\alpha = 1$. Let $\alpha > 1$ and suppose that α is not a limit ordinal. Thus $\alpha = \beta + 1$ for some ordinal β and $G_\alpha = OG(G_\beta, C)$. If H is a proper finitely generated subgroup of G_α then either H is cyclic or H is contained in some subgroup conjugate to some G_β. By induction hypothesis G_β has special rank 2 so, in any case, H has at most two generators. It follows that G_α has special rank 2.

Suppose now that α is a limit ordinal. If H is a proper finitely generated subgroup of G_α, then, since $G_\alpha = \bigcup_{\beta < \alpha} G_\beta$, it follows that $H \leq G_\beta$ for some $\beta < \alpha$. By induction hypothesis G_β has special rank 2, so H has at most two generators. Thus, in this case also, G_α has special rank 2. Setting $\alpha = \Omega$ we see that G has special rank 2, as required. $\qquad\square$

It seems to be unknown if a torsion-free group of finite rank can be uncountable.

CHAPTER 7

The Relationship Between the Factors of the Upper Central Series and the Nilpotent Residual

This chapter is dedicated to the following general problem.

Let G be a group. For which classes of groups $\mathfrak{X}$ is it the case that $G/\zeta_\infty(G) \in \mathfrak{X}$ always implies that the hypercentral residual, or more generally the locally nilpotent residual, of G also belongs to $\mathfrak{X}$?

This problem has already been discussed in Chapter 1 when $\mathfrak{X}$ is the class $\mathfrak{F}$ of finite groups. Furthermore, the case when $\mathbf{zl}(G) = n$ is finite can be stated more explicitly and is based on Baer's Theorem (Theorem 1.5.19).

For which classes of groups $\mathfrak{X}$ is it the case that $G/\zeta_n(G) \in \mathfrak{X}$ always implies that $\gamma_{n+1}(G) \in \mathfrak{X}$?

An important special case here is the case $n = 1$ and, in this case, our question asks for those classes of groups $\mathfrak{X}$ for which $G/\zeta(G) \in \mathfrak{X}$ implies that $\gamma_2(G) = G' \in \mathfrak{X}$.

In this chapter, we consider this topic in the case when $\mathfrak{X}$ is a class of groups satisfying this or that finiteness condition on ranks and related classes of groups. In presenting these results, we will not follow the chronological order in which they have appeared. Where possible we gather more general results and use these to obtain other interesting and important results as corollaries.

7.1. Hypercentral Extensions by Groups of Finite 0-Rank

We start with two results giving generalizations of Corollary 1.5.17. Indeed a number of the results in this section are generalizations of those occurring in Sections 1.5 and 1.6.

7.1.1. LEMMA. *Let G be a group and let A be an abelian normal subgroup of G. If $A \cap \zeta(G)$ contains a subgroup C such that A/C is periodic, then $[A, G]$ is also periodic and $\Pi([A, G]) \subseteq \Pi(A/C)$.*

EXERCISE 7.1. *Prove Lemma 7.1.1.*

Ranks of Groups: The Tools, Characteristics, and Restrictions
By Martyn R. Dixon, Leonid A. Kurdachenko and Igor Ya Subbotin
Copyright © 2017 John Wiley & Sons, Inc.

EXERCISE 7.2. *Prove that if G is a nilpotent group and if G/G' is periodic, then G is also periodic and $\Pi(G) = \Pi(G/G')$.*

A special case of Lemma 7.1.1 is the following corollary, which appears in [**18**, Proposition 3.2].

7.1.2. COROLLARY. *Let G be a group and let A be a normal subgroup of G. If $A \cap \zeta(G)$ contains a subgroup C such that A/C is locally finite, then $[A, G]$ is also locally finite and $\Pi([A, G]) \subseteq \Pi(A/C)$.*

PROOF. Let $\pi = \Pi(A/C)$. By Corollary 1.5.17 A' is a locally finite π-group. Lemma 7.1.1 shows that $[A/A', G/A']$ is also a locally finite π-group. However, $[A/A', G/A'] = [A, G]/A'$ and hence $[A, G]$ is also a locally finite π-group.   □

The result of Corollary 1.5.17 cannot be extended to the case when $G/\zeta(G)$ is periodic since S. I. Adyan [**1**] has constructed a torsion-free group G in which the central factor group $G/\zeta(G)$ is an infinite finitely generated p-group of finite exponent, for some prime p. In particular, G' is a nontrivial, non-periodic group.

Our first candidate for the class $\mathfrak{X}$ in our general problem of this chapter is the class of groups of finite 0-rank. We first give some results that will be useful for the other classes we consider. A number of the following results are analogous to Lemma 1.5.20. We recall that if G is a finitely generated group then $\mathbf{d}(G)$ represents the minimal number of generators of G.

7.1.3. LEMMA. *Let G be a group and let A be an abelian normal subgroup of G. Suppose that G satisfies the following conditions:*

(i) *$G/C_G(A) = \langle x_1 C_G(A), \dots, x_m C_G(A) \rangle$, where $x_1, \dots, x_m \in G$;*
(ii) *$A \cap \zeta(G)$ contains a subgroup C such that A/C is finitely generated and $\mathbf{d}(A/C) = d$.*

Then $[A, G]$ is finitely generated and $\mathbf{d}([A, G]) \leq dm$.

EXERCISE 7.3. *Prove Lemma 7.1.3.*

Our next two results also appear in [**18**].

7.1.4. COROLLARY. *Let G be a group and let A be an abelian normal subgroup of G. Suppose that G satisfies the following conditions:*

(i) *$G/C_G(A) = \langle x_1 C_G(A), \dots, x_m C_G(A) \rangle$, where $x_1, \dots, x_m \in G$;*
(ii) *$A \cap \zeta(G)$ contains a subgroup C such that A/C has finite special rank r.*

Then $[A, G]$ has finite special rank at most rm.

PROOF. Let B/C be a subgroup of A/C which has exactly r generators, say $B/C = \langle d_1 C, \ldots, d_r C \rangle$, and let

$$D/C = \langle d_j^{x_s} C | 1 \leq j \leq r, 1 \leq s \leq m \rangle.$$

Then D is G-invariant and D/C is a finitely generated subgroup of A/C, an abelian group of special rank r. Since $B/C \leq D/C$ it follows that D/C has exactly r generators. Furthermore, A/C is countable since it is abelian of finite special rank. These arguments imply that we can construct an ascending series of subgroups

$$D = D_1 \leq D_2 \leq \cdots \leq D_j \leq D_{j+1} \leq \cdots$$

such that each D_j is G-invariant, D_j/C has exactly r generators for each $j \in \mathbb{N}$ and $A = \cup_{j \in \mathbb{N}} D_j$. Clearly $[D_j, G] \leq [D_{j+1}, G]$, for all $j \in \mathbb{N}$, and $[A, G] = \cup_{j \in \mathbb{N}} [D_j, G]$. Lemma 7.1.3 shows that $[D_j, G]$ has at most rm generators for all $j \in \mathbb{N}$. Now let F be an arbitrary finitely generated subgroup of $[A, G]$. Then $F \leq [D_t, G]$ for some natural number t and since $[D_t, G]$ is abelian it follows that $\mathbf{d}(F) \leq \mathbf{d}([D_t, G]) \leq rm$, which proves the result.   □

7.1.5. COROLLARY. *Let G be a group and let A be an abelian normal subgroup of G such that $G/C_G(A)$ has finite special rank m. If $A \cap \zeta(G)$ contains a subgroup C such that A/C has finite special rank r, then $[A, G]$ has finite special rank at most rm.*

PROOF. Let F be an arbitrary finitely generated subgroup of $[A, G]$. Then there is a finitely generated subgroup H such that $F \leq [A, H]$. We let $K = HC_G(A)$. Clearly, $F \leq [A, K]$ and $K/C_G(A)$ is finitely generated so the hypotheses imply that $K/C_G(A)$ is at most m-generator. It follows from Corollary 7.1.4 that $[A, K]$ has finite special rank at most rm, which implies that F is at most rm-generator. Consequently, $[A, G]$ has finite special rank at most rm. This completes the proof.   □

We leave the following analogue of Corollary 7.1.4 as an exercise for the reader.

7.1.6. COROLLARY. *Let G be a group and let A be an abelian normal subgroup of G. Suppose that G satisfies the following conditions:*

(i) *$G/C_G(A) = \langle x_1 C_G(A), \ldots, x_m C_G(A) \rangle$, where $x_1, \ldots, x_m \in G$;*
(ii) *$A \cap \zeta(G)$ contains a subgroup C such that A/C has finite 0-rank r.*

Then $[A, G]$ has finite 0-rank at most rm.

EXERCISE 7.4. *Prove Corollary 7.1.6.*

7.1.7. COROLLARY. *Let G be a group and let A be an abelian normal subgroup of G such that $G/C_G(A)$ has finite special rank m. If $A \cap \zeta(G)$ contains a subgroup C such that A/C has finite 0-rank r, then $[A, G]$ has finite 0-rank at most rm.*

The proof of this corollary follows exactly as in the proof of Corollary 7.1.5 but instead of using Corollary 7.1.4 we use Corollary 7.1.6.

7.1.8. LEMMA. *Let G be a group and let C be a subgroup of $\zeta(G)$ such that G/C is abelian of finite 0-rank r. Then G' has finite 0-rank at most $r(r-1)/2$.*

EXERCISE 7.5. *Prove Lemma 7.1.8.*

The following corollaries are analogous to several results occurring in [**66**].

7.1.9. COROLLARY. *Let G be a group containing normal subgroups A, B such that the following conditions hold:*
   (i) $A \leq B \cap \zeta(G)$ and $B/A \leq \zeta(G/A)$;
   (ii) B/A has finite 0-rank r and G/B has finite special rank k.
Then $[B, G]$ has finite 0-rank at most $r(r + 2k - 1)/2$.

PROOF. Applying Lemma 7.1.8 to B we see that $\mathbf{r}_0(B') \leq r(r-1)/2$. Of course, B/B' is abelian and $(G/B')/C_{G/B'}(B/B')$ has finite special rank k. Applying Corollary 7.1.7 to G/B' we deduce that $[B/B', G/B'] = [B, G]/B'$ has finite 0-rank at most rk. Finally, by Lemma 2.2.3, we have
$$\mathbf{r}_0([B, G]) = \mathbf{r}_0(B') + \mathbf{r}_0([B, G]/B') \leq r(r-1)/2 + rk = r(r + 2k - 1)/2,$$
as required. □

7.1.10. COROLLARY. *Let G be a group containing normal subgroups A, B such that the following conditions hold:*
   (i) $A \leq B \cap \zeta(G)$ and $B \leq \zeta_m(G)$, for some natural number m;
   (ii) B/A has finite 0-rank r and G/B has finite special rank k.
Then $[B, G]$ has finite 0-rank at most $r(2r + 2k - 1)/2$.

PROOF. Since B is nilpotent, Proposition 1.2.11 implies that B has a torsion subgroup $\mathbf{Tor}(B)$ consisting of the elements of B of finite order. We first suppose that B is torsion-free and let
$$1 = Z_0 \leq Z_1 \leq \cdots \leq Z_m = B$$
be the upper G-central series of B. Since $A \leq B \cap \zeta(G) = Z_1$ it follows that B/Z_1 has finite 0-rank at most r. As in Corollary 1.2.9 it is possible

to prove that the factors Z_j/Z_{j-1} are torsion-free, for $1 \le j \le m$, and hence $m - 1 \le r$.

Let $\mathbf{r}_0(Z_{j+1}/Z_j) = r_{j+1}$, for $1 \le j \le m - 1$. By Lemma 2.2.3 we have $\mathbf{r}_0(B/Z_2) = r - r_2$. Proposition 6.2.4 shows that the special rank of B/Z_2 is also $r - r_2$ and hence G/Z_2 has special rank at most $k + r - r_2$. Then Corollary 7.1.9 implies that $C_2 = [Z_2, G]$ has finite 0-rank at most $r_2(r_2 + 2k + 2r - 2r_2 - 1)/2 = r_2(2k + 2r - r_2 - 1)/2$.

The centre of G/C_2 contains Z_2/C_2 and we now apply the above argument to G/C_2. We have $\mathbf{r}_0((Z_3/C_2)/(Z_2/C_2)) = \mathbf{r}_0(Z_3/Z_2) = r_3$. Also G/Z_3 has special rank at most $k + r - r_2 - r_3$, so at most $k + r - r_3$. Then, as above, $C_3/C_2 = [Z_3/C_2, G/C_2]$ has finite 0-rank at most $r_3(2k + 2r - r_3 - 1)/2$. Since $C_2 = [Z_2, G] \le [Z_3, G]$ we have $[Z_3/C_2, G/C_2] = [Z_3, G]/C_2$, and we deduce that

$$\mathbf{r}_0([Z_3, G]) \le r_2(2k + 2r - r_2 - 1)/2 + r_3(2k + 2r - r_3 - 1)/2$$
$$= (r_2 + r_3)(2k + 2r - 1)/2 - (r_2^2 + r_3^2)/2.$$

We now repeat this argument and after finitely many steps we deduce that

$$\mathbf{r}_0([B, G]) \le r(2k + 2r - 1)/2 - (r_2^2 + r_3^2 + \ldots r_m^2)/2$$
$$\le r(2k + 2r - 1)/2.$$

Finally, we note that if B is not torsion-free then we can apply the argument above to the group $G/\mathbf{Tor}\,(B)$ to deduce that

$$\mathbf{r}_0([B/\mathbf{Tor}\,(B), G/\mathbf{Tor}\,(B)] \le r(2k + 2r - 1)/2.$$

Since $\mathbf{r}_0([B, G]) = \mathbf{r}_0([B/\mathbf{Tor}\,(B), G/\mathbf{Tor}\,(B)])$ the result follows. $\square$

Although we are here concerned with the 0-rank of a group, a crucial role is still played by our knowledge of finite groups and the next few results illustrate how the special rank is affected.

7.1.11. LEMMA. *Let F be a field and let G be a finite abelian subgroup of $GL_n(F)$. Suppose also that if F has characteristic $p > 0$ then $p \notin \Pi(G)$. Then $\mathbf{d}(G) \le n$.*

EXERCISE 7.6. *Prove Lemma 7.1.11.*

EXERCISE 7.7. *Let G be a finite group of order n. Prove that G has special rank at most $\log_2(n)$ and hence $\mathbf{d}(G) \le \log_2(|G|)$.*

7.1.12. LEMMA. *Let p be a prime and let G be a finite p-group. If every abelian subgroup of G has finite special rank at most r, then G has special rank at most $r(5r + 1)/2$.*

PROOF. Let A be a maximal normal abelian subgroup of G. Since G is nilpotent, $A = C_G(A)$. Let m be a natural number such that $A = \Omega_m(A)$. Then we can think of G/A as a subgroup of $GL_r(\mathbb{Z}/p^m\mathbb{Z})$. In Example 6.2.5 we showed that the Sylow p-subgroup of $GL_r(\mathbb{Z}/p^m\mathbb{Z})$ has special rank at most $r(5r-1)/2$, so that G/A also has special rank at most $r(5r-1)/2$. It follows that G has special rank at most

$$r + r(5r-1)/2 = r(5r+1)/2,$$

as required. □

7.1.13. PROPOSITION. *Let G be a finite group. If every abelian subgroup of G has special rank at most r, then G has special rank at most $r(5r+3)/2$.*

PROOF. Let H be an arbitrary subgroup of G. If p is a prime, then Lemma 7.1.12 shows that every Sylow p-subgroup of H has special rank at most $r(5r+1)/2$. In particular, every Sylow p-subgroup of H is generated by at most $r(5r+1)/2$ elements. Since this is true for each prime p, Corollary 6.3.15 implies that H is generated by at most

$$r(5r+1)/2 + 1 = r(5r+3)/2$$

elements. This means that G has special rank at most $r(5r+3)/2$, which completes the proof. □

7.1.14. COROLLARY. *Let F be a field and let G be a finite subgroup of $GL_n(F)$. If F has characteristic 0, then G has special rank at most $n(5n+3)/2$.*

PROOF. By Lemma 7.1.11 every abelian subgroup of G has special rank at most n, so we can apply Proposition 7.1.13 to G to obtain the result. □

7.1.15. COROLLARY. *Let R be an integral domain and let G be a finite subgroup of $GL_n(R)$. If R has characteristic 0, then G has special rank at most $n(5n+3)/2$.*

PROOF. Since R is an integral domain it has a field of fractions F and the group $GL_n(R)$ can be embedded in the group $GL_n(F)$. Hence G is a subgroup of $GL_n(F)$ and we may apply Corollary 7.1.14 to G in this case. □

7.1.16. COROLLARY. *Let A be a torsion-free abelian group with finite 0-rank r and let G be a finite subgroup of $\mathbf{Aut}(A)$. Then G has special rank at most $r(5r+3)/2$.*

PROOF. Let D be the divisible envelope of A. Every automorphism of A extends to an automorphism of D. Also $D = \oplus_{j=1}^{r} D_j$ where $D_j \cong \mathbb{Q}$, for $1 \leq j \leq r$ and $\mathbf{Aut}\,(D) \cong GL_r(\mathbb{Q})$. The result now follows by Corollary 7.1.14.   □

7.1.17. COROLLARY. *Let G be a group and let A be a torsion-free abelian normal subgroup of G. Let C be a subgroup of $A \cap \zeta(G)$ such that A/C is torsion-free of finite 0-rank r. Then every finite subgroup of $G/C_G(A)$ has special rank at most $r(5r+3)/2$.*

PROOF. Let $H/C_G(A)$ be a finite subgroup of $G/C_G(A)$ and let $Z = C_H(A/C)$. If $z \in Z, a \in A$, then $a^z = ac$, for some element $c \in C$ and for every natural number m we have $z^{-m}az^m = ac^m$. Since $H/C_G(A)$ is finite there is a natural number t such that $z^t \in C_G(A)$, so we have $a = z^{-t}az^t = ac^t$ and hence $c^t = 1$. However C is torsion-free, so $c = 1$ and hence $z \in C_H(A)$. Consequently, $C_H(A/C) = C_H(A)$. Corollary 7.1.16 implies that $H/C_H(A/C)$ has special rank at most $r(5r+3)/2$, as required.   □

We may now prove the main result of this section.

7.1.18. THEOREM. *Let G be a locally generalized radical group and suppose that C is a subgroup of $\zeta(G)$ such that G/C has finite 0-rank r. Then G' has finite 0-rank at most $r(5r^2 + 5r - 1)/2$.*

PROOF. Let $R = \mathbf{Tor}\,(G)$ and note that $\mathbf{r}_0(G) = \mathbf{r}_0(G/R)$. Hence it is sufficient to prove the theorem for G/R and in so doing we may suppose that $R = 1$.

By Theorem 2.4.13, G has a series of normal subgroups $C \leq T \leq L \leq K \leq G$ such that T/C is locally finite, L/T is a torsion-free nilpotent group, K/L is a finitely generated torsion-free abelian group and G/K is finite. By Corollary 7.1.2, $[T, G]$ is locally finite, so $[T, G] = 1$, by our assumption and hence we may assume that $T = C$.

Let $\mathbf{r}_0(L/C) = r_1$ and $\mathbf{r}_0(K/L) = r_2$, so that $r_1 + r_2 = r$. Corollary 7.1.10 applied to L implies that L' has 0-rank at most $r_1(2r_1 - 1)/2$. Applying Corollary 7.1.7 to K/L' shows that $[L/L', K/L'] = [L, K]/L'$ has finite 0-rank at most $r_1 r_2$ and we deduce that

$$\mathbf{r}_0([L, K]) \leq r_1(2r_1 - 1)/2 + r_1 r_2.$$

Next we note that $L/[L, K]$ lies in the centre of $K/[L, K]$ so we may apply Lemma 7.1.8 to $K/[L, K]$ to deduce that $(K/[L, K])' = K'/[L, K]$

has finite 0-rank at most $r_2(r_2 - 1)/2$. Hence we have,

$$\mathbf{r}_0(K') \leq r_1(2r_1 - 1)/2 + r_1 r_2 + r_2(r_2 - 1)/2$$
$$= r_1^2 + r_1 r_2 + r_2^2/2 - r_1/2 - r_2/2 = r^2 - r/2 - r_2(r - r_2/2)$$
$$\leq r(2r - 1)/2.$$

Let $S/K' = \mathbf{Tor}\,(K/K')$ and note that $\mathbf{r}_0(S) = \mathbf{r}_0(K')$. Furthermore, K/S is a torsion-free abelian group. Let $P/CS = \mathbf{Tor}\,(K/CS)$. Lemma 7.1.1 implies that $[P/S, G/S] = [P, G]S/S$ is periodic, so we deduce that $[P, G] \leq S$. Consequently, $P/S \leq \zeta(G/S)$. The factor group K/P is torsion-free and has 0-rank at most r. Applying Corollary 7.1.17 to G/S, we see that the finite group $(G/S)/C_{G/S}(K/S)$ has special rank at most $r(5r + 3)/2$. Corollary 7.1.7 can be applied again and this now implies that $[K/S, G/S] = [K, G]S/S$ has 0-rank at most $r^2(5r + 3)/2$. We now have

$$\mathbf{r}_0([K, G]) \leq r^2(5r + 3)/2 + r(2r - 1)/2 = r(5r^2 + 5r - 1)/2.$$

The centre of $G/[K, G]$ contains $K/[K, G]$, which is of finite index in $G/[K, G]$. Consequently we may apply Theorem 1.5.12 to deduce that $G'/[K, G] = (G/[K, G])'$ is finite and hence $\mathbf{r}_0(G') = \mathbf{r}_0([K, G])$. Finally we have $\mathbf{r}_0(G') = r(5r^2 + 5r - 1)/2$, as required. $\qquad\square$

Our next goal is to establish an analogue of Theorem 1.5.21 for groups of finite 0-rank. Thus we shall be considering groups in which $G/\zeta_k(G)$ has finite 0-rank for some natural number k and we shall be interested in obtaining information concerning the 0-rank of $\gamma_{k+1}(G)$. There are two stages required to establish this goal, the first being the important special case when $G/\zeta_k(G)$ is locally finite.

We consider a slightly more general situation. We let A be a normal subgroup of the group G. Write $A = \gamma_1(A, G), \gamma_2(A, G) = [A, G]$ and recursively $\gamma_{\alpha+1}(A, G) = [\gamma_\alpha(A, G), G]$ for all ordinals α and $\gamma_\lambda(A, G) = \cap_{\mu<\lambda}\gamma_\mu(A, G)$ for all limit ordinals λ.

Our next result is really a generalization of [18, Proposition 3.2].

7.1.19. THEOREM. *Let G be a group and let A be a normal subgroup of G. If $A/(A \cap \zeta_k(G))$ is locally finite, then $\gamma_{k+1}(A, G)$ is also locally finite and $\Pi(\gamma_{k+1}(A, G)) \subseteq \Pi(A/A \cap \zeta_k(G))$.*

PROOF. Let $Z_j = \zeta_j(G)$ and let

$$1 = Z_0 \leq Z_1 \leq \cdots \leq Z_{k-1} \leq Z_k = Z$$

be a part of the upper central series of G. We use induction on k. If $k = 1$ then $A/(A \cap Z_1)$ is locally finite and an application of Corollary 7.1.2

shows that $\gamma_2(A, G) = [A, G]$ is locally finite and that $\Pi(\gamma_2(A, G)) \subseteq \Pi(A/A \cap Z_1)$.

Assume now that $k > 1$ and consider G/Z_1. Suppose we have already proved that $\gamma_k(AZ_1/Z_1, G/Z_1)$ is locally finite and that

$$\Pi(\gamma_k(AZ_1/Z_1, G/Z_1)) \subseteq \Pi(A/A \cap Z_k).$$

Let $K/Z_1 = \gamma_k(AZ_1/Z_1, G/Z_1)$ and apply Corollary 7.1.2 to deduce that $[K, G]$ is locally finite and $\Pi([K, G]) \subseteq \Pi(K/Z_1)$. We note that $\gamma_k(AZ_1/Z_1, G/Z_1) = \gamma_k(A, G)Z_1/Z_1$ and hence $\gamma_k(A, G) \leq K$. Then $\gamma_{k+1}(A, G) = [\gamma_k(A, G), G] \leq [K, G]$. It follows that $\gamma_{k+1}(A, G)$ is locally finite and $\Pi(\gamma_{k+1}(A, G)) \subseteq \Pi(K/Z_1) \subseteq \Pi(A/(A \cap Z_k))$, as required. □

7.1.20. COROLLARY. *Let G be a group and suppose that $G/\zeta_k(G)$ is locally finite for some natural number k. Then $\gamma_{k+1}(G)$ is also locally finite and $\Pi(\gamma_{k+1}(G)) \subseteq \Pi(G/\zeta_k(G))$.*

Next we obtain not only a generalization of this result but a generalization of Theorem 1.5.22. A slightly different version of this result occurs in [**77**].

7.1.21. THEOREM. *Let G be group and suppose that $G/\zeta_\infty(G)$ is locally finite. Then G contains a normal locally finite subgroup R such that G/R is locally nilpotent. Furthermore $\Pi(R) \subseteq \Pi(G/\zeta_\infty(G))$. In particular, the locally nilpotent residual $G^{\mathrm{L}\mathfrak{N}}$ of G is locally finite and $G/G^{\mathrm{L}\mathfrak{N}}$ is locally nilpotent.*

PROOF. Let $Z = \zeta_\infty(G)$ and let $\mathfrak{T}$ be the family of all finitely generated subgroups of G. If $U \in \mathfrak{T}$, then the upper hypercentre of U contains the U-invariant subgroup $U \cap Z = Z_U$ and U/Z_U is a finitely generated, locally finite group. Hence U/Z_U is finite. Furthermore, since $Z_U \leq \zeta_\infty(G)$ we have $Z_U \leq \zeta_\infty(U)$. We note that $\zeta_\infty(U)$ is finitely generated by Proposition 1.2.13 and that $\zeta_\infty(U) = \zeta_k(U)$, for some natural number k, by Proposition 1.2.5. Theorem 1.5.19 shows that the nilpotent residual $U^\mathfrak{N}$ of U is finite and clearly $U/U^\mathfrak{N}$ is nilpotent. Corollary 7.1.20 implies that $\Pi(U^\mathfrak{N}) \subseteq \Pi(U/Z_U) \subseteq \Pi(G/Z)$.

Now let V be another subgroup in $\mathfrak{T}$. Then there exists $W \in \mathfrak{T}$ such that $\langle U, V \rangle \leq W$. Of course $W^\mathfrak{N} \cap U \lhd U$ and $U/(W^\mathfrak{N} \cap U)$ is nilpotent. Hence $U^\mathfrak{N} \leq W^\mathfrak{N}$ and similarly $V^\mathfrak{N} \leq W^\mathfrak{N}$. It follows that $R = \bigcup_{U \in \mathfrak{T}} U^\mathfrak{N}$ is a locally finite subgroup of G such that $\Pi(R) \subseteq \Pi(G/Z)$. Also, R is a normal subgroup of G, since $U^\mathfrak{N}$ is a normal subgroup of U for each $U \in \mathfrak{T}$.

Let F/R be a finitely generated subgroup of G/R. Then there is a finitely generated subgroup V of G such that $F = VR$. Of course

$V^{\mathfrak{N}} \leq R \cap V$ and hence $V/(R \cap V) \cong VR/R = F/R$ is nilpotent. Consequently, G/R is locally nilpotent.

Next we let Y be a normal subgroup of G such that G/Y is locally nilpotent. If $U \in \mathfrak{T}$ then UY/Y is nilpotent and hence $U^{\mathfrak{N}} \leq U \cap Y$. In particular, $U^{\mathfrak{N}} \leq Y$ and we have $R = \bigcup_{U \in \mathfrak{T}} U^{\mathfrak{N}} \leq Y$. Since this is true for all such normal subgroups Y we have $R \leq G^{\mathbf{L}\mathfrak{N}}$, so $R = G^{\mathbf{L}\mathfrak{N}}$ and hence $G/G^{\mathbf{L}\mathfrak{N}}$ is locally nilpotent. This completes the proof. $\quad\square$

7.1.22. COROLLARY. *Let G be a group and let C be a G-invariant subgroup of $\zeta_\infty(G)$ such that G/C is locally finite. If $\mathbf{Tor}\,(G) = 1$, then G is a torsion-free hypercentral group.*

PROOF. By Proposition 1.2.11, $Z = \zeta_\infty(G)$ is torsion-free. Using Theorem 7.1.21 we deduce that G is locally nilpotent and hence is torsion-free, again by Proposition 1.2.11. Now Corollary 1.2.9 implies that G is hypercentral, as required. $\quad\square$

It is easy to see now that if G satisfies the hypotheses of Theorem 7.1.21, then $G/\mathbf{Tor}\,(G)$ is hypercentral and $\mathbf{Tor}\,(G)$ is locally finite, giving the result of [**77**]. To continue further we need to determine the connection between the 0-rank and the special rank.

7.1.23. LEMMA. *Let G be a nilpotent-by-finite group with finite 0-rank r. If $\mathbf{Tor}\,(G) = 1$ then G has special rank at most $5r^2(r + 1)/2$.*

PROOF. Let L be the maximal normal nilpotent subgroup of G and let

$$1 = Z_0 \leq Z_1 \leq \cdots \leq Z_k = L$$

be the upper central series of L. By Corollary 1.2.9, each factor Z_j/Z_{j-1} is torsion-free, for $1 \leq j \leq k$ and hence $k \leq \mathbf{r}_0(L)$. Furthermore, G/L is finite so $\mathbf{r}_0(G) = \mathbf{r}_0(L)$.

Let $C = Z_{k-1}$ and note that G/C is centre-by-finite. Let $V_k/C = C_{G/C}(Z_k/C)$. Since Z_k/C is torsion-free it follows that $T_{k-1}/C = \mathbf{Tor}\,(V_k/C)$ is finite. We may apply Theorem 1.5.12 to G/C and deduce that $(V_k/C)'$ is finite. Corollary 7.1.16 shows that G/V_k has finite special rank at most $r(5r + 3)/2$ so, by Lemma 6.1.2,

$$\mathbf{r}(G/T_{k-1}) \leq \mathbf{r}(V_k/T_{k-1}) + \mathbf{r}(G/V_k)$$
$$\leq r + r(5r + 3)/2 = r(5r + 5)/2.$$

Next we consider the factor group T_{k-1}/Z_{k-2}. By the same argument as that used above we deduce that there is a G-invariant subgroup T_{k-2} containing Z_{k-2} such that T_{k-2}/Z_{k-2} is finite and T_{k-1}/T_{k-2} has special rank at most $r(5r + 5)/2$. Then, by Lemma 6.1.2,

$$\mathbf{r}(G/T_{k-2}) \leq \mathbf{r}(T_{k-1}/T_{k-2}) + \mathbf{r}(G/T_{k-1}) \leq r(5r + 5)/2 + r(5r + 5)/2.$$

We may repeat these arguments to construct a sequence of G-invariant subgroups, T_j, such that $T_j \leq T_{j+1}$, T_j/Z_j is finite and $\mathbf{r}(T_{j+1}/T_j) \leq r(5r+5)/2$, for $2 \leq j \leq k-2$. This means that G/T_2 has special rank at most $(k-1)r(5r+5)/2$. Likewise, $T_2/\mathbf{Tor}\,(T_2)$ has special rank at most $r(5r+5)/2$. However $\mathbf{Tor}\,(T_2) = 1$ so G has special rank at most

$$(k-1)r(5r+5)/2 + r(5r+5)/2 = kr(5r+5)/2 \leq 5r^2(r+1)/2,$$

as required. □

We now generalize this to locally generalized radical groups.

7.1.24. THEOREM. *Let G be a locally generalized radical group. If G has finite 0-rank r, then $G/\mathbf{Tor}\,(G)$ has special rank at most $5r^2(r+1)/2$.*

PROOF. Since $\mathbf{Tor}\,(G/\mathbf{Tor}\,(G)) = 1$ and we are obtaining the result for $G/\mathbf{Tor}\,(G)$, we shall assume that $\mathbf{Tor}\,(G) = 1$. Then, by Theorem 2.4.13, G has normal subgroups $L \leq K$ such that L is torsion-free nilpotent, K/L is a finitely generated torsion-free abelian group and G/K is finite.

If $K = L$, then Lemma 7.1.23 shows that G has special rank at most $5r^2(r+1)/2$, so suppose that $K \neq L$. Let $\mathbf{r}_0(L) = r_1$ and $\mathbf{r}_0(K/L) = r_2$, so $r_1 + r_2 = r$. Let $T_1/L = \mathbf{Tor}\,(G/L)$. Applying Lemma 7.1.23 we deduce that G/T_1 has special rank at most $5r_2^2(r_2+1)/2$. We note that T_1/L is finite so we may apply Lemma 7.1.23 to T_1 to deduce that T_1 has special rank at most $5r_1^2(r_1+1)/2$. Hence G has special rank at most

$$\begin{aligned}
5r_1^2(r_1+1)/2 + 5r_2^2(r_2+1)/2 &= 5(r_1^3 + r_2^3 + r_1^2 + r_2^2)/2 \\
&= 5((r-r_2)^3 + r_2^3 + (r-r_2)^2 + r_2^2)/2 \\
&= 5(r^3 + r^2 + 3rr_2(r_2 - r) \\
&\quad + 2r_2(r_2 - r))/2 \\
&\leq 5r^2(r+1)2,
\end{aligned}$$

since $r \geq r_2$. In either case $\mathbf{r}(G) \leq 5r^2(r+1)/2$, as required. □

We may now obtain our analogue of Theorems 1.5.19 and 1.5.21.

7.1.25. THEOREM. *Let G be a locally generalized radical group and suppose that for some natural number k, $G/\zeta_k(G)$ has finite 0-rank r. Then $\gamma_{k+1}(G)$ has finite 0-rank and there is a function $\mathfrak{n}$ such that $r_0(\gamma_{k+1}(G)) \leq \mathfrak{n}(r, k)$.*

PROOF. Let $T = \mathbf{Tor}\,(G)$ and let

$$1 = Z_0/T \leq Z_1/T \leq \cdots \leq Z_{k-1}/T \leq Z_k/T = Z/T$$

be a part of the upper central series of G/T. By Theorem 2.4.13 the group G has a series $Z \leq P \leq Q \leq K \leq G$ of normal subgroups such that P/Z is locally finite, Q/P is torsion-free nilpotent, K/Q is finitely generated torsion-free abelian and G/K is finite. By Theorem 7.1.19 $S/T = \gamma_{k+1}(P/T, G/T)$ is locally finite and, since T is locally finite, Corollary 1.2.14 shows that S is locally finite. Hence $S = T$ and so $P/T \leq \zeta_k(G/T) = Z_k/T = Z/T$. Hence $P = Z$ and Q/Z is a torsion-free nilpotent group of finite 0-rank.

The remainder of the argument is to be applied to the group G/T. We shall show that $\gamma_{k+1}(G/T)$ has finite zero rank. We note that $\gamma_{k+1}(G/T) = \gamma_{k+1}(G)T/T$. Since T is periodic it follows that

$$\mathbf{r}_0(\gamma_{k+1}(G/T)) = \mathbf{r}_0(\gamma_{k+1}(G)T) = \mathbf{r}_0(\gamma_{k+1}(G)).$$

Thus in order to simplify our notation we may assume that $T = 1$ in the remainder of the argument and hence assume that Z is torsion-free. As we saw above, $\mathbf{Tor}\,(G/Z) = 1$ and Theorem 7.1.24 implies that $\mathbf{r}(G/Z) \leq 5r^2(r+1)/2$.

We use induction on k. If $k = 1$ then the central factor group G/Z_1 has finite 0-rank and then Theorem 7.1.18 implies that G' has finite 0-rank at most $r(5r^2 + 5r - 1)/2 = \mathfrak{n}(r, 1)$.

Assume now that $k > 1$ and that there is a function $\mathfrak{n}$, depending on r and k such that $\gamma_k(G/Z_1)$ has finite 0-rank at most $\mathfrak{n}(r, k-1)$. Set $V/Z_1 = \gamma_k(G/Z_1)$, let $L = \gamma_k(G)$ and note that $V = LZ_1$. Then $V' = [LZ_1, LZ_1] = L'$, since $Z_1 = \zeta_1(G)$. Also $[V, G] = [L, G] = \gamma_{k+1}(G)$. An application of Theorem 7.1.18 shows that L' has finite 0-rank at most $\mathfrak{n}(r, k-1)(5\mathfrak{n}(r, k-1)^2 + 5\mathfrak{n}(r, k-1) - 1)$.

The factor group L/L' is abelian and we have

$$(L/L')/(L/L' \cap Z_1L'/L') \cong (L/L')(Z_1L'/L')/(Z_1L'/L')$$
$$\cong LZ_1/L'Z_1 \cong L/(L \cap Z_1L'),$$

an epimorphic image of $L/L \cap Z_1$. Certainly,

$$\mathbf{r}_0(L/L \cap Z_1) = \mathbf{r}_0(LZ_1/Z_1) = \mathbf{r}_0(V/Z_1) \leq \mathfrak{n}(r, k-1)$$

and hence

$$\mathbf{r}_0((L/L')/(L/L' \cap Z_1L'/L')) \leq \mathfrak{n}(r, k-1).$$

By Proposition 1.1.6, $[L, Z] = 1$ and hence $G/C_G(L)$ is an image of G/Z. It follows that $G/C_G(L)$ has special rank at most $5r^2(r+1)/2$. We apply Corollary 7.1.7 to G/L' and the work above shows that $[L/L', G/L'] = [L, G]/L' = \gamma_{k+1}(G)/L'$ has finite 0-rank at most $5\mathfrak{n}(r, k-1)r^2(r+1)/2$.

It follows that $\gamma_{k+1}(G)$ has finite 0-rank and that

$$\mathbf{r}_0(\gamma_{k+1}(G)) \leq \mathbf{r}_0(\gamma_{k+1}(G)/L') + \mathbf{r}_0(L')$$
$$\leq 5\mathfrak{n}(r, k-1)r^2(r+1)/2$$
$$+ \mathfrak{n}(r, k-1)(5\mathfrak{n}(r, k-1)^2 + 5\mathfrak{n}(r, k-1) - 1).$$

Consequently, we let

$$\mathfrak{n}(r, k) = 5\mathfrak{n}(r, k-1)r^2(r+1)/2 + \mathfrak{n}(r, k-1)(5\mathfrak{n}(r, k-1)^2 + 5\mathfrak{n}(r, k-1) - 1).$$

The result follows.

$\square$

The previous result shows that $\gamma_{k+1}(G)$ has finite 0-rank and this implies that the nilpotent residual of G also has finite 0-rank. However the bound for the 0-rank of the nilpotent residual turns out to be simpler than that for $\gamma_{k+1}(G)$ as we shall now see, although it requires some further ideas. As in Section 1.6 a key role here will be played by the Z-decomposition of certain normal abelian subgroups. However, for torsion-free subgroups these Z-decompositions can exist only in certain extensions. It is convenient to use the language of modules over group rings.

The following lemma is the module theoretic version of Lemma 1.6.1 and its proof is identical to the proof of that lemma so is omitted.

7.1.26. LEMMA. *Let G be a hypercentral group, let R be an integral domain and let A be an RG-module. Suppose that A contains an RG-submodule C such that $C \leq \zeta_{RG}(A)$. Suppose further that A/C is a simple RG-module and that $G \neq C_G(A/C)$. Then A contains a simple RG-submodule D such that $A = C \oplus D$.*

7.1.27. COROLLARY. *Let G be a hypercentral group, let R be an integral domain and let A be an RG-module. Suppose that A contains an RG-nilpotent submodule C such that A/C is a simple RG-module and that $G \neq C_G(A/C)$. Then A contains a simple RG-submodule D such that $A = C \oplus D$.*

The following lemma and its subsequent corollaries are also the module theoretic versions of results from Section 1.6. In this case Lemma 7.1.28 is analogous to Lemma 1.6.3, as is the proof, which is omitted.

7.1.28. LEMMA. *Let G be a hypercentral group, let R be an integral domain and let A be an RG-module. Suppose that A contains a simple*

RG-submodule C such that $G \neq C_G(C)$ and that $A/C \leq \zeta_{RG}(A/C)$. Then A contains an RG-submodule D such that $A = C \oplus D$.

7.1.29. COROLLARY. Let G be a hypercentral group, let R be an integral domain and let A be an RG-module. Suppose that A contains an RG-submodule C satisfying the following conditions:

(i) $A/C \leq \zeta_{RG}(A/C)$;

(ii) C has a finite series of RG-submodules whose factors are RG-simple and G-eccentric.

Then A contains an RG-submodule D such that $A = C \oplus D$.

7.1.30. COROLLARY. Let G be a hypercentral group, let R be an integral domain and let A be an RG-module. Suppose that A has a finite RG-composition series. Then A has a Z-decomposition.

In the following result, we let $\mathbf{cl}_{RG}(L)$ denote the RG-composition length of a module L.

7.1.31. COROLLARY. Let G be a hypercentral group, let R be an integral domain and let A be an RG-module. Suppose that A contains an RG-nilpotent submodule C such that A/C has a finite RG-composition series. Then A has a Z-decomposition, $A = Y \oplus L$, where Y is the upper RG-hypercentre of A and L is a hypereccentric RG-submodule. Furthermore, $\mathbf{cl}_{RG}(L) \leq \mathbf{cl}_{RG}(A/C)$.

7.1.32. COROLLARY. Let G be a group and let A be a torsion-free normal abelian subgroup of G. Suppose that $A/(A \cap \zeta_k(G))$ has finite 0-rank r, for some natural number k. If $G/C_G(A)$ is hypercentral, then A contains a pure G-invariant subgroup D satisfying the following conditions:

(i) D has a finite series of G-invariant pure subgroups

$$1 = D_0 \leq D_1 \leq \cdots \leq D_n = D$$

such that D_j/D_{j-1} is rationally irreducible, for $1 \leq j \leq n$;

(ii) A/D is G-nilpotent and $r_0(D) \leq r$.

PROOF. The group G acts on A by conjugation, and this action can be extended naturally to the action of the group ring $\mathbb{Z}G$ on A, making A into a $\mathbb{Z}G$-module. Let B denote the divisible envelope of A. The action of G on A can be naturally extended to an action of G on B and B can then be made into a $\mathbb{Q}G$-module. Let C be the divisible envelope of $A \cap \zeta_k(G)$ in B, so that C is a $\mathbb{Q}G$-submodule of B. Furthermore, $\dim_{\mathbb{Q}}(B/C) = r_0(A/(A \cap \zeta_k(G))) = r$ is finite, so that B/C has a finite

$\mathbb{Q}G$-composition series. It follows from Corollary 7.1.31 that B has a Z-decomposition, $B = U \oplus E$, where U is the upper $\mathbb{Q}G$-hypercentre of B and E is a hypereccentric $\mathbb{Q}G$-submodule of B. Furthermore, $\mathbf{cl}_{\mathbb{Q}G}(E) \leq \mathbf{cl}_{\mathbb{Q}G}(B/C) \leq r$. Thus E has a finite series

$$1 = E_0 \leq E_1 \leq \cdots \leq E_n = E$$

of $\mathbb{Q}G$-submodules whose factors are simple $\mathbb{Q}G$-modules. Let $D = A \cap E$ and $D_j = A \cap E_j$, for $0 \leq j \leq n$, so that every subgroup D_j is pure and G-invariant. Also, each factor D_j/D_{j-1} is G-rationally irreducible for $1 \leq j \leq n$. Finally, $\mathbf{r}_0(D) = \dim_{\mathbb{Q}}(E) \leq \mathbf{cl}_{\mathbb{Q}G}(E) \leq \mathbf{cl}_{\mathbb{Q}G}(B/C) \leq r$. Since B/E is G-nilpotent, A/D is also G-nilpotent. $\square$

7.1.33. THEOREM. *Let G be a locally generalized radical group and suppose that $G/\zeta_k(G)$ has finite 0-rank r, for some natural number k. Then the nilpotent residual $G^{\mathfrak{N}}$ of G has finite 0-rank and $\mathbf{r}_0(G^{\mathfrak{N}}) \leq r(5r^2 + 5r + 1)/2$.*

PROOF. As in the proof of Theorem 7.1.25, we may assume that $\mathbf{Tor}\,(G) = 1$. Then the group G has a series $\zeta_k(G) = Z \leq V \leq K \leq G$ of normal subgroups such that V/Z is a torsion-free nilpotent group, K/V is finitely generated torsion-free abelian and G/K is finite. Let

$$1 = Z_0 \leq Z_1 \leq \cdots \leq Z_{k-1} \leq Z_k = Z$$

be a part of the upper central series of G. By Theorem 1.2.22, $G/C_G(Z)$ is nilpotent of class at most $k - 1$. Let $C = C_G(Z)$. Since $Z \leq C_G(C)$ it follows that $G/C_G(C)$ has finite 0-rank at most r. Clearly $C \cap Z \leq \zeta(C)$ and $C/(Z \cap C) \cong CZ/Z$ has finite 0-rank at most r. Using Theorem 7.1.18 we deduce that C' has finite 0-rank at most $r(5r^2 + 5r - 1)/2$. Let $D/C' = \mathbf{Tor}\,(C/C')$ and note that $\mathbf{r}_0(D) = \mathbf{r}_0(C')$. The subgroup D is G-invariant and C/D is a torsion-free abelian group. Hence $(G/D)/C_{G/D}(C/D)$ is nilpotent and has finite 0-rank. We have $(C \cap Z)D/D \leq \zeta^{\infty}_{G/D}(C/D)$ and $(C/D)/((C \cap Z)D/D) \cong C/(C \cap Z)D$ has finite 0-rank at most r. By Corollary 7.1.32, C/D contains a pure G-invariant subgroup V/D of 0-rank at most r such that $(C/D)/(V/D)$ is G-nilpotent. Since G/C is nilpotent, G/V is also nilpotent, so $G^{\mathfrak{N}} \leq V$. Finally,

$$\mathbf{r}_0(V) = \mathbf{r}_0(D) + \mathbf{r}_0(V/D) \leq r(5r^2 + 5r - 1)/2 + r = r(5r^2 + 5r + 1)/2,$$

as required. $\square$

7.2. Central Extensions by Groups of Finite Section Rank

In Section 1.5 we discussed groups G in which $G/\zeta(G)$ is finite and in Theorem 1.5.18 obtained bounds for the order of G' in terms of the order of $G/\zeta(G)$. In this section, we obtain bounds for the section p-rank and the special rank of G' in terms of $\mathbf{r}(G/\zeta(G))$ or $\mathbf{sr}_p(G/\zeta(G))$.

7.2.1. LEMMA. *Let G be a group and let C be a subgroup of $\zeta(G)$ such that G/C is a finitely generated abelian group. Suppose that $\mathbf{d}(G/C) = d$. Then G' is finitely generated and $\mathbf{d}(G') \leq d(d-1)/2$.*

EXERCISE 7.8. *Prove Lemma 7.2.1.*

The finite p-group case of the result we are going to establish plays a special role in this work. In Section 4.3 we presented some of the results in the work of A. Lubotzky and A. Mann [**179**]. We could mention here another result of this work (without proof) which is concerned with the special rank of the Schur multiplier and immediately obtain an estimate for the special rank of the derived subgroup, as was done in [**18**]. However, we prefer to accompany these results with some reasoning to show how the concept of special (or section) rank works in finite groups. This will allow us to demonstrate some of the traditional technical aspects of the theory of finite groups.

We shall need some further results concerned with powerful p-groups, which were introduced in Section 4.3. If G is a p-group for some prime p and n is a natural number, then let

$$\mho_n(G) = \langle g^{p^n} | g \in G \rangle.$$

Furthermore, a normal subgroup H of G is called *powerfully embedded in G* if

$$[H, G] \leq \mho_2(H) \text{ whenever } p = 2;$$
$$[H, G] \leq \mho_1(H) \text{ whenever } p > 2.$$

Thus a finite p-group is powerful if and only if it is powerfully embedded in itself. If a subgroup H is powerfully embedded in G, then H is powerful and, moreover, if K is a subgroup of G containing H such that K/H is cyclic, then $K' = [H, K]$, so that K is also powerful. We continue our discussion by proving the following trio of results from [**179**].

7.2.2. LEMMA. *Let p be a prime and let G be a finite p-group. Suppose that H is a normal subgroup of G which is powerfully embedded in G. If S is a subset of H such that $H = \langle S \rangle^G$, then $H = \langle S \rangle$.*

PROOF. Let $K = \mho_1(H)$. We know that $\mathbf{Frat}(H) = H'K$. If $p = 2$, then $\mathbf{Frat}(H) = K$, whereas if $p > 2$, then $H' \leq [H, G] \leq \mho_1(H) = K$, since H is powerfully embedded in G, so $\mathbf{Frat}(H) = K$ in this case as well. Since H is powerfully embedded in G, we have $[H, G] \leq K$. Hence $H/K \leq \zeta(G/K)$ and we have $H/K = \langle S \rangle^G K/K = \langle S \rangle K/K$. Therefore, $H = \langle S \rangle K$ and the non-generator property of the Frattini subgroup implies that $H = \langle S \rangle$.   □

We shall require the result of the next exercise during the proof of Lemma 7.2.3.

EXERCISE 7.9. *Let G be a nilpotent group and let g be an arbitrary element of G. Then $\boldsymbol{ncl}(\langle g, G' \rangle) < \boldsymbol{ncl}(G)$.*

7.2.3. LEMMA (Lubotzky and Mann [**179**]). *Let p be an odd prime and let G be a finite p-group. Suppose that C is a subgroup of $\zeta(G)$ such that G/C is a powerful p-group and $\boldsymbol{d}(G/C) = d$. Then G' is powerfully embedded in G and $\boldsymbol{d}(G') \leq d(d-1)/2$.*

PROOF. Let H be a subgroup of G such that $G' \leq H \leq \mathbf{Frat}(G)$. We show that H is powerfully embedded in G. To this end, we may assume that $\mho_1(H) = 1$ and we then show that $H \leq \zeta(G)$. In this case, since $G' \leq H$ we also have $\mho_1(G') = 1$. Suppose that $H \nleq \zeta(G)$ and consider $\bar{G} = G/[H, G, G]$. In this factor group $[\bar{H}, \bar{G}, \bar{G}] = 1$, so to keep our notation simple we now assume that in our group G we have $[H, G, G] = 1$, and note that, as we shall see later, this does not affect the generality of our reasoning.

Since $\gamma_2(G) \leq H$, we have $\gamma_3(G) \leq [H, G]$ and $\gamma_4(G) = [\gamma_3(G), G] \leq [H, G, G] = 1$. Consequently, G is nilpotent of class at most 3. Let a, b be arbitrary elements of G. Then $\langle a, a^b \rangle = \langle a, [a, b] \rangle \leq \langle a, G' \rangle$, which shows that $\langle a, a^b \rangle$ is nilpotent of class at most 2, by Exercise 7.9. It follows that for all elements $x, y \in \langle a, a^b \rangle$ we have $(xy)^p = x^p y^p z^p$, where $z = [y^{-1}, x^{-1}]$. However, $\mho_1(G') = 1$ so $z^p = 1$ and we deduce that $(xy)^p = x^p y^p$. In particular,

$$b^{-1} a^p b = (b^{-1} a b)^p = (a[a, b])^p = a^p [a, b]^p = a^p,$$

again since $\mho_1(G') = 1$. Hence $\mho_1(G) \leq \zeta(G)$. Since G/C is powerful we have

$$G'C/C = (G/C)' \leq \mho_1(G/C) = \mho_1(G)C/C$$

and hence $G'C \leq \mho_1(G)C \leq \zeta(G)$. Therefore $G' \leq \zeta(G)$ so that $\gamma_3(G) = 1$. Since $H \leq \mathbf{Frat}(G)$ it follows that

$$[H, G] \leq [\mathbf{Frat}(G), G] = [G'\mho_1(G), G] = \gamma_3(G)[\mho_1(G), G] = 1,$$

so that $H \leq \zeta(G)$. With this contradiction, we have proved that H is powerfully embedded in G.

Finally, since G/C is generated by d elements, G' is the normal closure of the commutators of these elements and the bound for $\mathbf{d}(G')$ now follows from Lemma 7.2.2, as required. ☐

We next obtain the analogue of Lemma 7.2.3 for $p = 2$, the proof being similar, but in need of some slight adjustments.

7.2.4. LEMMA (Lubotzky and Mann [**179**]). *Let G be a finite 2-group. Suppose that C is a subgroup of $\zeta(G)$ such that G/C is a powerful 2-group and $\mathbf{d}(G/C) = d$. Then G' is powerfully embedded in G and $\mathbf{d}(G') \le d(d-1)/2$.*

PROOF. In this case we let H be a subgroup of G such that $G' \le H \le G'\mho_2(G)$ and again show that H is powerfully embedded in G. As in the proof of Lemma 7.2.3 we may assume that $\mho_2(H) = 1$ and we shall show that $H \le \zeta(G)$. To this end, we note that, since $G' \le H$, it follows that $\mho_2(G') = 1$. Suppose for a contradiction that $H \nleq \zeta(G)$ and consider the factor group $G/[H, G, G]$. Again we may suppose that $[H, G, G] = 1$ and $\mho_1([H, G]) = 1$. As before, it follows that $\gamma_4(G) = 1$ and $\mho_1(\gamma_3(G)) = 1$. Then $G' \le \zeta_2(G)$.

Let a, b be arbitrary elements of G. As before, it is possible to show that $\langle a, a^b \rangle$ is nilpotent of class at most 2. We have

$$b^{-1}a^4b = (b^{-1}ab)^4 = (a[a,b])^4 = a^4[a,b]^4[a,b,a]^6 = a^4,$$

which shows that $\mho_2(G) \le \zeta(G)$. The fact that G/C is powerful implies that

$$G'C/C = (G/C)' \le \mho_2(G/C) = \mho_2(G)C/C$$

and hence $G'C \le \mho_2(G)C$. Since $\mho_2(G) \le \zeta(G)$ we obtain that $G' \le \zeta(G)$ and $H \le G'\mho_2(G) \le \zeta(G)$. With this contradiction it follows that H is powerfully embedded in G.

As before, G/C is generated by d elements, G' is the normal closure of the commutators of these elements and the bound for $\mathbf{d}(G')$ now follows from Lemma 7.2.2. ☐

7.2.5. COROLLARY. *Let p be a prime and let G be a finite p-group. Suppose that C is a subgroup of $\zeta(G)$ such that G/C is a powerful p-group and $\mathbf{d}(G/C) = d$. Then G' has special rank at most $d(d-1)/2$.*

PROOF. By Lemmas 7.2.3 and 7.2.4, $\mathbf{d}(G') \le d(d-1)/2$. These lemmas also imply that G' is powerfully embedded in G and, in particular, G' is powerful. If H is a subgroup of G', then Theorem 4.3.14 shows that $\mathbf{d}(H) \le \mathbf{d}(G')$ and the result now follows. ☐

This allows us to obtain the result we are seeking, at least for finite p-groups.

7.2.6. PROPOSITION. *Let p be a prime and let G be a finite p-group. Suppose that C is a subgroup of $\zeta(G)$ such that $\mathbf{r}(G/C) = d$. Then $\mathbf{r}(G') \leq m(m-1)/2 + d(3d-1)/2 = \mathfrak{s}_1(d)$, where $m = d(\iota(\log_2 d) + 1)$.*

PROOF. By Theorem 4.3.12, G/C contains a characteristic powerful subgroup H/C such that $|G/H| \leq p^m$, where $m = d(\iota(\log_2 d) + 1)$. Using Corollary 7.2.5 we deduce that $\mathbf{r}(H') \leq d(d-1)/2$. We apply Lemma 7.1.3 to the abelian normal subgroup H/H' of G/H' and from this we have

$$\mathbf{r}([H/H', G/H']) = \mathbf{d}([H/H', G/H']) \leq d^2.$$

However, $[H/H', G/H'] = [H, G]/H'$ so $\mathbf{r}([H, G]) \leq d^2 + d(d-1)/2 = d(3d-1)/2$.

Clearly, $H/[H, G] \leq \zeta(G/[H, G])$ and Theorem 1.5.18 implies that

$$|G'/[H, G]| = |(G/[H, G])'| \leq p^k,$$

where $k = m(m-1)/2$. Therefore,

$$\mathbf{r}(G') \leq m(m-1)/2 + d(3d-1)/2 = \mathfrak{s}_1(d).$$

$\square$

We note that $\iota(\log_2 d) \leq d$ so this last formula has the rough polynomial bound

$$\mathbf{r}(G') \leq \mathfrak{s}_1(d) \leq d(d^3 + 2d^2 + 3d - 2)/2.$$

The following corollary is immediate from Lemma 6.1.8 stating that the section p-rank of a finite p-group coincides with its special rank.

7.2.7. COROLLARY. *Let p be a prime and let G be a finite p-group. Suppose that C is a subgroup of $\zeta(G)$ and that $\mathbf{sr}_p(G/C) = d$. Then $\mathbf{sr}_p(G') \leq \mathfrak{s}_1(d)$.*

7.2.8. COROLLARY. *Let p be a prime and let G be a finite group. Suppose that C is a subgroup of $\zeta(G)$ such that G/C is a p-group and $\mathbf{sr}_p(G/C) = d$. Then G' is a p-group and $\mathbf{sr}_p(G') \leq \mathfrak{s}_1(d)$.*

PROOF. It is clear, from Theorem 1.5.18, that G' is a p-group. Let Q be a Sylow p'-subgroup of G. Since G/C is a p-group, $Q \leq C$ and hence Q is normal in G. Consequently $G = QP$, where P is a Sylow p-subgroup of G, using [118, Hauptsatz I.18.1]. Since $Q \leq \zeta(G)$ we have $G = Q \times P$ and hence $G' = P'$. The result now follows using Corollary 7.2.7. $\square$

7.2.9. LEMMA. *Let G be a group, let p be a prime and let A be an abelian normal p-subgroup of G. Suppose that C is a subgroup of $A \cap \zeta(G)$. If $p \notin \Pi(G/C_G(A))$, then $C_G(A) = C_G(\mathbf{\Omega}_1(A/C))$.*

EXERCISE 7.10. *Prove Lemma 7.2.9.*

In this section we shall prove several results which have appeared in [**18**]. Here is one such result.

7.2.10. LEMMA. *Let G be a group, let p be a prime and let A be an abelian normal p-subgroup of G. Suppose that G satisfies the following conditions:*

(i) *$G/C_G(A)$ is a periodic abelian p'-group;*
(ii) *$A \cap \zeta(G)$ contains a subgroup C such that $sr_p(A/C) = r$.*

Then $G/C_G(A)$ is finite and $d(G/C_G(A)) \le r$.

PROOF. Let $B/C = \Omega_1(A/C)$. Then $|B/C| \le p^r$ and we can think of $G/C_G(B/C)$ as a subgroup of $GL_r(\mathbb{F}_p)$. Lemma 7.1.11 shows that $d(G/C_G(B/C)) \le r$. By Lemma 7.2.9, $C_G(A) = C_G(B/C)$, which proves the result. □

7.2.11. COROLLARY. *Let p be a prime and let A be an abelian p-group. Suppose that $sr_p(A) = r$. If Q is a Sylow q-subgroup of $\textbf{Aut}(A)$, for some prime $q \ne p$, then Q is finite and has special rank at most $r(5r+1)/2$.*

PROOF. Consider the natural semidirect product $G = A \rtimes Q$. Lemma 7.2.9 shows that $C_G(A) = C_G(\Omega_1(A))$ and it follows that $1 = C_Q(A) = C_Q(\Omega_1(A))$. Since $\Omega_1(A)$ is finite, Q must be finite. By Lemma 7.2.10, every abelian subgroup of Q has special rank at most r. We now use Lemma 7.1.12 to deduce that Q has special rank at most $r(5r+1)/2$. □

7.2.12. PROPOSITION. *Let p be a prime and let A be an abelian p-group. Suppose that G is a periodic subgroup of $\textbf{Aut}(A)$. If $sr_p(A) = r$, then G is finite and has special rank at most $r(5r+3)/2$.*

PROOF. By Proposition 3.2.3, A is a Chernikov group so $A = D \times K$, where $D = C_1 \times \cdots \times C_t$ for certain Prüfer p-groups C_j, $1 \le j \le t$ and K is finite. Since G is periodic, Theorem 3.2.17 implies that $G/C_G(D)$ is finite. Since K is finite, so is $G/C_G(K)$ and, since $1 = C_G(A) = C_G(D) \cap C_G(K)$, Remak's Theorem shows that G is finite. Consequently, there is a natural number m such that $C_G(\Omega_m(A)) = C_G(A) = 1$ and hence G can be thought of as a subgroup of $GL_n(\mathbb{Z}/p^m\mathbb{Z})$.

Let H be an arbitrary subgroup of G. By Example 6.2.5, each Sylow p-subgroup of $GL_n(\mathbb{Z}/p^m\mathbb{Z})$ has special rank at most $r(5r-1)/2$, so the same is true for each Sylow p-subgroup of H. If $q \ne p$ is prime and Q is a Sylow q-subgroup of H, then Corollary 7.2.11 shows that $\mathbf{r}(Q) \leq r(5r + 1)/2$. Thus, for every prime s, the Sylow s-subgroup of H is generated by at most $r(5r + 1)/2$ elements. Corollary 6.3.15 implies that H is generated by at most $r(5r + 1)/2 + 1 = r(5r + 3)/2$ elements. It follows that G has special rank at most $r(5r + 3)/2$.   □

7.2.13. LEMMA. *Let G be a group and let C be a subgroup of $\zeta(G)$ such that G/C is finite. Let p be a prime and let $\boldsymbol{sr}_p(G/C) = r$. If A is an abelian normal p-subgroup of G, then $G/C_G(A)$ has special rank at most $r(5r + 3)/2$.*

EXERCISE 7.11. *Prove Lemma 7.2.13.*

For the finite group G and prime p the *lower p-series of G*,

$$1 = L_0 \leq L_1 \leq L_2 \leq \cdots \leq L_{n-1} \leq L_n,$$

is defined as follows: Let $L_1 = \mathbf{O}_{p'}(G), L_2/L_1 = \mathbf{O}_p(G/L_1)$ and if L_j has been defined, then $L_{j+1}/L_j = \mathbf{O}_p(G/L_j)$, if L_j/L_{j-1} is a p'-group and $L_{j+1}/L_j = \mathbf{O}_{p'}(G/L_j)$, if L_j/L_{j-1} is a p-group. Finally, the series terminates when $\mathbf{O}_p(G/L_n) = 1 = \mathbf{O}_{p'}(G/L_n)$.

A finite group is called *p-soluble* if $L_n = G$, for some natural number n, and the *p-length $l_p(G)$* of a p-soluble group G is the number of factors of the lower p-series that are p-groups. In the following proposition we shall require an important result, the Hall–Higman Theorem (see [110]), which allows us to deduce the p-soluble case of the result we want.

7.2.14. PROPOSITION. *Let p be a prime and let G be a finite p-soluble group. Suppose C is a subgroup of $\zeta(G)$ such that $\boldsymbol{sr}_p(G/C) = r$. Then $\boldsymbol{sr}_p(G') \leq \mathfrak{s}_1(r) \cdot r + r^2(5r^2 - 2r - 3)/2 = \mathfrak{s}_2(r)$.*

PROOF. Let P/C be a Sylow p-subgroup of G/C. By Lemma 6.1.8 it follows that $r = \mathbf{sr}_p(G/C) = \mathbf{r}(P/C)$ so $\mathbf{d}(P/C) \leq r$. Let

$$1 = L_0/C \leq L_1/C \leq L_2/C \leq \cdots \leq L_{n-1}/C \leq L_n/C = G/C$$

be the lower p-series of G/C.

The factor L_1/C is a p'-group so $[L_1, G]$ is a p'-group, using Corollary 7.1.2. Clearly, $L_1/[L_1, G] \leq \zeta(G/[L_1, G])$ and L_2/L_1 is a p-group, so applying Corollary 7.2.8 to $L_2/[L_1, G]$ we see that $\mathbf{sr}_p((L_2/[L_1, G])') \leq \mathfrak{s}_1(r) = k_1$. Now $L_2'/[L_1, G] = (L_2/[L_1, G])'$ and since $[L_1, G]$ is a p'-group we have $\mathbf{sr}_p(L_2'/[L_1, G]) = \mathbf{sr}_p(L_2') \leq k_1$. Let $Q/L_2' = \mathbf{O}_{p'}(L_2/L_2')$, so that $\mathbf{sr}_p(Q) = \mathbf{sr}_p(L_2') \leq k_1$. Also, L_2/Q is an abelian p-group and $(L_2/Q)/(CQ/Q)$ has special rank at most r. It follows, by Lemma 7.2.13 that $G/C_G(L_2/Q)$ has special rank at most $r(5r + 3)/2$. Corollary 7.1.4

implies that $[L_2/Q, G/Q] = [L_2, G]Q/Q$ has special rank at most $r^2(5r + 3)/2$, so

$$\mathbf{sr}_p([L_2, G]Q) \leq k_1 + r^2(5r + 3)/2 = k_2.$$

Clearly, $L_2/[L_2, G]Q \leq \zeta(G/[L_2, G]Q)$ and the number of p-factors in the lower p-series of G/L_2 is $\mathbf{l}_p(G/C) - 1$. Therefore in the remainder of the proof we may use the previous arguments and induction. By the Hall–Higman Theorem (see [**118**, Hauptsatz VI.6.6] or [**110**]) $\mathbf{l}_p(G/C) \leq \mathbf{d}(P/C) \leq r$ and so we obtain

$$\mathbf{sr}_p(G') \leq (r - 1)k_2 + k_1 = (r - 1)(k_1 + r^2(5r + 3)/2) + k_1$$
$$= rk_1 - k_1 + r^2(5r + 3)(r - 1)/2$$
$$= k_1 r + r^2(5r^2 - 2r - 3)/2,$$

as required. $\qquad\square$

7.2.15. LEMMA. *Let G be a finite group and suppose that $G/\zeta(G)$ has special rank r. If $G/\zeta(G) = S_1 \times \cdots \times S_m$ for certain non-abelian simple groups S_j, $1 \leq j \leq m$, then $\mathbf{r}(G') \leq 3m + r$.*

PROOF. Let $C = \zeta(G)$. By Theorem 1.5.13, $G' \cap C$ is an epimorphic image of the Schur Multiplier, $M(G/C)$. We have

$$M(G/C) = M(S_1 \times \cdots \times S_m) = M(S_1) \times \cdots \times M(S_m) \times \mathop{\mathrm{Dr}}_{1 \leq j < k \leq m} (S_j \otimes S_k),$$

by [**262**, p. 148]. Since $S_j \otimes S_k = S_j/S_j' \otimes S_k/S_k'$ and S_j, S_k are non-abelian simple groups, it follows that $S_j \otimes S_k = 1$, for all j, k such that $1 \leq j < k \leq m$ and hence $M(G/C) = M(S_1) \times \cdots \times M(S_m)$. For $1 \leq j \leq m$, $\mathbf{r}(M(S_j)) \leq 3$ (see [**97**, pp.302–303]), so it is evident that $\mathbf{r}(M(G/C)) \leq 3m$ and, since $G'/G' \cap C \cong G'C/C$, we deduce that $\mathbf{r}(G') \leq 3m + r$, as required. $\qquad\square$

If p is a prime, then clearly a product of two normal p-soluble subgroups is p-soluble. Thus every finite group G has a p-soluble radical R, the largest normal p-soluble subgroup of G. Then every normal subgroup of G/R has elements of order p and furthermore G/R contains no nontrivial normal abelian subgroups. The proof of Proposition 4.1.11 can be used to obtain the following result.

7.2.16. PROPOSITION. *Let p be a prime and let G be a finite group of section p-rank r. Then G has normal subgroups $R \leq S \leq V \leq G$ such that R is the p-soluble radical of G, S/R is a direct product of at most r non-abelian finite simple groups, V/S is soluble and $|G/V| \leq r!$.*

We can now generalize Proposition 7.2.6 to all finite groups.

7.2.17. THEOREM. *Let p be a prime and let G be a finite group. Suppose that C is a subgroup of $\zeta(G)$ such that $\mathbf{sr}_p(G/C) = r$. Then there exists a function $\mathfrak{s}_3$ such that $\mathbf{sr}_p(G') \leq \mathfrak{s}_3(r)$.*

PROOF. By Proposition 7.2.16, G has normal subgroups $C \leq R \leq S \leq V \leq G$ such that R is p-soluble, V/S is soluble, S/R is a direct product of at most r non-abelian finite simple groups and $|G/V| \leq r!$.

By Proposition 7.2.14, $\mathbf{sr}_p(R') \leq \mathfrak{s}_2(r)$. Let $Q/R' = \mathbf{O}_{p'}(R/R')$, so that also $\mathbf{sr}_p(Q) = \mathbf{sr}_p(R') \leq \mathfrak{s}_2(r)$ and of course R/CQ is an abelian p-group, of special rank at most r. By Lemma 7.2.13, $(G/Q)/C_{G/Q}(R/Q)$ has special rank at most $r(5r + 3)/2$ and Corollary 7.1.4 allows us to deduce that $[R/Q, G/Q] = [R, G]Q/Q$ has special rank at most $r^2(5r + 3)/2$. Consequently,

$$\mathbf{sr}_p([R, G]Q) \leq \mathfrak{s}_2(r) + r^2(5r + 3)/2.$$

We let $Y = [R, G]Q$ and note that R/Y lies in the centre of G/Y. By Lemma 7.2.15, $X/Y = (S/Y)'$ has special rank at most $4r$. Hence X has section p-rank at most $\mathfrak{s}_2(r) + r^2(5r+3)/2 + 4r$. Now S/X is abelian so that V/X is soluble. Let $U/X = (V/X)'$. Then Proposition 7.2.14 implies that $\mathbf{sr}_p(U/X) \leq \mathfrak{s}_2(r)$, so that

$$\mathbf{sr}_p(U) \leq \mathfrak{s}_2(r) + \mathfrak{s}_2(r) + r^2(5r + 3)/2 + 4r = 2\mathfrak{s}_2(r) + r^2(5r + 3)/2 + 4r.$$

Let $D/U = \mathbf{O}_{p'}(V/U)$ so that $\mathbf{sr}_p(D) = \mathbf{sr}_p(U)$ and V/D is an abelian p-group. By Lemma 7.2.13, $(G/D)/C_{G/D}(V/D)$ has special rank at most $r(5r + 3)/2$. Using Corollary 7.1.4 we deduce that $W/D = [V/D, G/D] = [V, G]D/D$ has special rank at most $r^2(5r + 3)/2$. Then

$$\mathbf{sr}_p(W) \leq 2\mathfrak{s}_2(r) + r^2(5r+3)/2 + 4r + r^2(5r+3)/2 = 2\mathfrak{s}_2(r) + r^2(5r+3) + 4r.$$

The centre of G/W contains V/W and $|G/V| \leq r! = t$, say. It follows from Theorem 1.5.18 that $K/W = (G/W)'$ has order at most t^m, where $m = (\log_2 t - 1)/2$. Exercise 7.7 shows that in this case the special rank of K/W is at most

$$\log_2 t^m = m \log_2 t = \log_2 t(\log_2 t - 1)/2 = \log_2 r!(\log_2 r! - 1)/2.$$

Hence,

$$\mathbf{sr}_p(K) \leq 2\mathfrak{s}_2(r) + r^2(5r + 3) + 4r + \log_2 r!(\log_2 r! - 1)/2 = \mathfrak{s}_3(r)$$

and since G/K is abelian we have $\mathbf{sr}_p(G') \leq \mathfrak{s}_3(r)$, as required.   □

We can deduce a number of different interesting corollaries from this one result and we now list these. They are very easy consequences of Theorem 7.2.17.

7.2.18. COROLLARY. *Let p be a prime and let G be a group. Suppose that C is a subgroup of $\zeta(G)$ such that G/C is finite and let $\mathbf{sr}_p(G/C) = r$. Then G' is finite and $\mathbf{sr}_p(G') \leq \mathfrak{s}_3(r)$.*

EXERCISE 7.12. *Prove Corollary 7.2.18.*

7.2.19. COROLLARY. *Let p be a prime and let G be a group. Suppose that C is a subgroup of $\zeta(G)$ such that G/C is locally finite and let $\mathbf{sr}_p(G/C) = r$. Then G' is locally finite and $\mathbf{sr}_p(G') \leq \mathfrak{s}_3(r)$.*

EXERCISE 7.13. *Prove Corollary 7.2.19.*

7.2.20. COROLLARY. *Let p be a prime and let G be a group. Suppose that C is a subgroup of $\zeta(G)$ such that G/C is locally finite and that the Sylow p-subgroups of G/C are Chernikov. Then G' is locally finite and the Sylow p-subgroups of G' are Chernikov.*

PROOF. By Theorem 3.3.5, G/C has finite section p-rank satifsfying $\mathbf{sr}_p(G/C) = \mathbf{sr}_p(W/C)$, where W/C is a Wehrfritz p-subgroup of G/C. Corollary 7.2.19 shows that $\mathbf{sr}_p(G')$ is finite. Again using Theorem 3.3.5 we deduce that the Sylow p-subgroups of G' are Chernikov. $\square$

7.2.21. COROLLARY. *Let G be a group and let C be a subgroup of $\zeta(G)$ such that G/C is finite. Let $\mathbf{bs}(G/C) = b$. Then G' is finite and $\mathbf{bs}(G') \leq \mathfrak{s}_3(b)$.*

7.2.22. COROLLARY. *Let G be a group and let C be a subgroup of $\zeta(G)$ such that G/C is locally finite. Suppose that $\mathbf{bs}(G/C) = b$ is finite. Then G' is locally finite and $\mathbf{bs}(G') \leq \mathfrak{s}_3(b)$.*

7.2.23. COROLLARY. *Let G be a group and let C be a subgroup of $\zeta(G)$ such that G/C is finite. Suppose that $\mathbf{r}(G/C) = r$. Then G' is finite and $\mathbf{r}(G') \leq \mathfrak{s}_3(r) + 1$.*

PROOF. By Theorem 6.3.17, $\mathbf{bs}(G/C) \leq r$ and Corollary 7.2.21 implies that $\mathbf{bs}(G') \leq \mathfrak{s}_3(r)$. Corollary 6.3.16 now allows us to deduce that $\mathbf{r}(G') \leq \mathfrak{s}_3(r) + 1$. $\square$

7.2.24. COROLLARY. *Let G be a group and let C be a subgroup of $\zeta(G)$ such that G/C is locally finite. Suppose that $\mathbf{r}(G/C) = r$ is finite. Then G' is locally finite and $\mathbf{r}(G') \leq \mathfrak{s}_3(r) + 1$.*

7.2.25. COROLLARY. *Let G be a group and let C be a subgroup of $\zeta(G)$ such that G/C is locally finite. Suppose that G/C has finite section rank σ. Then G' is locally finite and has finite section rank σ_1 where $\sigma_1(p) \leq \mathfrak{s}_3(\sigma(p))$.*

7.2.26. COROLLARY. *Let G be a group and let C be a subgroup of $\zeta(G)$ such that G/C is locally finite with Chernikov Sylow p-subgroups for all primes p. Then G' is locally finite and its Sylow p-subgroups are Chernikov for all primes p.*

We next deduce a result due to A. Schlette [**231**].

7.2.27. COROLLARY. *Let G be a group and let C be a subgroup of $\zeta(G)$ such that G/C is Chernikov. Then G' is also Chernikov.*

PROOF. Of course Chernikov groups are locally finite and $\Pi(G/C)$ is finite. Then Corollary 1.5.17 shows that G' is locally finite and $\Pi(G') \subseteq \Pi(G/C)$. In particular, $\Pi(G')$ is finite and Corollary 7.2.26 shows that the Sylow p-subgroups of G' are Chernikov for all primes p. By Theorem 4.1.6, G' contains a normal divisible abelian subgroup D such that the Sylow p-subgroups of G'/D are finite for all primes p. Since $\Pi(G'/D)$ is finite, Lemma 4.1.8 shows that G'/D is finite. However, D is abelian, $\Pi(D)$ is finite and the Sylow p-subgroups of D are Chernikov for all primes p. These facts imply that D is Chernikov, as required. $\square$

As we have seen in previous chapters, Chernikov groups have several invariants associated with them which we now recall. We denote $|G/\mathbf{Div}(G)|$ by $\mathbf{o}(G)$ where, as before, we let $\mathbf{Div}(G)$, the divisible radical of the Chernikov group G, denote the largest normal divisible subgroup of G. As usual, $\mathbf{Div}(G) = \underset{p \in \mathbf{Sp}(G)}{\mathrm{Dr}} D_p$, where D_p is the unique Sylow p-subgroup of $\mathbf{Div}(G)$ and $\mathbf{Sp}(G)$ is the spectrum of G. The subgroup D_p is uniquely determined by it Zaitsev rank $\mathbf{r}_Z(D_p)$ and we set $\mathbf{r}_{Z,p}(G) = \mathbf{r}_Z(D_p)$, the Zaitsev rank of the Sylow p-subgroup of D. We note also that $\mathbf{sr}_p(\mathbf{Div}(G)) = \mathbf{r}_{Z,p}(G)$ and that $\mathbf{r}_Z(G) = \sum_{p \in \mathbf{Sp}(G)} \mathbf{r}_{Z,p}(G)$.

We now find the connection between the numerical invariants introduced above, pertaining to $G/\zeta(G)$ and G'.

7.2.28. LEMMA. *Let G be a group and let A be a subgroup of $\zeta(G)$. If G/A is a periodic divisible abelian group, then G is abelian.*

EXERCISE 7.14. *Prove Lemma 7.2.28.*

7.2.29. LEMMA. *Let G be a group and let A be an abelian normal subgroup of G. Suppose that G satisfies the following conditions:*
   (i) *$G/C_G(A) = \langle g_1 C_G(A), \ldots, g_m C_G(A) \rangle$, for elements $g_1, \ldots, g_m$ of G;*
   (ii) *$A \cap \zeta(G)$ contains a subgroup C such that A/C is a divisible Chernikov group.*

Then $[A, G]$ is a divisible Chernikov group. Furthermore, $\mathbf{Sp}([A, G]) \subseteq \mathbf{Sp}(A/C)$ and $\mathbf{r}_{Z,p}([A, G]) \leq m \cdot \mathbf{r}_{Z,p}(A/C)$, for all primes $p \in \mathbf{Sp}([A, G])$.

EXERCISE 7.15. Prove Lemma 7.2.29.

These results allow us to obtain the following quantitative version of Corollary 7.2.27.

7.2.30. THEOREM. Let G be a group and suppose that C is a subgroup of $\zeta(G)$ such that G/C is Chernikov. Then G' is also Chernikov. Furthermore, if $o(G/C) = k$, then $o(G') \leq k^m$, where $m = (\log_2 k - 1)/2$, $\mathbf{Sp}(G') \subseteq \mathbf{Sp}(G/C)$ and $\mathbf{r}_{Z,p}(G') \leq k \cdot \mathbf{r}_{Z,p}(G/C)$, for all primes $p \in \mathbf{Sp}(G')$.

PROOF. Let $D/C = \mathbf{Div}(G/C)$. Lemma 7.2.28 implies that D is abelian. Also Lemma 7.2.29 shows that $[D, G]$ is a Chernikov group such that $\mathbf{Sp}([D, G]) \subseteq \mathbf{Sp}(D/C)$ and $\mathbf{r}_{Z,p}([D, G]) \leq k \cdot \mathbf{r}_{Z,p}(D/C)$, for each prime $p \in \mathbf{Sp}([D, G])$. Clearly $D/[D, G] \leq \zeta(G/[D, G])$ and G/D is a finite group of order k. So Theorem 1.5.18 applies and we deduce that $G'/[D, G] = (G/[D, G])'$ is finite of order at most k^m, where $m = (\log_2 k - 1)/2$, which proves the result. $\square$

Theorem 7.2.30 is a minor refinement of [151, Corollary 1].

7.2.31. LEMMA. Let G be a group and let C be a subgroup of $\zeta(G)$ such that G/C is a polyrational group of special rank r. Then G' has special rank at most $r(r-1)/2$.

EXERCISE 7.16. Prove Lemma 7.2.31.

The following theorem is the main result of the paper [18]. We here give a different proof of this result and, of course, the resulting bound is also different. We recall that the function f_5 was introduced in Theorem 2.5.7.

7.2.32. THEOREM. Let G be a locally generalized radical group and let p be a prime. Suppose that C is a subgroup of $\zeta(G)$ such that G/C has finite section p-rank r. Then G' has finite section p-rank. Moreover, there exists a function $\mathfrak{s}_4$ such that $\mathbf{sr}_p(G') \leq \mathfrak{s}_4(r)$.

PROOF. By Theorem 3.4.2, G contains normal subgroups $C \leq T \leq L \leq K \leq G$ such that T/C is locally finite, L/T is torsion-free nilpotent of 0-rank at most r, K/L is a finitely generated torsion-free abelian group of 0-rank at most r and G/K is finite of order at most $f_5(r)$. Note that Theorem 6.2.2 shows that K/T has special rank at most $2r$ and Theorem 7.1.24 shows that G/T has special rank at most $5(2r)^2(2r + 1)/2 = 10r^2(2r + 1)$.

By Corollary 7.2.19, T' is locally finite and has section p-rank at most $\mathfrak{s}_3(r)$. The factor T/T' is abelian and T/CT' is periodic. Let $Q/CT' = \mathbf{O}_{p'}(T/CT')$. By Lemma 7.1.1 we deduce that $[Q/T', G/T'] = [Q, G]/T'$ is a periodic p'-group and hence $\mathbf{sr}_p([Q, G]) = \mathbf{sr}_p(T') \leq \mathfrak{s}_3(r)$.

It is clear that $Q/[Q, G] \leq \zeta(G/[Q, G])$ and that T/Q is an abelian p-group having section p-rank at most r. Theorem 6.1.10 implies that T/Q has special rank at most r also, so we may apply Corollary 7.1.5 to deduce that $R/[Q, G] = [T/[Q, G], K/[Q, G]]$ has special rank at most $2r^2$. By Theorem 6.3.17, $\mathbf{sr}_p(R/[Q, G]) \leq 2r^2$ so that $\mathbf{sr}_p(R) \leq \mathfrak{s}_3(r) + 2r^2$.

For the factor K/R we note that $T/R \leq \zeta(K/R)$ and Lemma 7.2.31 implies that $V/R = (K/R)'$ has special rank at most $2r(2r - 1)/2 = r(2r - 1)$. Theorem 6.3.17 shows that $\mathbf{sr}_p(V/R) \leq r(2r - 1)$ so that $\mathbf{sr}_p(V) \leq \mathfrak{s}_3(r) + 2r^2 + r(2r - 1) = \mathfrak{s}_3(r) + r(4r - 1)$.

The factor K/V is abelian. Let $S/CV = \mathbf{O}_{p'}(K/CV)$. Using Lemma 7.1.1 we see that $U/V = [S/V, G/V]$ is a periodic p'-group and it follows that $\mathbf{sr}_p(U) = \mathbf{sr}_p(V) \leq \mathfrak{s}_3(r) + r(4r - 1)$. Clearly $S/U \leq \zeta(G/U)$ and $\mathbf{Tor}\,(K/S)$ is a p-group. By Theorem 6.1.10, $\mathbf{r}(\mathbf{Tor}\,(K/S)) = \mathbf{sr}_p(\mathbf{Tor}\,(K/S)) \leq r$. The group $(K/S)/\mathbf{Tor}\,(K/S)$ is a p-group and has section p-rank at most r so it has special rank at most r. Hence K/S has special rank at most $2r$. Furthermore, G/K has special rank at most $10r^2(2r + 1)$ and Corollary 7.1.5 shows that $W/U = [K/U, G/U]$ has special rank at most $2r \cdot 10r^2(2r + 1) = 20r^3(2r + 1)$, since K/U is abelian. Then $\mathbf{sr}_p(W/U) \leq 20r^3(2r + 1)$ and $\mathbf{sr}_p(W) \leq 20r^3(2r + 1) + \mathfrak{s}_3(r) + r(4r - 1)$.

The centre of G/W contains K/W and $|G/K| \leq f_5(r) = t$. Theorem 1.5.18 shows that $(G/W)' = G'/W$ is finite of order at most t^m, where $m = (\log_2 t - 1)/2$. Hence G'/W has finite special rank at most $\log_2 t^m = m \log_2 t = (\log_2 t - 1)(\log_2 t)/2 = (\log_2 f_5(r))(\log_2 f_5(r) - 1)/2$. Finally,

$$\mathbf{sr}_p(G') \leq \mathbf{sr}_p(G'/W) + \mathbf{sr}_p(W) \leq \mathbf{r}(G'/W) + \mathbf{sr}_p(W)$$
$$\leq \log_2 f_5(r)(\log_2 f_5(r) - 1)/2 + 20r^3(2r + 1) + r(4r - 1)$$
$$+ \mathfrak{s}_3(r) = \mathfrak{s}_4(r).$$

$\square$

We shall again separate out some interesting important corollaries.

7.2.33. COROLLARY. *Let G be a locally generalized radical group and let C be a subgroup of $\zeta(G)$ such that G/C has finite section rank σ. Then G' has finite section rank σ_1, where $\sigma_1(p) \leq \mathfrak{s}_4(\sigma(p))$.*

For soluble groups, this result was obtained in the paper [**86**], where the bounding function there is also dependent upon $\mathbf{dl}(G/C)$.

7.2.34. COROLLARY. *Let G be a locally generalized radical group and let C be a subgroup of $\zeta(G)$ such that $\boldsymbol{bs}(G/C) = b$ is finite. Then $\boldsymbol{bs}(G')$ is finite and $\boldsymbol{bs}(G') \leq \mathfrak{s}_4(b)$.*

We can also deduce the following result of L. A. Kurdachenko and P. Shumyatsky [**161**].

7.2.35. COROLLARY. *Let G be a locally generalized radical group and let C be a subgroup of $\zeta(G)$ such that G/C has finite special rank r. Then G' has finite special rank at most $3\mathfrak{s}_4(r) + 1$.*

PROOF. By Theorem 6.3.17, $\mathbf{bs}(G/C) \leq r$ and Corollary 7.2.34 shows that $\mathbf{bs}(G') \leq \mathfrak{s}_4(r)$. We apply Theorem 6.3.17 again to deduce that $\mathbf{r}(G') \leq 3\mathfrak{s}_4(r) + 1$, as required. $\square$

Finally we note that the existence of a function giving a bound for the special rank of G' was proved in [**161**], but the function itself was not described. For soluble groups Corollary 7.2.35 was obtained in [**86**], where the bound obtained depended also on the derived length of G/C.

7.3. Hypercentral Extensions by Groups of Finite Section p-Rank

In Section 7.2 we obtained an analogue of Theorem 1.5.12 for the class of groups of finite section p-rank. A natural next step is to obtain analogues of Theorems 1.5.19 and 1.5.22 for this class of groups. There are various results that we obtain here, depending on which of the ranks we are discussing. We begin, once again, with finite groups.

7.3.1. LEMMA. *Let G be a group, let p be a prime and let A be an abelian normal p-subgroup of G. Suppose that G satisfies the following conditions:*

(i) *$G/C_G(A)$ is periodic and $\boldsymbol{sr}_p(G/C_G(A)) = m$;*
(ii) *$A \cap \zeta(G)$ contains a subgroup C such that $\boldsymbol{sr}_p(A/C) = r$.*

Then

(a) *$G/C_G(A)$ is Chernikov and $\boldsymbol{r}(G/C_G(A)) \leq r(5r+1)/2 + m + 1$;*
(b) *$[A, G]$ has finite section p-rank and $\boldsymbol{sr}_p([A, G]) \leq r(5r^2 + r + 2m + 2)/2$.*

EXERCISE 7.17. *Prove Lemma 7.3.1.*

Our first theorem of this section is a result of Kurdachenko, Semko and Pypka [160].

7.3.2. THEOREM. *Let p be a prime, let G be a finite group and let $sr_p(G/\zeta_k(G)) = r$, for some natural number k. Then there is a function $\mathfrak{b}_1(r, k)$ such that $sr_p(\gamma_{k+1}(G)) \leq \mathfrak{b}_1(r, k)$.*

PROOF. Let

$$1 = Z_0 \leq Z_1 \leq \cdots \leq Z_{k-1} \leq Z_k = Z$$

be part of the upper central series of G. We use induction on k.

If $k = 1$, then G/Z_1 has finite section p-rank r and Theorem 7.2.17 shows that $\gamma_2(G) = G'$ has finite section p-rank at most $\mathfrak{s}_3(r)$. Accordingly, we let $\mathfrak{b}_1(r, 1) = \mathfrak{s}_3(r)$.

Assume now that $k > 1$ and that we have already found a function $\mathfrak{b}_1(r, k - 1)$ such that $sr_p(\gamma_k(G/Z_1)) \leq \mathfrak{b}_1(r, k - 1)$. Let $K/Z_1 = \gamma_k(G/Z_1)$ and let $L = \gamma_k(G)$ so that $K = LZ_1$. By the induction hypothesis $sr_p(K/Z_1) \leq \mathfrak{b}_1(r, k - 1)$ and since $L/(L \cap Z_1) \cong LZ_1/Z_1 = K/Z_1$ we have $sr_p(L/(L \cap Z_1)) \leq \mathfrak{b}_1(r, k-1) = t$, also. Theorem 7.2.17 shows that L' has section p-rank at most $\mathfrak{b}_1(\mathfrak{b}_1(r, k - 1), 1)$.

The factor group L/L' is a finite abelian group, so $L/L' = P/L' \times Q/L'$, where P/L' is the Sylow p-subgroup of L/L' and Q/L' is the Sylow p'-subgroup of L/L'. Since $Q \triangleleft G$, $U/L' = [Q/L', G/L']$ is a p'-group. Proposition 1.1.6 shows that $[L, Z] = 1$ and hence $[P, Z] = 1$, so that $G/C_G(P)$ is a homomorphic image of G/Z. Thus $sr_p(G/C_G(P)) \leq r$ and $sr_p(G/C_G(P/L')) \leq r$. By Lemma 7.3.1, $V/L' = [P/L', G/L']$ has special rank at most $t(5t^2 + t + 2r + 2)/2$. However, $[L/L', G/L'] = [P/L', G/L'][Q/L', G/L']$, so that $sr_p([L/L', G/L']) \leq t(5t^2 + t + 2r + 2)/2$. Then

$$\begin{aligned} sr_p([L, G]) &\leq sr_p([L/L', G/L']) + sr_p(L') \\ &\leq \mathfrak{b}_1(r, k - 1)(5\mathfrak{b}_1(r, k - 1)^2 + \mathfrak{b}_1(r, k - 1) + 2r + 2)/2 \\ &\quad + \mathfrak{b}_1(\mathfrak{b}_1(r, k - 1), 1) \\ &= \mathfrak{b}_1(r, k). \end{aligned}$$

This proves the result since $\gamma_{k+1}(G) = [L, G]$.   □

This one result allows us to deduce a host of natural corollaries which we state and prove next.

7.3.3. COROLLARY. *Let p be a prime and let G be a periodic group. Suppose that $G/\zeta_k(G)$ is finite and $sr_p(G/\zeta_k(G)) = r$, for some natural number k. Then $\gamma_{k+1}(G)$ is finite and $sr_p(\gamma_{k+1}(G)) \leq \mathfrak{b}_1(r, k)$.*

PROOF. Since G is a periodic nilpotent-by-finite group it is locally finite and since $G/\zeta_k(G)$ is finite this means that there is a finite subgroup K of G such that $G = K\zeta_k(G)$. Corollary 1.1.8 shows that $\gamma_{k+1}(G) = \gamma_{k+1}(K)$. Furthermore, $G/\zeta_k(G) = K\zeta_k(G)/\zeta_k(G) \cong K/(K \cap \zeta_k(G))$ and $K \cap \zeta_k(G) \leq \zeta_k(K)$ so that $\mathbf{sr}_p(K/\zeta_k(K)) \leq r$. The result now follows by Theorem 7.3.2. $\square$

7.3.4. LEMMA. *Let G be a group such that $G/\zeta_k(G)$ is finite, for some natural number k. If $\mathbf{Tor}(G) = 1$, then $G = \zeta_k(G)$.*

PROOF. By Theorem 1.5.19, $\gamma_{k+1}(G)$ is finite and hence is a subgroup of $\mathbf{Tor}(G)$. Hence $\gamma_{k+1}(G) = 1$, which is equivalent to $G = \zeta_k(G)$. $\square$

7.3.5. COROLLARY. *Let p be a prime and let G be a finitely generated group. Suppose that $G/\zeta_k(G)$ is finite, for some natural number k, and that $\mathbf{sr}_p(G/\zeta_k(G)) \leq r$. Then $\gamma_{k+1}(G)$ is finite and $\mathbf{sr}_p(\gamma_{k+1}(G)) \leq \mathfrak{b}_1(r, k)$.*

PROOF. Theorem 1.5.19 implies that $\gamma_{k+1}(G)$ is finite. The group $G/\gamma_{k+1}(G)$ is a finitely generated nilpotent group, so $T/\gamma_{k+1}(G) = \mathbf{Tor}(G/\gamma_{k+1}(G))$ is finite, as is T. Moreover, G/T is torsion-free. By Corollary 2.4.5 G contains a torsion-free normal subgroup H such that G/H is finite and clearly $H \cap T = 1$. By Remak's Theorem we have the embedding $G \longrightarrow G/H \times G/T = V$. Since G/T is nilpotent of class at most k, $\gamma_{k+1}(G/T) = 1$ so that

$$\gamma_{k+1}(V) = \gamma_{k+1}(G/H) \times \gamma_{k+1}(G/T) = \gamma_{k+1}(G/H).$$

Applying Theorem 7.3.2 to G/H shows that $\mathbf{sr}_p(\gamma_{k+1}(G/H)) \leq \mathfrak{b}_1(r, k)$ and hence $\mathbf{sr}_p(\gamma_{k+1}(G)) \leq \mathbf{sr}_p(\gamma_{k+1}(V)) \leq \mathfrak{b}_1(r, k)$, as required. $\square$

7.3.6. COROLLARY. *Let p be a prime and let G be a group. Suppose that $G/\zeta_k(G)$ is finite and that $\mathbf{sr}_p(G/\zeta_k(G)) = r$, for some natural number k. Then $\gamma_{k+1}(G)$ is finite and $\mathbf{sr}_p(\gamma_{k+1}(G)) \leq \mathfrak{b}_1(r, k)$.*

PROOF. Since $G/\zeta_k(G)$ is finite, there is a finitely generated subgroup K such that $G = K\zeta_k(G)$. Then $G/\zeta_k(G) \cong K/(K \cap \zeta_k(G))$ and $K \cap \zeta_k(G) \leq \zeta_k(K)$. Corollary 1.1.8 shows that $\gamma_{k+1}(G) = \gamma_{k+1}(K)$ and application of Corollary 7.3.5 to K yields the result. $\square$

7.3.7. COROLLARY. *Let p be a prime and let G be a group. Suppose that $G/\zeta_k(G)$ is locally finite, for some natural number k, and that $\mathbf{sr}_p(G/\zeta_k(G)) = r$. Then $\gamma_{k+1}(G)$ is locally finite and $\mathbf{sr}_p(\gamma_{k+1}(G)) \leq \mathfrak{b}_1(r, k)$.*

PROOF. Corollary 7.1.20 shows that $\gamma_{k+1}(G)$ is locally finite. Let A/B be an arbitrary finite elementary abelian p-section of $\gamma_{k+1}(G)$ and choose a finite subgroup H of A such that $A = HB$. Then there is a finitely generated subgroup K of G such that $H \leq \gamma_{k+1}(K)$. Since $K \cap \zeta_k(G) \leq \zeta_k(K)$ it follows that $K/\zeta_k(K)$ is finite and Corollary 7.3.5 shows that $\mathbf{sr}_p(\gamma_{k+1}(K)) \leq \mathfrak{b}_1(r,k)$. It follows that A/B has order at most $p^{\mathfrak{b}_1(r,k)}$ and hence every elementary abelian p-section of $\gamma_{k+1}(G)$ is finite of order at most $p^{\mathfrak{b}_1(r,k)}$. Thus $\mathbf{sr}_p(\gamma_{k+1}(G)) \leq \mathfrak{b}_1(r,k)$, as required. $\square$

7.3.8. COROLLARY. *Let p be a prime and let G be a group. Suppose that $G/\zeta_k(G)$ is locally finite, for some natural number k, and that the Sylow p-subgroups of $G/\zeta_k(G)$ are Chernikov. Then $\gamma_{k+1}(G)$ is locally finite and the Sylow p-subgroups of $\gamma_{k+1}(G)$ are Chernikov.*

PROOF. By Theorem 3.3.5, $G/\zeta_k(G)$ has finite section p-rank and $\mathbf{sr}_p(G/\zeta_k(G)) = \mathbf{sr}_p(W/\zeta_k(G))$, whenever $W/\zeta_k(G)$ is a Wehrfritz p-subgroup of $G/\zeta_k(G)$. Corollary 7.3.7 shows that $\mathbf{sr}_p(\gamma_{k+1}(G))$ is finite and Theorem 3.3.5 can be used again to deduce that the Sylow p-subgroups of $\gamma_{k+1}(G)$ are Chernikov. $\square$

The next two corollaries are straightforward consequences.

7.3.9. COROLLARY. *Let G be a group and let k be a natural number such that $G/\zeta_k(G)$ is finite. Let $\mathbf{bs}(G/\zeta_k(G)) = b$. Then $\gamma_{k+1}(G)$ is finite and $\mathbf{bs}(\gamma_{k+1}(G)) \leq \mathfrak{b}_1(b,k)$.*

7.3.10. COROLLARY. *Let G be a group and let k be a natural number such that $G/\zeta_k(G)$ is locally finite. Suppose that $\mathbf{bs}(G/\zeta_k(G)) = b$ is finite. Then $\gamma_{k+1}(G)$ is locally finite and $\mathbf{bs}(\gamma_{k+1}(G)) \leq \mathfrak{b}_1(b,k)$.*

The finite special rank versions of these results can also be easily determined.

7.3.11. COROLLARY. *Let G be a group and let k be a natural number such that $G/\zeta_k(G)$ is finite. Let $\mathbf{r}(G/\zeta_k(G)) = r$. Then $\gamma_{k+1}(G)$ is finite and $\mathbf{r}(\gamma_{k+1}(G)) \leq \mathfrak{b}_1(r,k) + 1$.*

PROOF. By Theorem 6.3.17, $\mathbf{bs}(G/\zeta_k(G)) \leq r$ and Corollary 7.3.9 implies that $\mathbf{bs}(\gamma_{k+1}(G)) \leq \mathfrak{b}_1(r,k)$. Then Corollary 6.3.16 shows that $\mathbf{r}(\gamma_{k+1}(G)) \leq \mathfrak{b}_1(r,k) + 1$, as required. $\square$

In the case when G is a finite group, the result of Corollary 7.3.11 appears in a paper of N. Yu. Makarenko [**185**].

7.3.12. COROLLARY. *Let G be a group and let k be a natural number such that $G/\zeta_k(G)$ is locally finite. Suppose that $r(G/\zeta_k(G)) = r$ is finite. Then $\gamma_{k+1}(G)$ is locally finite and $r(\gamma_{k+1}(G)) \leq \mathfrak{b}_1(r, k) + 1$.*

7.3.13. COROLLARY. *Let G be a group and let k be a natural number such that $G/\zeta_k(G)$ is locally finite with finite section rank σ. Then $\gamma_{k+1}(G)$ is locally finite and has finite section rank σ_1, where $\sigma_1(p) \leq \mathfrak{b}_1(\sigma(p), k)$.*

7.3.14. COROLLARY. *Let G be a group and let k be a natural number such that $G/\zeta_k(G)$ is locally finite. Suppose that the Sylow p-subgroups of $G/\zeta_k(G)$ are Chernikov, for all primes p. Then $\gamma_{k+1}(G)$ is locally finite and its Sylow p-subgroups are Chernikov, for all primes p.*

The following result has appeared in the paper [213] of J. Otal and J. M. Peña. This result also appears as a special case of [225, Corollary to Theorem 4.23].

7.3.15. COROLLARY. *Let G be a group and let k be a natural number such that $G/\zeta_k(G)$ is Chernikov. Then $\gamma_{k+1}(G)$ is Chernikov.*

PROOF. The group $G/\zeta_k(G)$ is locally finite and $\Pi(G/\zeta_k(G))$ is finite. Corollary 7.1.20 shows that $\gamma_{k+1}(G)$ is also locally finite and $\Pi(\gamma_{k+1}(G)) \subseteq \Pi(G/\zeta_k(G))$, a finite set. Corollary 7.3.14 shows that the Sylow p-subgroups of $\gamma_{k+1}(G)$ are Chernikov for all primes p, so $\gamma_{k+1}(G)$ contains a normal divisible abelian subgroup D such that the Sylow p-subgroups of $\gamma_{k+1}(G)/D$ are finite for all primes p, by Theorem 4.1.6. Since $\Pi(\gamma_{k+1}(G)/D)$ is finite, Lemma 4.1.8 shows that $\gamma_{k+1}(G)/D$ is finite. Finally, $\Pi(D)$ is finite and the Sylow p-subgroups of the divisible abelian group D are Chernikov for all primes p, so D is Chernikov, whence so is $\gamma_{k+1}(G)$. $\qquad\square$

We note that it is possible to obtain the relationships between the invariants for $G/\zeta_k(G)$ and $\gamma_{k+1}(G)$, in the case above, as was done in Section 7.2. Such relationships can be obtained using the proofs above, but a result analogous to Theorem 7.2.30 is omitted. The next theorem, due to L. A. Kurdachenko and J. Otal [150], is left as an exercise.

7.3.16. THEOREM. *Let G be a locally generalized radical group and let k be a natural number such that $G/\zeta_k(G)$ has finite special rank r. Then $\gamma_{k+1}(G)$ has finite special rank and there is a function $\mathfrak{b}_2(r, k)$ such that $r(\gamma_{k+1}(G)) \leq \mathfrak{b}_2(r, k)$.*

EXERCISE 7.18. *Prove Theorem 7.3.16.*

We next prove a lemma occurring in [**66**].

7.3.17. LEMMA. *Let G be a group, let $\mathcal{L}$ be a local system for G and let p be a prime. If there exists a positive integer t such that the section p-rank of every subgroup $H \in \mathcal{L}$ is at most t, then G has section p-rank at most t.*

PROOF. Let U, V be subgroups of G such that $U \triangleleft V$ and V/U is a finite elementary abelian p-group. Then

$$V/U = \langle a_1 U \rangle \times \cdots \times \langle a_n U \rangle,$$

for certain elements $a_1, a_2, \ldots, a_n$ of V such that $a_i^p \in U$ for each i, where $1 \leq i \leq n$. Since $\mathcal{L}$ is a local system, there is subgroup $K_i \in \mathcal{L}$ such that $a_i \in K_i$, for $1 \leq i \leq n$ and, for the same reason, there is a subgroup $Y \in \mathcal{L}$ such that $K_i \leq Y$ for $1 \leq i \leq n$. Then $A = \langle a_1, \ldots, a_n \rangle \leq Y$. Now $V = AU$ so $V/U \cong A/A \cap U$. Since $Y \in \mathcal{L}$ it has section p-rank at most t and since $A \leq Y$ it follows that the elementary abelian p-sections of A have rank at most t. Hence $n \leq t$ and it follows that $\mathbf{sr}_p(G) \leq t$. $\square$

Our next result is the appropriate analogue of Corollary 7.1.4.

7.3.18. LEMMA. *Let G be a group, let p be a prime and let A be an abelian normal p-subgroup of G. Suppose that G satisfies the following conditions:*

(i) *$G/C_G(A) = \langle x_1 C_G(A), \ldots, x_m C_G(A) \rangle$, where $x_1, \ldots, x_m \in G$;*
(ii) *$\zeta(G) \cap A$ contains a subgroup B such that A/B has finite section p-rank t.*

Then $[A, G]$ has finite section p-rank at most tm.

EXERCISE 7.19. *Prove Lemma 7.3.18.*

This yields the following result, which is analogous to Corollary 7.1.5.

7.3.19. COROLLARY. *Let G be a group, let p be a prime and let A be an abelian normal p-subgroup of G. Suppose that G satisfies the following conditions:*

(i) *$G/C_G(A)$ is a group of finite rank t;*
(ii) *$\zeta(G) \cap A$ contains a subgroup B such that A/B has finite section p-rank r.*

Then $[A, G]$ has finite section p-rank at most tr.

PROOF. Let $\mathcal{L}$ be the local family consisting of all the finitely generated subgroups of $G/C_G(A)$. If $F/C_G(A) \in \mathcal{L}$ then

$$F/C_G(A) = \langle x_1 C_G(A), \ldots, x_m C_G(A) \rangle,$$

for certain elements $x_1, \ldots, x_m$ of G such that $m \leq t$. Lemma 7.3.18 implies that $[A, F]$ has finite section p-rank at most $rm \leq rt$. Of course $\{[A, F] | F/C_G(A) \in \mathcal{L}\}$ is a local system for $[A, G]$ and Lemma 7.3.17 implies that $\mathbf{sr}_p([A, G]) \leq tr$. $\square$

We require further information concerning normal abelian subgroups.

7.3.20. LEMMA. *Let G be a group, let p be a prime and let A be a normal torsion-free abelian subgroup of G. Suppose that*

(i) $G/C_G(A)$ *is locally finite;*
(ii) $\zeta(G) \cap A$ *contains a subgroup B such that A/B has finite section p-rank r.*

Then $G/C_G(A)$ is finite and has special rank at most $r(5r + 3)/2$.

PROOF. Let T/B be the torsion subgroup of A/B and let $a \in T$, so there is a natural number k such that $a^k \in B$. If $g \in G$ then $(a^g)^k = (a^k)^g = a^k$. Since A is torsion-free it follows that $a^g = a$ and hence $T \leq \zeta(G)$. Now, A/T is torsion-free and, by Proposition 3.2.3, it has finite 0-rank at most r, by our hypotheses. Let $U = C_G(A/T)$ and note that we may think of G/U as a periodic subgroup of $GL_r(\mathbb{Q})$, so Proposition 1.4.2 shows that G/U is finite. If $a \in A$ and $u \in U$, then $a^u = ac$, for some element $c \in T$. Since $G/C_G(A)$ is periodic there is a natural number m such that $u^m \in C_G(A)$ so $a = u^{-m} a u^m = ac^m$. It follows that $c^m = 1$ and since A is torsion-free we deduce that $c = 1$. Hence $U = C_G(A/T) = C_G(A)$ so $G/C_G(A)$ is finite. Using Corollary 7.1.17 we see that $G/C_G(A)$ has special rank at most $r(5r + 3)/2$. $\square$

7.3.21. LEMMA. *Let G be a group, let p be a prime and let A be an abelian normal subgroup of G. Suppose that*

(i) $G/C_G(A)$ *is a locally finite group of finite section p-rank t;*
(ii) $\zeta(G) \cap A$ *contains a subgroup B such that A/B has finite section p-rank r;*
(iii) $\mathbf{Tor}\,(A)$ *is a p-group.*

Then

(a) $G/C_G(A)$ *is Chernikov and has finite special rank at most $r(5r + 2) + 2t + 1$;*
(b) $[A, G]$ *has finite special rank at most $r^2(5r + 2) + r(2t + 1)$.*

PROOF. (a) Let $T = \mathbf{Tor}\,(A)$. By Lemma 7.3.1, $G/C_G(T)$ is Chernikov and has finite special rank at most $r(5r + 1)/2 + t + 1$. Lemma 7.3.20 shows that $G/C_G(A/T)$ is finite and has special rank at most $r(5r + 3)/2$. Remak's Theorem gives the embedding

$$G/(C_G(T) \cap C_G(A/T)) \longrightarrow G/C_G(T) \times G/C_G(A/T)),$$

which shows that $G/(C_G(T) \cap C_G(A/T))$ is Chernikov and has finite special rank at most $r(5r+1)/2+t+1+r(5r+3)/2 = r(5r+2)+t+1$.

Let $a \in A$ and $x \in C_G(T) \cap C_G(A/T)$. Then $a^x = ac$, for some element $c \in T$. Also, there is a natural number m such that $x^m \in C_G(A)$, so $a = x^{-m}ax^m = ac^m$ and it follows that $c^m = 1$. Since $\Pi(A) = \{p\}$ we deduce that $m = p^s$ for some non-negative integer s and hence $(C_G(T) \cap C_G(A/T))/C_G(A)$ is a p-group. Condition (i) and Theorem 6.1.10 together imply that $(C_G(T) \cap C_G(A/T))/C_G(A)$ is Chernikov and has finite special rank at most t. By our work above $G/C_G(A)$ is Chernikov and has finite special rank at most $r(5r + 2) + t + 1 + t = r(5r + 2) + 2t + 1$. . (b) Let Q/B be the Sylow p'-subgroup of A/B and let $a \in Q$. Then there is a p'-number k such that $a^k \in B$. If $g \in G$ then $(a^g)^k = (a^k)^g = a^k$ and since A is abelian we have $(a^{-1}a^g)^k = 1$. However, the hypothesis on A implies that $a^g = a$ so that $Q \leq \zeta(G)$. Clearly, A/Q contains no p'-elements and since it has p-rank at most r, Corollary 6.2.10 shows that A/Q has finite special rank at most r. Corollary 7.1.5 now shows that $[A, G]$ has finite special rank at most $r(r(5r + 2) + 2t + 1) = r^2(5r + 2) + r(2t + 1)$. □

7.3.22. LEMMA. *Let G be a locally generalized radical group, let p be a prime and let A be an abelian normal subgroup of G. Suppose that*

(i) *$G/C_G(A)$ has finite section p-rank t;*
(ii) *$\zeta(G) \cap A$ contains a subgroup B such that A/B has finite section p-rank r;*
(iii) *$\mathbf{Tor}\,(A)$ is a p-group.*

Then $[A, G]$ has finite special rank at most $5r^3 + 2r^2 + r(20t^3 + 10t^2 + 2t + 1)$.

PROOF. Let $T = \mathbf{Tor}\,(A)$ and let $R/C_G(A) = \mathbf{Tor}\,(G/C_G(A))$. By Lemma 7.3.21, $R/C_G(A)$ is Chernikov and has finite special rank at most $r(5r + 2) + 2t + 1$. By Theorem 3.4.2, $G/C_G(A)$ has 0-rank at most $2t$ and Theorem 7.1.24 shows that G/R has finite special rank at most $5(2t)^2(2t + 1)/2 = 10t^2(2t + 1)$. It follows that $G/C_G(A)$ has finite special rank at most $r(5r + 2) + 2t + 1 + 10t^2(2t + 1)$.

Let Q/B be the Sylow p'-subgroup of A/B and let $a \in Q$. As in the proof of Lemma 7.3.21 $Q \leq \zeta(G)$ and A/Q has finite special rank r.

Corollary 7.1.5 then shows that $[A, G]$ has finite special rank at most

$$r(r(5r + 2) + 2t + 1 + 10t^2(2t + 1))$$
$$= 5r^3 + 2r^2 + r(20t^3 + 10t^2 + 2t + 1).$$

$\square$

7.3.23. COROLLARY. *Let G be a locally generalized radical group, let p be a prime and let A be an abelian normal subgroup of G. Suppose that G satisfies the following conditions:*

(i) *$G/C_G(A)$ has finite section p-rank t;*
(ii) *$\zeta(G) \cap A$ contains a subgroup B such that A/B has finite section p-rank r.*

Then $[A, G]$ has finite section p-rank at most $5r^3 + 2r^2 + r(20t^3 + 10t^2 + 2t + 1)$.

PROOF. Let Q be the Sylow p'-subgroup of A. By Lemma 7.3.22, $[A/Q, G/Q] = V/Q$ has finite special rank at most $5r^3 + 2r^2 + r(20t^3 + 10t^2 + 2t + 1)$ and hence V has finite section p-rank at most $5r^3 + 2r^2 + r(20t^3 + 10t^2 + 2t + 1)$. However, $[A, G] \leq V$ so $[A, G]$ also has finite section p-rank at most $5r^3 + 2r^2 + r(20t^3 + 10t^2 + 2t + 1)$. $\square$

7.3.24. THEOREM. *Let G be a locally generalized radical group and let p be a prime. Suppose that $G/\zeta_k(G)$ has finite section p-rank at most t. Then $\gamma_{k+1}(G)$ has finite section p-rank and there exists a function $\mathfrak{b}_3(t, k)$ such that $\mathbf{sr}_p(\gamma_{k+1}(G)) \leq \mathfrak{b}_3(t, k)$.*

PROOF. Let

$$1 = Z_0 \leq Z_1 \leq \cdots \leq Z_{k-1} \leq Z_k = Z$$

be part of the upper central series of G. We use induction on k. If $k = 1$ then the central factor group G/Z_1 has finite section p-rank r and Theorem 7.2.32 implies that $\gamma_2(G) = G'$ has finite section p-rank and that there is a function $\mathfrak{s}_4$ such that $\mathbf{sr}_p(\gamma_2(G)) \leq \mathfrak{s}_4(t)$. We let $\mathfrak{b}_3(t, 1) = \mathfrak{s}_4(t)$.

Assume now that $k \geq 1$ and apply our induction hypothesis to G/Z_1 so that there is a function $\mathfrak{b}_3(t, k-1)$ such that $\mathbf{sr}_p(\gamma_k(G/Z_1)) \leq \mathfrak{b}_3(t, k-1)$. Set $K/Z_1 = \gamma_k(G/Z_1)$, let $L = \gamma_k(G)$ and note that $K = LZ_1$. Then $K' = [LZ_1, LZ_1] = L'$, since $Z_1 = Z(G)$. Also $[K, G] = [L, G] = \gamma_{k+1}(G)$. Since $L/(L \cap Z_1) \cong LZ_1/Z_1 = K/Z_1$ we deduce that $\mathbf{sr}_p(L/(L \cap Z_1)) \leq \mathfrak{b}_3(t, k-1)$. Theorem 7.2.32 shows that L' has finite section p-rank and

$$\mathbf{sr}_p(L') \leq \mathfrak{s}_4(\mathfrak{b}_3(t, k-1)).$$

Proposition 1.1.6 shows that $[L, Z] = 1$ so $G/C_G(L)$ is an image of G/Z and hence $\mathbf{sr}_p(G/C_G(L/L')) \leq t$. Corollary 7.3.23 shows that $V/L' = [L/L', G/L']$ has special rank at most $5r^3 + 2r^2 + r(20t^3 + 10t^2 + 2t + 1)$. Then

$$\mathbf{sr}_p(\gamma_{k+1}(G)) = \mathbf{sr}_p([L, G]) \leq \mathbf{sr}_p([L/L', G/L']) + \mathbf{sr}_p(L')$$
$$\leq 5r^3 + 2r^2 + r(20t^3 + 10t^2 + 2t + 1) + \mathfrak{s}_3(\mathfrak{b}_3(t, k-1))$$
$$= 5\mathfrak{b}_3(t, k-1)^3 + 2\mathfrak{b}_3(t, k-1)^2$$
$$+ \mathfrak{b}_3(t, k-1)(20t^3 + 10t^2 + 2t + 1) + \mathfrak{s}_3(\mathfrak{b}_3(t, k-1))$$
$$= \mathfrak{b}_3(t, k).$$

$\square$

That $\gamma_{k+1}(G)$ has finite section p-rank has been proved in [66], but there a different bound was obtained. There are two very nice subsidiary results.

7.3.25. COROLLARY. *Let G be a locally generalized radical group. Suppose that $G/\zeta_k(G)$ has finite section rank σ, for some natural number k. Then $\gamma_{k+1}(G)$ has finite section rank σ_1, where $\sigma_1(p) \leq \mathfrak{b}_3(\sigma(p), k)$, for every prime p.*

7.3.26. COROLLARY. *Let G be a locally generalized radical group. Suppose that $\mathbf{bs}(G/\zeta_k(G)) = b$ is finite, for some natural number k. Then $\gamma_{k+1}(G)$ has finite bounded section rank and $\mathbf{bs}(\gamma_{k+1}(G)) \leq \mathfrak{b}_3(b, k)$.*

We now turn our attention to groups in which $G/\zeta_\infty(G)$ has finite section p-rank, or finite rank. The following example, from [66], shows that further hypotheses are needed in order to obtain analogues of Theorems 1.5.22 and 1.6.8.

7.3.27. EXAMPLE. Let $G = \langle a \rangle \rtimes \langle d \rangle$ be an infinite dihedral group, where a is an element of infinite order, d has order 2 and $a^d = a^{-1}$. Let $A_0 = \langle a \rangle$ and, for each $n \geq 0$, let $A_{n+1} = A_n^2$. Each of the subgroups A_n is clearly normal in G and G/A_n is a finite dihedral group of order 2^{n+1}. In the group $D_n = G/A_n$ let $a_n = aA_n$ and $v_n = dA_n$. Let b be an element of order 2 and let $K_n = \langle b \rangle \wr D_n$, the wreath product of $\langle b \rangle$ by D_n. Let B_n denote the base group of this wreath product, so that $K_n = B_n \rtimes D_n$. Let $B = \mathrm{Dr}_{n \in \mathbb{N}} B_n$ denote the direct product of the groups B_n, let $D = \mathrm{Cr}_{n \in \mathbb{N}} D_n$ denote the Cartesian product of the groups D_n and define an action of D on B as follows:

If $b = (b_n) \in \mathrm{Dr}_{n \in \mathbb{N}} B_n$ and $y = (d_n) \in \mathrm{Cr}_{n \in \mathbb{N}} D_n$ let $b^y = (b_n^{d_n})$.

Let $H = B \rtimes D$ be the corresponding semidirect product defined by this action.

The mapping $f : G \longrightarrow D$ defined by $f(x) = (xA_n)_{n \in \mathbb{N}}$ is clearly a homomorphism and since $\cap_{n \in \mathbb{N}} A_n = 1$, Remak's Theorem implies that this homomorphism is an embedding of G into D. Thus we may form the subgroup of H defined by $L = B \rtimes f(G)$.

A short calculation shows that $B = \zeta_\infty(L)$ and that the length of the upper central series of L is the first infinite ordinal, ω. Hence $L/\zeta_\infty(L) \cong f(G) \cong G$ is infinite dihedral, so is of finite special rank.

We suppose, for a contradiction, that L contains a normal subgroup P such that P has finite special rank and L/P is locally nilpotent. Since B is an elementary abelian 2-group and L/B is polycyclic, it follows that P is also polycyclic. If P is periodic, then P is finite, so $f(G) \cap P$ is a finite normal subgroup of the infinite dihedral group $f(G)$. This implies that $f(G) \cap P = 1$ and hence $f(G)P/P \cong f(G)$ is locally nilpotent, which is a contradiction. Hence P is non-periodic and since L has torsion-free rank 1, it follows that P has torsion-free rank 1. Since P is polycyclic it contains an infinite cyclic normal subgroup of finite index and hence a characteristic such subgroup $\langle u \rangle$. Since $L = B \cdot f(G)$ and $f(G)$ is dihedral, we have $u = xf(a^s)$, for some element $x \in B$ and some integer s. Since $x \in \zeta_k(L)$, for some natural number k, it is easy to see that $u^l = f(a^{sl})$, for some positive power l. However, u^l has infinitely many conjugates in L whereas, by construction, $\langle u^l \rangle$ is a normal subgroup of L. This gives us the required contradiction and it follows that L has no normal subgroup of finite special rank with locally nilpotent factor group.

This example shows that even if $G/\zeta_\infty(G)$ is polycyclic and even has 0-rank 1, then there may not exist a normal subgroup N, even of finite special rank, such that G/N is locally nilpotent. In the example $G/\zeta_\infty(G)$ is non-periodic, so now we discuss the periodic case, beginning with the following result.

7.3.28. LEMMA. *Let p be a prime, let G be a finite group, let Z be a G-invariant subgroup of the hypercentre of G and let $G^{\mathfrak{N}}$ be the nilpotent residual of G. Suppose that $\boldsymbol{sr}_p(G/Z) = r$. Then $\Pi(G^{\mathfrak{N}}) \subseteq \Pi(G/Z)$ and $\boldsymbol{sr}_p(G^{\mathfrak{N}}) \leq r + \boldsymbol{s}_3(r)$.*

PROOF. Note that Theorem 7.1.21 shows that $\Pi(G^{\mathfrak{N}}) \subseteq \Pi(G/Z) = \pi$. There is a series of G-invariant subgroups

$$1 = Z_0 \leq Z_1 \leq \cdots \leq Z_n \leq Z_{n+1} = Z$$

whose factors are G-central. Let $C = C_G(Z)$ and note that, by Theorem 1.2.22, G/C is nilpotent. Clearly $Z \leq C_G(C)$, so that $G/C_G(C)$ has section p-rank at most r. We have $C \cap Z \leq \zeta(C)$ and $C/(Z \cap C) \cong CZ/Z$ so $\mathbf{sr}_p(C/(C \cap Z)) \leq r$ and $\Pi(C/(C \cap Z)) \subseteq \pi$. Applying Theorem 7.2.17 to C shows that $\mathbf{sr}_p(C') \leq \mathfrak{s}_3(r)$. Of course, C' is G-invariant and C/C' is abelian, so $C/C' \leq C_{G/C'}(C/C')$.

Since G/C is nilpotent so is $(G/C')/C_{G/C'}(C/C')$ and Corollary 1.6.5 shows that C/C' has a Z-decomposition, say

$$C/C' = \zeta_{\mathbb{Z}G}^{\infty}(C/C') \oplus E_{\mathbb{Z}G}^{\infty}(C/C').$$

Clearly $(C \cap Z)C'/C' \leq \zeta_{\mathbb{Z}G}^{\infty}(C/C')$ from which it follows that $R/C' = E_{\mathbb{Z}G}^{\infty}(C/C')$ has section p-rank at most r. Furthermore, the factor $C/R \cong (C/C')/(R/C')$ is G-hypercentral and hence $C/R \leq \zeta_{\infty}(G/R)$. Since G/C is nilpotent, G/R is therefore nilpotent and hence $G^{\mathfrak{N}} \leq R$. Finally

$$\mathbf{sr}_p(R) \leq \mathbf{sr}_p(C') + \mathbf{sr}_p(R/C') \leq \mathfrak{s}_3(r) + r,$$

whence $\mathbf{sr}_p(G^{\mathfrak{N}}) \leq r + \mathfrak{s}_3(r)$. $\qquad\square$

The lemma above appears in [160] and [66], where different bounds for the section p-rank are given.

7.3.29. COROLLARY. *Let p be a prime, let G be a finitely generated group and let Z be a G-invariant subgroup of the hypercentre of G. Suppose that G/Z is finite and $\mathbf{sr}_p(G/Z) = r$. Then $G^{\mathfrak{N}}$ is finite and $\Pi(G^{\mathfrak{N}}) \subseteq \Pi(G/Z)$. Also $\mathbf{sr}_p(G^{\mathfrak{N}}) \leq r + \mathfrak{s}_3(r)$ and $G/G^{\mathfrak{N}}$ is nilpotent.*

PROOF. By Propositions 1.2.13 and 1.2.5, Z is a finitely generated nilpotent group and hence satisfies the maximal condition on subgroups. Consequently, there is a natural number k such that $Z \leq \zeta_k(G)$, so $G/\zeta_k(G)$ is finite. Theorem 7.1.21 shows that $\Pi(G^{\mathfrak{N}}) \subseteq \Pi(G/Z)$. Theorem 1.5.19 implies that $\gamma_{k+1}(G)$ is finite, so $G/G^{\mathfrak{N}}$ is nilpotent. Since $G/\gamma_{k+1}(G)$ is a finitely generated nilpotent group of class at most k, its torsion subgroup $T/\gamma_{k+1}(G)$ is finite, so G/T is a finitely generated torsion-free nilpotent group of class at most k.

Corollary 2.4.5 implies that G contains a torsion-free normal subgroup H such that G/H is finite and clearly $T \cap H = 1$. Remak's Theorem implies the embedding

$$G \longrightarrow G/H \times G/T = V$$

and of course $\gamma_{k+1}(G/T) = 1$. Let U/H be the nilpotent residual of G/H, so that U/H coincides with the nilpotent residual of V. Clearly the upper hypercentre of G/H contains ZH/H and

$$\mathbf{sr}_p((G/H)/(ZH/H)) \leq \mathbf{sr}_p(G/Z).$$

Also, G/U is nilpotent so $G^{\mathfrak{N}} \leq G \cap U$ and Lemma 7.3.28, applied to G/H, shows that $\mathbf{sr}_p(G^{\mathfrak{N}}) \leq \mathbf{sr}_p(U) \leq r + \mathfrak{s}_3(r)$. $\qquad\qquad\square$

7.3.30. THEOREM. *Let G be a group and let p be a prime. Suppose that the upper hypercentre of G contains a G-invariant subgroup Z such that G/Z is locally finite and has finite section p-rank r. Then the locally nilpotent residual $G^{\mathrm{L}\mathfrak{N}}$ of G is locally finite, $\Pi(G^{\mathrm{L}\mathfrak{N}}) \subseteq \Pi(G/Z)$ and $\mathbf{sr}_p(G^{\mathrm{L}\mathfrak{N}}) \leq r + \mathfrak{s}_3(r)$. Moreover $G/G^{\mathrm{L}\mathfrak{N}}$ is locally nilpotent.*

PROOF. The facts that $G^{\mathrm{L}\mathfrak{N}}$ is locally finite and $G/G^{\mathrm{L}\mathfrak{N}}$ is locally nilpotent follow from Theorem 7.1.21. This same theorem also implies that $\Pi(G^{\mathrm{L}\mathfrak{N}}) \subseteq \Pi(G/\zeta_\infty(G)) \subseteq \Pi(G/Z) = \pi$, say.

Let $\mathcal{L}$ be the family of all finitely generated subgroups of G. If $U \in \mathcal{L}$ then the hypercentre of U contains the U-invariant subgroup $U \cap Z = Z_U$ and U/Z_U is a finitely generated, locally finite group, so is finite. Also, since $U/Z_U \cong UZ/Z \leq G/Z$, we have $\mathbf{sr}_p(U/Z_U) \leq r$ and $\Pi(U/Z_U) \subseteq \pi$. Corollary 7.3.29 shows that $U^{\mathfrak{N}}$ is finite and $\mathbf{sr}_p(U^{\mathfrak{N}}) \leq r + \mathfrak{s}_3(r)$. Furthermore $U/U^{\mathfrak{N}}$ is nilpotent.

Now let V be another subgroup in $\mathcal{L}$. Then there exists $W \in \mathcal{L}$ such that $\langle U, V \rangle \leq W$. Of course $W^{\mathfrak{N}} \cap U \vartriangleleft U$ and $U/(W^{\mathfrak{N}} \cap U)$ is nilpotent, so $U^{\mathfrak{N}} \leq W^{\mathfrak{N}}$. Similarly $V^{\mathfrak{N}} \leq W^{\mathfrak{N}}$ and it follows, as in the proof of Theorem 7.1.21, that $R = \bigcup_{U \in \mathcal{L}} U^{\mathfrak{N}}$ is a normal locally finite subgroup of G such that G/R is locally nilpotent. Then $G^{\mathrm{L}\mathfrak{N}} \leq R$ and Lemma 7.3.17 shows that $\mathbf{sr}_p(G^{\mathrm{L}\mathfrak{N}}) \leq \mathbf{sr}_p(R) \leq r + \mathfrak{s}_3(r)$ as required. $\qquad\qquad\square$

This result and some of its corollaries below have been proved in the paper [**66**], where different bounds were obtained for the section p-ranks of the locally nilpotent residual. We list many of the corollaries that can be obtained quite specifically, since they are important special cases.

7.3.31. COROLLARY. *Let G be a group and let p be a prime. Suppose that the upper hypercentre of G contains a G-invariant subgroup Z such that G/Z is locally finite and has Chernikov Sylow p-subgroups. Then the locally nilpotent residual $G^{\mathrm{L}\mathfrak{N}}$ of G is locally finite and its Sylow p-subgroups are Chernikov.*

PROOF. Theorem 3.3.5 implies that G/Z has finite section p-rank and $\mathbf{sr}_p(G/Z) = \mathbf{sr}_p(W/Z)$, where W/Z is a Wehrfritz p-subgroup of G/Z. Theorem 7.3.30 implies that $\mathbf{sr}_p(G^{\mathrm{L}\mathfrak{N}})$ is finite and, using Theorem 3.3.5 again, we deduce that the Sylow p-subgroups of $G^{\mathrm{L}\mathfrak{N}}$ are Chernikov. $\qquad\qquad\square$

7.3.32. COROLLARY. *Let G be a group. Suppose that the upper hypercentre of G contains a G-invariant subgroup Z such that G/Z is locally finite and has finite section rank σ. Then the locally nilpotent residual $G^{\mathrm{L}\mathfrak{N}}$ of G is locally finite and there is a function σ_1 such that $G^{\mathrm{L}\mathfrak{N}}$ has finite section rank σ_1. Furthermore, $\Pi(G^{\mathrm{L}\mathfrak{N}}) \subseteq \Pi(G/Z)$, $\sigma_1(p) \le \sigma(p) + \mathfrak{s}_3(\sigma(p))$, for all primes p, and $G/G^{\mathrm{L}\mathfrak{N}}$ is hypercentral.*

PROOF. A locally nilpotent group of finite section rank is hypercentral by Theorem 4.2.12, so $G/G^{\mathrm{L}\mathfrak{N}}$ is an extension of a subgroup of the upper hypercentre by a hypercentral group and therefore is hypercentral. □

7.3.33. COROLLARY. *Let G be a group. Suppose that the upper hypercentre of G contains a G-invariant subgroup Z such that G/Z is locally finite and has Chernikov Sylow p-subgroups for all primes p. Then the locally nilpotent residual $G^{\mathrm{L}\mathfrak{N}}$ of G is locally finite and its Sylow p-subgroups are Chernikov for all primes p. Furthermore, $\Pi(G^{\mathrm{L}\mathfrak{N}}) \subseteq \Pi(G/Z)$ and $G/G^{\mathrm{L}\mathfrak{N}}$ is hypercentral.*

7.3.34. COROLLARY. *Let G be a group. Suppose that the upper hypercentre of G contains a G-invariant subgroup Z such that G/Z is locally finite and has finite bounded section rank b. Then the locally nilpotent residual $G^{\mathrm{L}\mathfrak{N}}$ of G is locally finite and has finite bounded section rank at most $b + \mathfrak{s}_3(b)$. Furthermore, $\Pi(G^{\mathrm{L}\mathfrak{N}}) \subseteq \Pi(G/Z)$ and $G/G^{\mathrm{L}\mathfrak{N}}$ is hypercentral.*

7.3.35. COROLLARY. *Let G be a group. Suppose that the upper hypercentre of G contains a G-invariant subgroup Z such that G/Z is locally finite and has finite special rank r. Then the locally nilpotent residual $G^{\mathrm{L}\mathfrak{N}}$ of G is locally finite and has finite special rank at most $r + \mathfrak{s}_3(r)+1$. Furthermore, $\Pi(G^{\mathrm{L}\mathfrak{N}}) \subseteq \Pi(G/Z)$ and $G/G^{\mathrm{L}\mathfrak{N}}$ is hypercentral.*

PROOF. By Theorem 6.3.17, $\mathbf{bs}(G/Z) \le r$ and Corollary 7.3.34 shows that $\mathbf{bs}(G^{\mathrm{L}\mathfrak{N}}) \le r + \mathfrak{s}_3(r)$. Using Corollary 6.3.16 we deduce that $\mathbf{r}(G^{\mathrm{L}\mathfrak{N}}) \le \mathfrak{s}_3(r) + r + 1$, as required. □

7.3.36. COROLLARY. *Let G be a group. Suppose that the upper hypercentre of G contains a G-invariant subgroup Z such that G/Z is Chernikov. Then the locally nilpotent residual $G^{\mathrm{L}\mathfrak{N}}$ of G is Chernikov and $G/G^{\mathrm{L}\mathfrak{N}}$ is hypercentral.*

PROOF. By Corollary 7.3.33, $G^{\mathrm{L}\mathfrak{N}}$ is locally finite and its Sylow p-subgroups are Chernikov, for all primes p. Furthermore, $\Pi(G^{\mathrm{L}\mathfrak{N}})$ is finite, so that $G^{\mathrm{L}\mathfrak{N}}$ is Chernikov. □

CHAPTER 8

Finitely Generated Groups of Finite Section Rank

One of the first natural objectives in the study of a class of groups is to clarify the structure of finitely generated groups in this class. In Chapter 5 the structure of finitely generated locally generalized radical groups G of finite 0-rank such that $\mathbf{Tor}\,(G)$ is trivial was described (see Theorem 5.2.8). Such groups are minimax and, as we shall see, this situation is also typical for the other ranks. The structure of soluble-by-finite groups of finite special rank was obtained in the paper [**224**] of D. J. S. Robinson and he also described the structure of soluble-by-finite groups whose elementary abelian sections are finite (see [**226**]). This enabled him to determine the structure of soluble-by-finite groups having finite section rank, a result whose initial proof was homological. A purely group theoretic proof of this latter result is given in [**156**]. As with Robinson's proof, the proof given in [**156**] is based on the search for supplements to certain normal subgroups. The search for such supplements is carried out by Robinson using homological techniques; in [**156**] this is done by purely group theoretical methods, and clarifies certain important features of the structure of the normal subgroups.

In this chapter, the basis for our presentation is the approach used in [**156**]. The main results of this paper will be presented, including extensions based on work in the earlier chapters. We also present some results on the existence of supplements, used in other areas of group theory. Their proofs are given for a comprehensive treatment of this topic .

8.1. The $Z(G)$-Decomposition in Some Abelian Normal Subgroups

In this section we return to the $Z(G)$-decomposition, introduced in Chapter 1. We begin with the following useful property of nilpotent minimax groups.

8.1.1. LEMMA. *Let G be a torsion-free nilpotent minimax group. Then G has a finite subnormal series*

$$H_0 \lhd H_1 \lhd \ldots \lhd H_{k-1} \lhd H_k = G,$$

where H_0 is finitely generated and H_j/H_{j-1} is divisible Chernikov, for $1 \leq j \leq k$.

PROOF. Let

$$1 = Z_0 \leq Z_1 \leq \cdots \leq Z_{n-1} \leq Z_n = G$$

be the upper central series of G. By Corollary 1.2.9, the factors of this series are torsion-free. We use induction on n and first let $n = 1$, so that G is an abelian minimax group. In this case G contains a finitely generated subgroup H such that G/H is a divisible Chernikov group and the result holds.

Next we let $n > 1$ and suppose inductively that G/Z_1 has a subnormal series

$$H_1/Z_1 \vartriangleleft H_2/Z_1 \vartriangleleft \ldots \vartriangleleft H_{k-1}/Z_1 \vartriangleleft H_k/Z_1 = G/Z_1,$$

where H_1/Z_1 is finitely generated and H_j/H_{j-1} is divisible Chernikov, for $1 \leq j \leq k$. Since Z_1 is a torsion-free abelian minimax group, it contains a finitely generated subgroup F such that Z_1/F is a divisible Chernikov group. Then H_1/F has the central divisible Chernikov subgroup Z_1/F and H_1/Z_1 is finitely generated. Hence H_1/F contains a finitely generated subgroup H/F such that $H_1/F = (Z_1/F)(H/F)$. Clearly $H/F \vartriangleleft H_1/F$ and we now have

$$H_1/H \cong (H_1/F)/(H/F) = (Z_1/F)(H/F)/(H/F)$$
$$\cong (Z_1/F)/(H/F \cap Z_1/F).$$

Since Z_1/F is a divisible Chernikov group, each homomorphic image is likewise, and hence H_1/H is divisible Chernikov. However H/F and F are finitely generated, so H is also finitely generated. Now let $H_0 = H$ and the factors H_j/H_{j-1} are divisible Chernikov groups, for $1 \leq j \leq k$. □

8.1.2. LEMMA. *Let G be a hypercentral group and suppose that there is a periodic abelian normal subgroup T such that G/T is finitely generated. Then the upper G-central series of T has length at most ω.*

EXERCISE 8.1. *Prove Lemma 8.1.2.*

8.1.3. LEMMA. *Let G be a hypercentral group and let T be a periodic abelian normal subgroup of G. Suppose that G/T contains a finitely generated normal subgroup L/T such that G/L is a periodic divisible group. Then every L-invariant subgroup of T is also G-invariant.*

EXERCISE 8.2. *Prove Lemma 8.1.3.*

The following corollary is easy to deduce.

8.1.4. COROLLARY. *Let G be a hypercentral group and let T be a periodic abelian normal subgroup of G. Suppose that G has a finite subnormal series*

$$T = L_0 \vartriangleleft L = L_1 \vartriangleleft L_2 \vartriangleleft \ldots \vartriangleleft L_n = G$$

such that L_1/T is finitely generated and L_{j+1}/L_j is periodic and divisible, for $1 \leq j \leq n-1$. Then every L-invariant subgroup of T is also G-invariant.

We shall require more further technical results.

8.1.5. LEMMA. *Let G be a group and let A, B normal subgroups of G satisfying the following conditions:*
   (i) $A \leq B \cap \zeta(G)$ *and* $B/A \leq \zeta(G/A)$;
   (ii) G/A *is minimax and* $\Pi(A)$ *is infinite.*
Then $\Pi(B/[B,G])$ is infinite.

PROOF. Since B/A is minimax there is a finite subset $\{b_1, \ldots, b_n\}$ of B such that $G/\langle b_1, \ldots, b_n, A \rangle$ is periodic and minimax, so is Chernikov. Lemma 3.2.5 shows that for each $b \in B$ the mapping ξ_b defined by $\xi_b(x) = [b,x]$, for $x \in G$, is an endomorphism of G and $[b,G] = \mathrm{Im}(\xi_b) \cong G/\mathrm{ker}(\xi_b) = G/C_G(b)$. Of course $A \leq C_G(b)$, so that $G/C_G(b)$ is minimax, as is $[b,G]$. Hence $[b_1, G] \ldots [b_n, G] = C$ is a minimax subgroup of $[B,G]$. Then $A \cap C$ is abelian minimax and hence $\Pi(A \cap C)$ is finite, so that $\Pi(AC/C)$ is infinite since $AC/C \cong A/(A \cap C)$. Let $D = \langle C, b_1, b_2, \ldots, b_n \rangle$, so that $D/C \leq \zeta(G/C)$ and $(B/C)/(D/C)$ is Chernikov. Hence $\pi = \Pi(B/D)$ is finite. If $g \in B$, then $(gC)^k \leq D/C \leq \zeta(G/C)$ for some π-number k and, by Lemma 3.2.5, we deduce that $[gC, G/C]^k = 1$. Hence $[gC, G/C]$ is a periodic π-group, a fact which is valid for each $g \in B$ and hence $[B/C, G/C]$ is also a periodic π-group. Since $C \leq [B,G]$ we have $[B/C, G/C] = [B,G]/C$ and since $\Pi(AC/C)$ is infinite, it also follows that $\Pi(A[B,G]/[B,G])$ is infinite. Finally, this means that $\Pi(B/[B,G])$ is infinite. $\square$

8.1.6. COROLLARY. *Let G be a nilpotent group and let T be a periodic subgroup of $\zeta(G)$ such that G/T is minimax. If $\Pi(T)$ is infinite, then $\Pi(G/G')$ is also infinite.*

PROOF. Let

$$1 = Z_0 \leq Z_1 \leq \cdots \leq Z_n = G$$

be the upper central series of G. We use induction on n, the case $n = 2$ being Lemma 8.1.5.

Suppose that $n > 2$ and that our assertion has been proved for groups of nilpotency class less than n. By Lemma 8.1.5, $\Pi(Z_2/[Z_2, G])$

is infinite and since $Z_2/[Z_2, G] \leq \zeta(G/[Z_2, G])$ we have $\mathbf{ncl}\,(G/[Z_2, G]) < \mathbf{ncl}\,(G)$. The induction hypothesis can be applied to give the result.   □

8.1.7. COROLLARY. *Let G be a nilpotent group and let T be a periodic subgroup of $\zeta(G)$ such that G/T is minimax. If $\Pi(T)$ is infinite, then G contains a subgroup K such that G/K is a periodic abelian group and $\Pi(G/K)$ is infinite.*

PROOF. Let $P/G' = \mathbf{Tor}\,(G/G')$ and note that $\Pi(P/G')$ is infinite, by Corollary 8.1.6. Since G/P is minimax there is a finite maximal $\mathbb{Z}$-independent subset, $\{g_1 G', \ldots, g_n G'\}$, of G/G' and we let $K/G' = \langle g_1 G', \ldots, g_n G' \rangle$, a torsion-free group. Also $(G/G')/(K/G')$ is periodic, so $P/G' \cap K/G' = 1$. Hence

$$P/G' \cong (P/G')(K/G')/(K/G'),$$

which proves that $\Pi(G/K) = \Pi((G/G')/(K/G'))$ is infinite.   □

We next need the following generalization of Lemma 7.2.29.

8.1.8. LEMMA. *Let G be a group and let A be an abelian normal subgroup of G. Suppose that G satisfies the following conditions:*
  (i) *$G/C_G(A)$ has finite special rank t;*
  (ii) *$A \cap \zeta(G)$ contains a subgroup C such that A/C is a divisible Chernikov group.*
Then $[A, G]$ is a divisible Chernikov group. Furthermore, $\Pi([A, G]) \subseteq \Pi(A/C)$ and $\mathbf{sr}_p([A, G]) \leq m \cdot \mathbf{sr}_p(A/C)$, for all primes $p \in \Pi([A, G])$.

EXERCISE 8.3. *Prove Lemma 8.1.8.*

8.1.9. LEMMA. *Let G be a group and let A, B be normal subgroups of G satisfying the following conditions:*
  (i) *$A \leq B \cap \zeta(G)$ and $B/A \leq \zeta(G/A)$;*
  (ii) *G/A is minimax and A is a divisible p-subgroup having infinite section p-rank, for some prime p.*
Then $[B, G]$ is minimax.

PROOF. Since B/A is minimax there is a finite subset $\{b_1, \ldots, b_n\}$ of B such that $G/\langle b_1, \ldots, b_n, A \rangle$ is periodic and minimax, so Chernikov. We may assume that this factor group is divisible. Lemma 3.2.5 shows that for each $b \in B$ the mapping ξ_b defined by $\xi_b(x) = [b, x]$, for $x \in G$, is an endomorphism of G and $[b, G] = \mathbf{Im}(\xi_b) \cong G/\mathbf{ker}(\xi_b) = G/C_G(b)$. Of course $A \leq C_G(b)$, so that $G/C_G(b)$ is minimax, as is $[b, G]$. Hence $[b_1, G] \ldots [b_n, G] = C$ is a minimax subgroup of $[B, G]$. Then $A \cap C$ is abelian minimax and hence $\mathbf{sr}_p(A \cap C)$ is finite. Therefore $\mathbf{sr}_p(AC/C)$

is infinite since $AC/C \cong A/(A \cap C)$. By Lemma 7.2.28, B/C is abelian. Since $AC/C \le \zeta(G/C)$ and $(B/C)/(AC/C)$ is divisible, Lemma 8.1.8 implies that $[B/C, G/C] = [B, G]/C$ is divisible Chernikov. Then, since C is minimax, $[B, G]$ is also minimax. $\square$

8.1.10. LEMMA. *Let G be a nilpotent group and let T be a divisible p-subgroup of $\zeta(G)$, for some prime p. Suppose that G/T is minimax and that T has infinite section p-rank. Then there is a subgroup K such that $G' \le K$ and G/K is a divisible p-group having infinite section p-rank.*

PROOF. Since G is nilpotent it has a central series

$$1 = Z_0 \le Z_1 = T \le \cdots \le Z_n = G$$

in which Z_j/Z_{j-1} is minimax, for $2 \le j \le n$. By Lemma 8.1.9, $C_2 = [Z_2, G]$ is minimax. Then $C_2 \cap T$ is Chernikov and hence $TC_2/C_2 \cong T/(C_2 \cap T)$ has infinite section p-rank. Clearly $Z_2/C_2 \le \zeta(G/C_2)$. Repeating this argument enough times we see that, after finitely many steps, TG'/G' has infinite section p-rank. Since TG'/G' is divisible, $G/G' = TG'/G' \times K/G'$, for some subgroup K (see [**88**, Theorem 21.2], for example). The result follows. $\square$

If G is a finitely generated group, then G is a homomorphic image of a free group F, having a finite set of free generators, and hence F contains a normal subgroup K such that $G \cong F/K$. If there is a finite subset M of K such that $K = \langle M \rangle^F$, then G is *finitely presented*. Each word $w \in M$ determines an identity $w = 1$, which all elements of the group G satisfy, and G is said to be generated by a finite set and satisfies a finite set of defining relations. Finitely presented groups form a much narrower class of groups than the class of finitely generated groups. For example, the class of finitely generated groups forms a set of uncountable cardinality, whereas the set of finitely presented groups is easily shown by simple set theoretic considerations to be only countable. Finitely presented groups arise in the consideration of various issues, not only in algebra, but also in other branches of mathematics, perhaps most significantly in many geometric and topological problems. Clearly finite groups and finitely generated abelian groups are examples of finitely presented groups and we shall see other examples later.

We shall also need the following important idea connected with finitely presented groups. If G is a group and H is a normal subgroup of G then, following the work of R. Baer [**11**], we say that G/H is *finitely presented relative to G* if G contains a subgroup U with the property that $G = HU$ and $H \cap U$ contains a finite subset M such that $H \cap U = \langle M \rangle^U$. We state next the following result of Baer [11].

8.1.11. LEMMA. *Let G be a finitely generated group and let H be a normal subgroup of G. If G/H is finitely presented relative to G, then H has a finite subset M such that $H = \langle M \rangle^G$.*

EXERCISE 8.4. *Prove Lemma 8.1.11.*

We obtain several more results due to R. Baer [11].

8.1.12. PROPOSITION. *Let G be a finitely generated group. Then the following are equivalent:*

(i) *G is finitely presented;*
(ii) *If V is a group and U is a normal subgroup of G such that $V/U \cong G$, then V/U is finitely presented relative to V;*
(iii) *If V is a finitely generated group and U is a normal subgroup of V such that $V/U \cong G$, then U has a finite subset M such that $U = \langle M \rangle^V$.*

PROOF. To show that (i) implies (ii), let G be finitely presented. Then there is a free group F and a finite subset S of F such that $G \cong F/\langle S \rangle^F$ and we let $R = \langle S \rangle^F$. Let U, V be as stated in (ii). Then $V/U \cong F/R$ and, since F is free, there is a homomorphism $\phi : F \longrightarrow V$ inducing the isomorphism between V/U and F/R. It follows that $V = U\phi(F)$ and $\phi(R) = U \cap \phi(F)$. We now have $\phi(R) = \langle \phi(S) \rangle^{\phi(F)}$, so that V/U is finitely presented relative to V. Hence (ii) follows.

Lemma 8.1.11 shows that (ii) implies (iii).

Finally suppose that V is a finitely generated group having a normal subgroup U such that $V/U \cong G$ and $U = \langle M \rangle^V$, for some finite subset M. Since G is an epimorphic image of a finitely generated group, it is also finitely generated and hence there is a free group F of finite free rank such that $G \cong F/R$ for some normal subgroup R of F. By (iii) R has a finite subset S such that $R = \langle S \rangle^F$. This shows that G is finitely presented and (i) follows from (iii). □

8.1.13. LEMMA (Baer [11]). *Let G be a finitely generated group, let H be a normal subgroup of G and let U be a finitely generated subgroup of G such that $HU \lhd G$. Then H contains a finite subset M such that $\langle M \rangle^G U$ is a normal subgroup of G.*

EXERCISE 8.5. *Prove Lemma 8.1.13.*

8.1.14. PROPOSITION (Baer [11]). *Let G be a finitely generated group and let H be a normal subgroup of G such that G/H is finitely* *presented relative to G. If K is a G-invariant subgroup of H such that H/K is finitely generated and finitely presented relative to H, then K contains a finite subset M such that $K = \langle M \rangle^G$.*

PROOF. By Lemma 8.1.11, H contains a finite subset S such that $H = \langle S \rangle^G$. Since H/K is finitely presented relative to H, there is a subgroup U of H such that $H = UK$ and $K \cap U$ contains a finite subset S_1 such that $K \cap U = \langle S_1 \rangle^U$. Now H/K is finitely generated and $H/K = UK/K \cong U/(K \cap U)$ which implies that there is a finitely generated subgroup V such that $U = (K \cap U)V$. Then $H = KU = K(K \cap U)V = KV$. Since G, V are both finitely generated and $H, K \vartriangleleft G$, Lemma 8.1.13 implies that H has a finite subset S_2 such that $\langle S_2 \rangle^G V$ is G-invariant. Let $D = \langle S_1 \rangle^G$ and $L = \langle S_2 \rangle^G$ and note that $K \cap U \leq D$. The subgroup DL is G-invariant and clearly $DL = \langle S_1 \cup S_2 \rangle^G$. Since LV is G-invariant, DLV is also G-invariant. Since $DL \leq K$ we have $H = KV = K(DLV)$ and hence $K \cap DLV = DL(K \cap V) = DL$. This shows that H/DL is the direct product of K/DL and DLV/DL. Since $H = \langle S \rangle^G$, H/DL is the normal closure in G/DL of some finite subset and there is a finite subset S_3 such that the direct factor K/DL is the normal closure in G/DL of the natural image of S_3. Also, $DL = \langle S_1 \cup S_2 \rangle^G$ implies that $K = DL = \langle S_1 \cup S_2 \cup S_3 \rangle^G$, so we set $M = S_1 \cup S_2 \cup S_3$. □

The next proposition, showing that the class of finitely presented groups is closed under taking extensions, appeared in the paper of P. Hall [105], where only an outline of the proof was given. Therefore we give the proof due to Baer [11].

8.1.15. PROPOSITION. *Let G be a group and let H be a normal subgroup of G. If H and G/H are finitely presented, then G is also finitely presented.*

PROOF. Clearly G is finitely generated, so there is a free group F of finite free rank and a normal subgroup R of F such that $G \cong F/R$. Let E be a preimage of H in F. Since H is finitely generated, we may suppose that E is also finitely generated and hence $H \cong E/R$. By Proposition 8.1.12, F/E is finitely presented relative to F. Since E/R is finitely presented, Proposition 8.1.12 shows that E/R is finitely presented relative to E. By Proposition 8.1.14, R contains a finite subset M such that $R = \langle M \rangle^F$. Hence G is finitely presented. □

8.1.16. COROLLARY. *Every finitely generated abelian-by-finite group or finite-by-abelian group is finitely presented.*

8.1.17. COROLLARY. *Polycyclic-by-finite groups are finitely presented. In particular, finitely generated nilpotent groups are finitely presented.*

We shall now consider a further generalization of hypercentral groups. If G is a group, then let

$$\mathbf{FC}(G) = \{x \in G | x^G \text{ is finite}\}.$$

It is easy to see that $\mathbf{FC}(G)$ is a characteristic subgroup of G, which is called the *FC-centre* of G. Starting with the FC-centre we can construct the upper FC-central series of a group G,

$$1 = F_0 \leq F_1 \leq \ldots F_\alpha \leq F_{\alpha+1} \leq \ldots F_\gamma,$$

defined by setting $F_1 = \mathbf{FC}(G), F_{\alpha+1}/F_\alpha = \mathbf{FC}(G/F_\alpha)$, for all $\alpha < \gamma$, $F_\lambda = \bigcup_{\beta<\lambda} F_\beta$, for all limit ordinals λ and $\mathbf{FC}(G/F_\gamma) = 1$. The last term, F_γ, of this series is called the upper FC-hypercentre (or sometimes just the FC-hypercentre) of G. If $F_\gamma = G$ then G is called FC-hypercentral, unless γ is finite in which case G is called FC-nilpotent.

8.1.18. LEMMA. *Let G be a finitely generated FC-nilpotent group. Then G is nilpotent-by-finite.*

EXERCISE 8.6. *Prove Lemma 8.1.18.*

8.1.19. PROPOSITION. *Let G be a finitely generated FC-hypercentral group. Then G is FC-nilpotent and also nilpotent-by-finite.*

EXERCISE 8.7. *Prove Proposition 8.1.19.*

If G is a group and H is a normal subgroup of G, then we say that H is *G-hyperfinite* if H has a series of G-invariant subgroups,

$$1 = F_0 \leq F_1 \leq \ldots F_\alpha \leq F_{\alpha+1} \ldots F_\gamma = H,$$

whose factors are all finite.

8.1.20. PROPOSITION. *Let G be a group and let T be an abelian normal subgroup of G such that G/T is finitely generated nilpotent. If T is G-hyperfinite then T has a $Z(G)$-decomposition.*

PROOF. Since T is G-hyperfinite and G/T is nilpotent it follows that G is FC-hypercentral. Let A be a finite subgroup of T. Since G/T is finitely generated, there is a finitely generated subgroup L such that $G = TL$. Then $\langle A, L \rangle$ is also finitely generated, so Proposition 8.1.19 implies that it is nilpotent-by-finite. It follows that every periodic subgroup of $\langle A, L \rangle$ is finite. Hence $A^L = A^G$ is finite and it follows that T contains a local system $\mathcal{L}$ of finite G-invariant subgroups. If B

is a finite G-invariant subgroup of T, then Corollary 1.6.5 shows that B has a $Z(G)$-decomposition and $B = C(B) \times E(B)$, where $C(B)$ is the upper G-hypercentre of B and $E(B)$ is a maximal G-invariant, G-hypereccentric subgroup of B. Let D be a finite G-invariant subgroup of T containing B. We note that $E(D) \cap B$ is G-invariant and G-hypereccentric. As we observed in Section 1.6 we have $E(D) \cap B \leq E(B)$. On the other hand, $E(B)$ is a G-invariant, G-hypereccentric subgroup of D and the work of Section 1.6 shows that $E(B) \leq E(D)$ which means that $E(B) = E(D) \cap B$. Clearly $C(B) \leq C(D)$ and we have

$$B = C(B) \times E(B) = (C(D) \cap B) \times (E(D) \cap B).$$

Consequently $\{C(Y)|Y \in \mathcal{L}\}$ and $\{E(Y)|Y \in \mathcal{L}\}$ are local families. Furthermore, $T = C(T) \times E(T)$, where $C(T) = \cup_{Y \in \mathcal{L}} C(Y)$ and $E(T) = \cup_{Y \in \mathcal{L}} E(Y)$. Clearly the direct decomposition for $T = C(T) \times E(T)$ is a $Z(G)$-decomposition of T. $\qquad \square$

8.1.21. COROLLARY. *Let G be a group and let T be an abelian normal subgroup of G such that G/T is torsion-free, nilpotent minimax group. If T is G-hyperfinite then T has a $Z(G)$-decomposition.*

PROOF. By Lemma 8.1.1, G has a finite subnormal series

$$T = L_0 \lhd L = L_1 \lhd \ldots \lhd L_{n-1} \lhd L_n = G$$

such that L_1/T is finitely generated and, for $1 \leq j \leq n-1$, L_{j+1}/L_j is periodic divisible. By Proposition 8.1.20, T has a $Z(L)$-decomposition $T = C \times E$, where C is the upper L-hypercentre of T and E is a maximal L-invariant, L-hypereccentric subgroup of T. As we saw in Section 1.6, C and E are G-invariant.

Let U, V be L-invariant subgroups of C such that $U \leq V$ and V/U is L-central. By Corollary 8.1.4, U and V are G-invariant and Proposition 3.2.6 implies that V/U is G-central.

Let A, B be L-invariant subgroups of E such that $B \leq A$ and A/B is G-chief. Since T is G-hyperfinite, A/B is finite, so that A has a finite series of L-invariant subgroups

$$B = B_0 \leq B_1 \leq \cdots \leq B_m = A$$

whose factors are L-chief. Since $A \leq E$ we have $L \neq C_L(B_{j+1}/B_j)$, for $0 \leq j \leq m-1$, and it follows that $C_G(A/B) \neq G$. Hence A/B is G-eccentric. It follows that C is the upper G-hypercentre of T and E is a maximal G-hypereccentric subgroup of T. Hence T has a $Z(G)$-decomposition. $\qquad \square$

8.1.22. COROLLARY. *Let G be a group and let T be an abelian normal subgroup of G such that G/T is a torsion-free, nilpotent minimax group. If T is G-hyperfinite and C is the upper G-hypercentre of T, then T/C is G-hypereccentric.*

8.2. Splittings over Some Normal Subgroups

In this section we shall obtain and use several results concerned with n-fold commutators and we now recall the notation used. For the group G and elements x, g let

$$[x,_1 g] = [x, g] \text{ and } [x,_{n+1} g] = [[x,_n g], g] \text{ for all } n \in \mathbb{N}.$$

We require the following technical results for such commutators, the first of which appears in [**227**].

8.2.1. LEMMA. *Let G be a group and let $g \in G$ be such that $[G, g]$ is abelian. Given $x, y \in G$ and a natural number n, there exists $u_n = u_n(y, g) \in \langle y \rangle^{\langle g \rangle}$ such that*

$$[xy,_n g] = [x,_n g]^{u_n}[y,_n g].$$

PROOF. We have $[xy, g] = [x, g]^y[y, g]$, so for $n = 1$ we have $u_1 = y$ and the result holds. We use induction on n and, to this end, suppose that we have already proved

$$[xy,_{n-1} g] = [x,_{n-1} g]^{u_{n-1}}[y,_{n-1} g],$$

for some element $u_{n-1} \in \langle y \rangle^{\langle g \rangle}$. As above we have

$$[xy,_n g] = [[xy,_{n-1} g], g] = [[x,_{n-1} g]^{u_{n-1}}[y,_{n-1} g], g]$$

$$= [y,_{n-1} g]^{-1}[[x,_{n-1} g]^{u_{n-1}}, g][y,_{n-1} g] \cdot [[y,_{n-1} g], g].$$

We have $g = (u_{n-1}gu_{n-1}^{-1})^{u_{n-1}}$ and $u_{n-1}gu_{n-1}^{-1} = ag$, where $a \in [G, g^{-1}]$. It follows that

$$[[x,_{n-1} g]^{u_{n-1}}, g] = [[x,_{n-1} g]^{u_{n-1}}, (u_{n-1}gu_{n-1}^{-1})^{u_{n-1}}]$$

$$= [[x,_{n-1} g], u_{n-1}gu_{n-1}^{-1}]^{u_{n-1}} = [[x,_{n-1} g], ag]^{u_{n-1}}.$$

Now $[[x,_{n-1} g], ag] = [[x,_{n-1} g], g][[x,_{n-1} g], a]^g$. Since $1 = [x, g^{-1}g] = [x, g][x, g^{-1}]^g$ we have $[G, g^{-1}] = [G, g]$, so $a \in [G, g]$ and because $[G, g]$ is abelian, $[[x,_{n-1} g], a] = 1$. Hence $[[x,_{n-1} g, ag] = [[x,_{n-1}, g], g] = [x,_n g]$. Finally,

$$[xy,_n g] = ([y,_{n-1} g]^{-1}[[x,_{n-1} g]^{u_{n-1}}, g][y,_{n-1} g])[[y,_{n-1} g], g]$$

$$= ([y,_{n-1} g]^{-1}[x,_n g]^{u_{n-1}}[y,_{n-1} g])[y,_n g] = [x,_n g]^{u_n}[y,_n g],$$

where $u_n = u_{n-1}[y,_{n-1} g] \in \langle y \rangle^{\langle g \rangle}$.   □

8.2.2. LEMMA. *Let G be a group and let A be an abelian normal subgroup of G. If $C_G(A) \neq gC_G(A) \in \zeta(G/C_G(A))$, then $[xa,_n g] = [x,_n g][a,_n ,g]$, for all $x \in G, a \in A$ and $n \in \mathbb{N}$.*

PROOF. We use induction on n. When $n = 1$, we have $[xa, g] = [x, g]^a[a, g] = [x, g][a, g]$, since $[x, g] \in C_G(A)$, so we suppose that $n > 1$ and that we have already proved

$$[xa,_{n-1} g] = [x,_{n-1} g][a,_{n-1} ,g].$$

Applying a basic commutator identity and the induction hypothesis, we obtain

$$[xa,_n g] = [[xa,_{n-1} g], g] = [[x,_{n-1} g][a,_{n-1} g], g]$$

$$= ([a,_{n-1} g]^{-1}[[x,_{n-1} g], g][a,_{n-1} g])[a,_{n-1} g], g]$$

$$= [a,_{n-1} g]^{-1}[x,_n g][a,_{n-1} g][a,_n g].$$

Since $A \lhd G$ we have $[a,_{n-1} g] \in A$ and the choice of g implies that $[x, g] \in C_G(A)$. Hence $[x,_n g] \in C_G(A)$ also and hence

$$[a,_{n-1} g]^{-1}[x,_n g][a,_{n-1} g] = [x,_n g].$$

Therefore $[xa,_n g] = [x,_n g][a,_n g]$, as required. $\square$

The proof of the following result from [**227**] is now quite easy.

8.2.3. LEMMA. *Let G be a group and let g be an element of G such that $[G, g]$ is abelian. Given $x \in G$ and $n \in \mathbb{N}$, there exists $v_n = v_n(x, g) \in \langle x \rangle^{\langle g \rangle}$ such that*

$$[x^{-1},_n g] = ([x,_n g]^{-1})^{v_n}.$$

PROOF. We use the notation of Lemma 8.2.1 and let $y = x^{-1}$ in the statement of that lemma to deduce $1 = [x,_n g]^{u_n}[x^{-1},_n g]$. Then

$$[x^{-1},_n g] = ([x,_n g]^{u_n})^{-1} = ([x,_n g]^{-1})^{u_n},$$

where $u_n = u_n(x^{-1}, g) \in \langle x^{-1} \rangle^{\langle g \rangle} = \langle x \rangle^{\langle g \rangle}$. So now let $v_n = u_n(x^{-1}, g)$ to deduce the result. $\square$

8.2.4. LEMMA. *Let G be a group and let A be an abelian normal subgroup of G. If $g \in G, a \in A$ and $n \in \mathbb{N}$, then*

$$[a^{-1},_n g] = [a,_n g]^{-1}.$$

EXERCISE 8.8. *Prove Lemma 8.2.4.*

For the group G and natural number n and for each element $g \in G$ we define the *left n-Engelizer* $\mathbf{E}_{G,n}(g)$ by

$$\mathbf{E}_{G,n}(g) = \{x \in G | [x,_n g] = 1\}.$$

We note that $\mathbf{E}_{G,1}(g) = C_G(g)$, but in general $\mathbf{E}_{G,n}(g)$ need not be a group.

8.2.5. LEMMA. *Let G be a group and let $g \in G$ be such that $[G, g]$ is nilpotent of class k. Then, for each $n \in \mathbb{N}$, there exists a natural number $m = m(n, k)$, depending only on k, n, such that $\langle \mathbf{E}_{G,n}(g) \rangle \subseteq \mathbf{E}_{G,m}(g)$.*

PROOF. We shall apply the arguments of the proof of [**227**, Corollary 3*] which are relevant to our situation.

Let $x \in \langle \mathbf{E}_{G,n}(g) \rangle$ so that $x = y_1 \ldots y_t$, where $[y_j,_n g] = 1$, for $1 \le j \le t$. We set $L = \langle y_1, \ldots, y_t, g \rangle$ and $D = [L, g]$. By hypothesis, D is nilpotent of class at most k. Now D/D' is abelian, so Lemmas 8.2.2 and 8.2.4 show that $[v,_n g] \in D'$, for all $v \in L$. Therefore $\langle g \rangle^L / D'$ is nilpotent of class at most n. Since D is nilpotent of class at most k, [**106**, Theorem 7] shows that $\langle g \rangle^L$ is nilpotent of class at most $n \binom{k+1}{2} - \binom{k}{2}$. Since $[x, g] \in \langle g \rangle^L$ we have $[x,_m g] = 1$, where

$$m = m(n, k) = n \binom{k+1}{2} - \binom{k}{2} + 1,$$ and hence $x \in \mathbf{E}_{G,m}(g)$.   □

We are now in a position to prove the following result from [**154**].

8.2.6. PROPOSITION. *Let G be a group and let A be an abelian normal subgroup of G. Suppose there exists $g \in G$ such that $C_G(A) \neq gC_G(A) \in \zeta(G/C_G(A))$ and that G also satisfies the conditions:*
  (i) *G/A is nilpotent;*
  (ii) *$A = [A, g] \neq 1$.*
Then there exists a subgroup L of G such that $G = LA$ and $g \in L$. Furthermore, $L \cap A \subseteq \mathbf{E}_{G,m}(g)$, for some natural number m, so that $L \cap A$ is a proper G-invariant subgroup of A.

PROOF. Clearly G is not nilpotent since $A = [A, g]$. Let $x \in G$ be arbitrary so that, by hypothesis, $[x, g] \in C_G(A)$. By (i) there is a natural number n such that $a = [x,_n g] \in A$. Since $A = [A, g]$, there exists $a_1 \in A$ such that $a = [a_1, g]$. For the same reason there exists an element $a_2 \in A$ such that $a_1 = [a_2, g]$ so that $a = [a_2, g, g]$ and in this way we inductively construct elements $a_k \in A$, for $k \ge 1$, such that $a = [a_k,_k g]$. In particular

$$[x,_n g] = a = [a_n,_n g].$$

Applying Lemmas 8.2.2 and 8.2.4 we see that

$$[xa_n^{-1},_n g] = [x,_n g][a_n^{-1},_n g] = [x,_n g][a_n,_n g]^{-1} = 1.$$

It follows that $xa_n^{-1} \in \mathbf{E}_{G,n}(g)$. We let $L = \langle \mathbf{E}_{G,n}(g) \rangle$ so that $xa_n^{-1} \in L$ and hence $x \in LA$. Since x was an arbitrary element of G we deduce that $G = LA$ and, of course, $g \in L$.

By hypothesis, $[G,g] \le C_G(A)$ and, since $C_G(A)$ is nilpotent, $[G,g]$ is also nilpotent. By Lemma 8.2.5, there exists $m \in \mathbb{N}$ such that $L = \langle \mathbf{E}_{G,n}(g) \rangle \subseteq \mathbf{E}_{G,m}(g)$ and, in particular, $A \cap L \subseteq \mathbf{E}_{G,m}(g)$. If $A \le L$, then $[A,_m g] = 1$, so $A = 1$, since $A = [A,g]$, which is a contradiction. Thus $A \cap L \ne A$ and it now follows that $A \cap L$ is a proper G-invariant subgroup of A. $\square$

We require some more terminology concerning complements. For the group G with normal subgroup A, if all complements to A are conjugate, then we say that G *splits conjugately over* A. We are interested now in obtaining criteria for such a splitting to exist.

8.2.7. THEOREM. *Let G be a group and let A be an abelian normal subgroup of G. Let $g \in G$ be an element such that $C_G(A) \ne gC_G(A) \in \zeta(G/C_G(A))$ and suppose that G also satisfies the conditions:*

(i) *G/A is nilpotent;*
(ii) *$A = [A,g] \ne 1$;*
(iii) *$C_A(g) = 1$.*

Then G splits conjugately over A.

PROOF. Proposition 8.2.6 implies that there is a subgroup L and a least natural number s such that $G = AL$ and $L \subseteq \mathbf{E}_{G,s}(g)$. Also Lemma 8.2.5 shows that there is a natural number r such that $R = \langle \mathbf{E}_{G,s}(g) \rangle \subseteq \mathbf{E}_{G,r}(g)$. Let $D = A \cap R$. Since $D \subseteq \mathbf{E}_{G,r}(g)$ we have $[D,_r g] = 1$. If t is the least natural number such that $C = [D,_t g] \ne 1$, then $[C,g] = [D,_{t+1} g] = 1$ and hence $C \le C_A(g) = 1$, by (iii). In particular, it follows that $[D,g] = 1$ and hence $D = A \cap R \le C_A(g) = 1$. Since $L \le R$ the Dedekind Law implies that $R = R \cap AL = L$. It follows that A is complemented by L in G and also that $L = \mathbf{E}_{G,s}(g)$.

Let Y be another complement to A in G. Then $g \in AY$, so that $g = yb$ for some $b \in A$ and $y \in Y$. Hence $y = ga$, where $a = b^{-1}$. Since L, Y are isomorphic there is a natural isomorphism $\phi : L \longrightarrow Y$ satisfying $\phi(g) = y$. Then, as $L = \mathbf{E}_{G,s}(g)$, we have $[Y,_s y] = 1$, so that $Y \subseteq \mathbf{E}_{G,s}(y)$. By hypothesis, $A = [A,g] = [g,A]$, so $a = [g,c]$, for some $c \in A$. Then $y = ga = gg^{-1}c^{-1}gc = c^{-1}gc$ and hence $\mathbf{E}_{G,s}(y) = \mathbf{E}_{G,s}(g^c) = \mathbf{E}_{G,s}(g)^c$. Therefore $\mathbf{E}_{G,s}(y)$ is a subgroup of G and since $A \lhd G$, we have

$$\mathbf{E}_{G,s}(y) \cap A = \mathbf{E}_{G,s}(g^c) \cap A = \mathbf{E}_{G,s}(g)^c \cap A = 1.$$

Now $\mathbf{E}_{G,s}(y) = Y(\mathbf{E}_{G,s}(y) \cap A) = Y$, so $Y = \mathbf{E}_{G,s}(y) = \mathbf{E}_{G,s}(g)^c = L^c$, so that Y, L are conjugate, as required. $\qquad\square$

8.2.8. COROLLARY. *Let G be a group and let A be an abelian normal subgroup of G. Suppose that $g \in G$ and that G also satisfies the conditions:*

(i) *G/A is nilpotent;*

(ii) *A has a finite series of G-invariant subgroups*

$$1 = A_0 \leq A_1 \leq \cdots \leq A_n = A$$

such that $gC_G(A_{j+1}/A_j) \in \zeta(G/C_G(A_{j+1}/A_j))$, for $0 \leq j \leq n-1$;

(iii) *$A_{j+1}/A_j = [A_{j+1}/A_j, gA_j]$, for $0 \leq j \leq n-1$;*

(iv) *$C_{A_{j+1}/A_j}(g) = 1$, for $0 \leq j \leq n-1$.*

Then G splits conjugately over A.

PROOF. We use induction on n and note that the case $n = 1$, follows from Theorem 8.2.7. Let $B = A_1$ and suppose that we have already proved that G/B splits conjugately over A/B. Let L/B be a complement to A/B, so that $G = AL$. Then $g = xa$, for some $x \in L, a \in A$. Since $L/B \cong G/A$, we have $xC_L(B) \in \zeta(L/C_L(B))$ and also L/B is nilpotent. By hypothesis, $B = [B, g] = [B, x]$ and $C_B(g) = C_B(x) = 1$. Theorem 8.2.7 shows that $L = B \rtimes K$, for some subgroup K and every complement to B in L is conjugate to K. Then $G = AL = A(BK) = AK$ and

$$A \cap K = A \cap (L \cap K) = B \cap K = 1$$

and hence K is a complement to A in G.

Now let D be another complement to A in G. Then DB/B is a complement to A/B in G/B. By the induction hypothesis, there exists $v \in G$ such that $D^vB/B = (DB/B)^{vB} = L/B$ and hence $D^v \leq L$. Since $L = D^vB$ and $D^v \cap B = 1$, D^v is a complement to B in L. Hence there exists $u \in L$ such that $K = (D^v)^u$, so D is conjugate to K. This completes the proof. $\qquad\square$

8.2.9. LEMMA. *Let G be a group and let A be a periodic abelian normal subgroup of G. If A is G-hypereccentric, then $A = [A, G]$ and $C_A(G) = 1$.*

EXERCISE 8.9. *Prove Lemma 8.2.9.*

This result can be used to obtain further splitting criteria as follows.

8.2.10. LEMMA. *Let G be a group and let A be a periodic abelian normal subgroup of G. Suppose that A is G-hypereccentric and that G contains a normal subgroup H which satisfies the conditions:*

(i) *H is hypercentral;*
(ii) *$A \leq H$;*
(iii) *G/A is finitely generated nilpotent;*
(iv) *$G/H = \langle gH \rangle$ is cyclic.*

Then G splits conjugately over A.

PROOF. By Lemma 8.1.2, A has its upper H-central series

$$1 = C_0 \leq C_1 \leq \cdots \leq C_n \leq C_{n+1} \leq \ldots C_\omega = A,$$

of length at most ω and we note that each $C_j \lhd G$. By (iii), there is a finitely generated subgroup K such that $H = AK$. We put $L = \langle K, g \rangle$ and note that $G = AL$. Clearly $L/(L \cap A)$ is finitely generated nilpotent and hence is finitely presented by Corollary 8.1.17. Proposition 8.1.12 shows that $L \cap A$ contains elements $a_1, \ldots, a_t$ such that $L \cap A = \langle a_1 \rangle^L \ldots \langle a_t \rangle^L$ and hence there is a natural number k such that $a_1, \ldots, a_t \in C_k$. Therefore $L \cap A \leq C_k$, since each subgroup C_j is G-invariant.

Since $G = AL$ and A is abelian, each L-invariant subgroup of A is also G-invariant. Since A is G-hypereccentric, Lemma 8.2.9 shows that $[C_{j+1}/C_j, G/C_j] = C_{j+1}/C_j$ and $C_{j+1}/C_j \cap \zeta(G/C_j) = 1$. On the other hand, C_{j+1}/C_j is H-central, so (iv) implies that $C_{j+1}/C_j = [C_{j+1}/C_j, G/C_j] = [C_{j+1}/C_j, gC_j]$ and $C_{C_{j+1}/C_j}(g) = 1$, for all $j \in \mathbb{N}$.

By the above, LC_k satisfies the hypotheses of Corollary 8.2.8 and hence there is a subgroup D such that $LC_k = C_k \rtimes D$ and all complements to C_k in LC_k are conjugate to D. Hence

$$G = AL = A(C_k D) = AD \text{ and}$$
$$A \cap D = A \cap (LC_k \cap D) = (A \cap LC_k) \cap D = C_k \cap D = 1,$$

since $L \cap A \leq C_k$. Hence D is a complement to A in G.

Suppose that Y is another complement to A in G. Then $Y \cong G/A \cong D$. Since G/A is finitely generated, so is $Y = \langle y_1, \ldots, y_s \rangle$, for certain y_j, with $1 \leq j \leq s$. Since $G = AD$ we have $y_j = a_j d_j$, for certain $a_j \in A, d_j \in D$ and we let $D_1 = \langle d_1, \ldots, d_s \rangle$. Then $G = AY \leq AD_1$ and hence $D_1 = D$. There exists $m \in \mathbb{N}$ such that $a_1, \ldots, a_s \in C_m$, so that $Y \leq C_m D$ and hence $YC_m = D_1 C_m = DC_m$. Applying Corollary 8.2.8 to $C_m D$ we deduce that every complement to C_m in $C_m D$ (in particular Y) is conjugate to D. □

8.2.11. LEMMA. *Let G be a locally nilpotent group and suppose that G contains an abelian normal subgroup T such that G/T is finitely generated. Then G is hypercentral.*

EXERCISE 8.10. *Prove Lemma 8.2.11.*

We prove next a further splitting result that has appeared in [**227**].

8.2.12. PROPOSITION. *Let G be a group and let A, B be subgroups of G such that $A \leq B$. Suppose that*

(i) *A is an abelian normal subgroup of G and that B is ascendant in G;*

(ii) *B splits conjugately over A;*

(iii) *$C_A(B) = 1$.*

Then G splits conjugately over A.

PROOF. Let

$$B = B_0 \leq B_1 \leq \ldots B_\alpha \leq B_{\alpha+1} \leq \ldots B_\gamma = G$$

be an ascending series from B to G. We use transfinite induction on γ. If $\gamma = 0$, then the result is clear. So suppose that $\gamma > 0$ and the result is true for all $\beta < \gamma$.

If γ is a limit ordinal, then for each $\beta < \gamma$ there is a complement X_β for A in B_β. Then

$$B_\beta = B_\beta \cap X_{\beta+1}A = (B_\beta \cap X_{\beta+1})A,$$

whence $B_\beta \cap X_{\beta+1}$ is conjugate to X_β in B_β. Replacing $X_{\beta+1}$ by a suitable conjugate we may therefore suppose that $X_\beta \leq X_{\beta+1}$, for all $\beta < \gamma$. It is then easy to see that $X = \bigcup_{\beta<\gamma} X_\beta$ is a complement to A in B_γ. Next we suppose that Y is another complement to A in B_γ. Then, for all $\beta < \gamma$, $X \cap B_\beta$ and $Y \cap B_\beta$ are complements to A in B_β and hence there is an element $a_\beta \in A$ such that $(X \cap B_\beta)^{a_\beta} = Y \cap B_\beta$. If $\beta < \delta < \gamma$, then

$$(a_\beta a_\delta^{-1})^{-1}(X \cap B_\beta)a_\beta a_\delta^{-1} = a_\delta(Y \cap B_\beta)a_\delta^{-1}$$

$$\leq a_\delta(Y \cap B_\delta)a_\delta^{-1} = X \cap B_\delta \leq X \cap B_\gamma.$$

Then $[a_\beta a_\delta^{-1}, X \cap B_\beta] \leq A \cap (X \cap B_\beta) = 1$ and hence $a_\beta a_\delta^{-1} \in C_A(X \cap B_\beta)$. But $B_\beta = A(X \cap B_\beta)$ and A is abelian so $a_\beta a_\delta^{-1} \in C_A(B_\beta) \leq C_A(B) = 1$. We deduce that $a_\beta = a_\delta = a$, say, and then $X^a = Y$.

Now suppose that $\gamma - 1$ exists and we know $B_{\gamma-1} = A \rtimes X$, for some subgroup X. If $g \in B_\gamma$, then X^g is a complement to A in $B_{\gamma-1}$ and the induction hypothesis implies that $X^g = X^a$, for some $a \in A$. Hence

$ga^{-1} \in U = N_G(X) \cap B_\gamma$ and $B_\gamma = UA$. Also $U \cap A \le C_A(B_{\gamma-1}) = 1$, so U is a complement to A in B_γ.

Suppose that Y is also a complement to A in B_γ. Then $Y \cap B_{\gamma-1}$ is a complement to A in $B_{\gamma-1}$ and, by the induction hypothesis, there is an element $h \in B_{\gamma-1}$ such that $(Y \cap B_{\gamma-1})^h = X$. As above we have $N_G(Y \cap B_{\gamma-1}) \cap B_\gamma = Y$ and therefore

$$Y^h = (N_G(Y \cap B_{\gamma-1}) \cap B_\gamma)^h = N_G((Y \cap B_{\gamma-1})^h) \cap B_\gamma$$
$$= N_G(X) \cap B_\gamma = U.$$

The result now follows.   □

8.2.13. PROPOSITION. *Let G be a group and let A be a periodic abelian normal subgroup of G. Suppose that G/A is finitely generated nilpotent and that A is G-hypereccentric. Then G splits conjugately over A.*

PROOF. Let H be the Hirsch–Plotkin radical of G. By Lemma 8.2.11, H is hypercentral since $A \le H$. Also G/H is a finitely generated nilpotent group, so there is a finite subnormal series

$$H = H_0 \triangleleft H_1 \triangleleft \ldots \triangleleft H_{n-1} \triangleleft H_n = G$$

whose factors are cyclic. We use induction on n and note that the case $n = 1$ follows from Lemma 8.2.10. Let $L = H_{n-1}$. By Proposition 8.1.20, A has a $Z(L)$-decomposition, say $A = C \times E$, where C is the upper L-hypercentre of A and E is a maximal L-invariant, L-hypereccentric subgroup of A. As we saw above, C and E are G-invariant.

Consider the factor group G/C. By the induction hypothesis, L contains a subgroup D such that $L/C = A/C \rtimes D/C$ and every complement to A/C in L/C is conjugate to D/C. Proposition 8.2.12 implies that there is a subgroup K such that $G/C = A/C \rtimes K/C$ and every complement to A/C in G/C is conjugate to K/C. Clearly $K/C \cong G/A$. Let P be the Hirsch–Plotkin radical of K; by Lemma 8.2.11, P is hypercentral. Also K/P is cyclic, by the choice of C. Since $G = AK$, every K-invariant subgroup of C is G-invariant and hence C is K-hypereccentric. Lemma 8.2.10 shows that $K = C \rtimes V$, for some subgroup V and every complement to C in K is conjugate to V. Hence

$$G = AK = A(CV) = AV \text{ and}$$
$$A \cap V = A \cap (K \cap V) = (A \cap K) \cap V = C \cap V = 1,$$

so KV is a complement to A in G.

If U is another complement to A in G, then UC/C is a complement to A/C in G/C and hence there is an element $g \in G$ such that

$U^g C/C = (UC/C)^{gC} = K/C$. Hence $U^g \leq K$ and also U^g is a complement to C in K. Hence there is an element $x \in K$ such that $V = (U^g)^x$
and U is conjugate to V. This completes the proof. □

8.2.14. THEOREM. *Let G be a group and let A be a periodic abelian
normal subgroup of G. Suppose that there is an ascendant subgroup H
of G such that $A \leq H$ and H/A is finitely generated nilpotent. If A is
H-hypereccentric, then G splits conjugately over A.*

PROOF. Proposition 8.2.13 shows that H splits conjugately over A
and Lemma 8.2.9 shows that $C_A(H) = 1$. The result now follows by
Proposition 8.2.12. □

8.2.15. COROLLARY. *Let G be a group and let A be an abelian
normal G-hyperfinite subgroup of G. Suppose that G/A is a nilpotent minimax group such that $\mathbf{Tor}(G/A)$ is finite. If C is the upper
G-hypercentre of A, then G/C splits conjugately over A/C.*

PROOF. By Lemma 8.1.1 G has a finite subnormal series

$$A = L_0 \lhd L_1 \lhd \ldots \lhd L_{n-1} \lhd L_n = G$$

such that L_1/A is finitely generated and L_{j+1}/L_j is periodic divisible,
for $1 \leq j \leq n-1$. By Proposition 8.1.20 A has a $Z(L)$-decomposition,
say $A = C \times E$, where C is the upper L-hypercentre of A and E is a
maximal L-invariant, L-hypereccentric subgroup of A. Using the arguments of the proof of Corollary 8.1.21 we deduce that C is the upper
G-hypercentre of A and E is a maximal G-invariant, G-hypereccentric
subgroup of A. We may now apply Theorem 8.2.14 to G/C to obtain
the result. □

8.2.16. COROLLARY. *Let G be a group and let A be an abelian normal G-hyperfinite subgroup of G. Suppose that G contains a normal
subgroup L such that $A \leq L$ and L/A is a nilpotent minimax group
such that $\mathbf{Tor}(L/A)$ is finite. If C is the upper G-hypercentre of A,
then G/C splits conjugately over A/C.*

PROOF. Since L is a normal subgroup of G, it follows that C is
a G-invariant subgroup of A. Corollary 8.2.15 and Proposition 8.2.12
can now be applied to deduce the result. □

8.3. Residually Finite Groups Having Finite 0-Rank

In Theorem 5.2.8 we considered finitely generated groups of finite
0-rank and proved that a finitely generated generalized radical such
group G is minimax modulo its subgroup $\mathbf{Tor}(G)$. In this section

we discuss finitely generated residually finite groups of finite 0-rank. The main issue here is to determine conditions under which such a group will be minimax. Immediately we note that such groups are not minimax in general. If $G = \langle a \rangle \wr \langle g \rangle$, where a has prime order p and g is an element of infinite order, then G is finitely generated, $\mathbf{Tor}\,(G)$ is an infinite elementary abelian p-group and $G/\mathbf{Tor}\,(G)$ is infinite cyclic, so that $\mathbf{r}_0(G) = 1$. A classical result of P. Hall [**107**, Theorem 1] shows that G is residually finite, but clearly is not minimax.

In this section we also start to apply the results proved earlier concerning the existence of supplements to certain abelian normal subgroups.

8.3.1. PROPOSITION. *Let G be a finitely generated group and suppose that G contains a torsion-free nilpotent normal subgroup H such that G/H is polycyclic-by-finite. If H has finite 0-rank, then H, and likewise G, is minimax.*

PROOF. Since G/H is polycyclic-by-finite it is finitely presented, by Corollary 8.1.17, so Proposition 8.1.12 shows that H contains a finite subset M such that $\langle M \rangle^G = H$. Thus we can consider H/H' as a finitely generated $\mathbb{Z}(G/H)$-module. Then H/H' contains a free abelian subgroup C/H' such that H/C is periodic and $\Pi(H/C)$ is finite (see [**152**, Corollary 1.8], for example). Since H has finite 0-rank, C/H' is finitely generated and, for the same reason, the Sylow p-subgroups of H/C are Chernikov for each prime p. Hence H/C is Chernikov, since $\Pi(H/C)$ is finite, and consequently H/H' is minimax. That H is minimax now follows from Corollary 5.2.7. Clearly G is minimax. $\square$

8.3.2. COROLLARY. *Let G be a finitely generated group and suppose that G contains a locally nilpotent normal subgroup H of finite 0-rank such that G/H is polycyclic-by-finite. Then $H/\mathbf{Tor}\,(H)$ is a nilpotent minimax group.*

PROOF. Since G/H is polycyclic-by-finite it is finitely presented, by Corollary 8.1.17. Let $T = \mathbf{Tor}\,(H)$. Since H is locally nilpotent, Proposition 1.2.11 shows that T is a characteristic subgroup of G consisting of the set of elements of H of finite order and H/T is torsion-free nilpotent, by Corollary 2.3.4. Proposition 8.3.1 implies that H/T is minimax. $\square$

8.3.3. PROPOSITION. *Let G be a finitely generated group and suppose that G contains a nilpotent normal subgroup H satisfying the conditions:*

(i) H has finite 0-*rank;*

(ii) $\mathbf{Tor}\,(H)$ is G-*hyperfinite;*

(iii) G/H is polycyclic-by-finite.

Then H is minimax.

PROOF. Let $T = \mathbf{Tor}\,(H)$ and note that, since H is nilpotent, T is a characteristic subgroup of H and H/T is torsion-free nilpotent. By Corollary 8.3.2, H/T is minimax. Then H/H' is an extension of a periodic normal subgroup TH'/H' by the minimax group H/TH'. By hypothesis, TH'/H' is G-hyperfinite. Since G/H' is finitely generated, its abelian normal subgroup H/H' satisfies the condition max-G, by a classical result of P. Hall [**105**, Theorem 3]. Then, since TH'/H' is G-hyperfinite, it is finite and hence H/H' is minimax. That H is minimax now follows from Proposition 5.2.6.   □

8.3.4. LEMMA. *Let G be a group and let H be a subgroup of G such that $|G : H|$ is finite. If H is residually finite, then G is also residually finite.*

EXERCISE 8.11. *Prove Lemma 8.3.4.*

8.3.5. LEMMA. *Let G be an abelian minimax group. Then G is residually finite if and only if $\mathbf{Tor}\,(G)$ is finite.*

EXERCISE 8.12. *Prove Lemma 8.3.5.*

8.3.6. PROPOSITION. *Let G be a soluble-by-finite minimax group. If $\mathbf{Tor}\,(G)$ is finite, then G is residually finite.*

EXERCISE 8.13. *Prove Proposition 8.3.6.*

8.3.7. LEMMA. *Let G be a finitely generated group and let A be an abelian normal subgroup of G such that $G = A \rtimes K$, for some subgroup K. If A is G-hyperfinite, then A is finite.*

EXERCISE 8.14. *Prove Lemma 8.3.7.*

8.3.8. LEMMA. *Let G be a group and let A be an abelian normal subgroup of G. Suppose that F is a subgroup of A such that $A = F[A, G]$. Then $A = F^G[A, {}_n G]$, for every natural number n.*

EXERCISE 8.15. *Prove Lemma 8.3.8.*

8.3.9. LEMMA. *Let G be a finitely generated residually finite group and suppose that G contains a hypercentral normal subgroup H satisfying the following conditions:*

(i) H has finite 0-rank;

(ii) $\mathbf{Tor}\,(H)$ is G-hyperfinite;

(iii) $\mathbf{Tor}\,(H)$ contains a G-invariant abelian subgroup A such that $\mathbf{Tor}\,(H)/A$ is finite;

(iv) G/H is polycyclic-by-finite.

Then A is finite.

PROOF. Let $\mathcal{S}$ be the family of all normal subgroups of G having finite index in G and let $\mathcal{L} = \{U \cap A | U \in \mathcal{S}\}$. Clearly, if $V \in \mathcal{L}$, then A/V is finite. Let $A_V/V = [H/V, A/V] \neq A/V$, since H is hypercentral and let $C = \bigcap_{V \in \mathcal{L}} A_V$. If $a \in A, h \in H$, then $[a, h]V = [aV, hV] \in A_V/V$, for each $V \in \mathcal{L}$ and hence $[a, h] \in C$. Thus $A/C \le \zeta(H/C)$. By Corollary 8.3.2, $H/\mathbf{Tor}\,(H)$ is nilpotent. Then, since $\mathbf{Tor}\,(H)/A$ is finite and $A/C \le \zeta(H/C)$ we see that H/C is nilpotent. Using Proposition 8.3.3 we deduce that H/C is minimax and since A is periodic it follows that A/C is Chernikov. However, by the definition of C, A/C is residually finite so it is finite, and hence there is a finite subgroup F such that $A = CF$. Of course $C \le A_V$, for all $V \in \mathcal{L}$ and it follows that

$$A/V = CF/V = A_V F/V = (A_V/V)(FV/V) = [H/V, A/V](FV/V).$$

Since H is hypercentral and A/V is finite there is a natural number n such that $[A,_n H] \le V$ and Lemma 8.3.8 implies that $A/V = F^H V/V$.

By Lemma 8.1.1, H has a finite subnormal series

$$A = L_0 \lhd L_1 \lhd \ldots \lhd L_{k-1} \lhd L_k = G$$

such that L_1/A is finitely generated and L_{j+1}/L_j is periodic divisible, for $1 \le j \le k - 1$. Since L_1/A is finitely generated, $L_1 = LA$, for some finitely generated subgroup L. Then $\langle F, L \rangle$ is a finitely generated nilpotent group, so $\mathbf{Tor}\,(\langle F, L \rangle)$ is finite. However, $F^L \le \mathbf{Tor}\,(\langle F, L \rangle)$, so F^L is also finite. Since $F \le A$ and A is abelian, we have $F^L = F^{L_1}$. By Corollary 8.1.4, every L_1-invariant subgroup of A is H-invariant and hence $F^{L_1} = F^H$ is H-invariant. Thus F^H is finite. Since $A/V = F^H V/V$ we have $|A/V| \le |F^H|$, for each $V \in \mathcal{L}$. Since H is hypercentral and F^H is finite, there is a natural number t such that $[F^H,_t H] = 1$ and hence $[A,_t H] \le V$, for each $V \in \mathcal{L}$. Therefore $[A,_t H] = 1$, since $\bigcap_{V \in \mathcal{L}} V = 1$. Since H/A is nilpotent, it follows that H is nilpotent and Proposition 8.3.3 shows that H is minimax. Hence $\mathbf{Tor}\,(H)$ is Chernikov and since $\mathbf{Tor}\,(H)$ is residually finite, $\mathbf{Tor}\,(H)$ is finite and hence A is finite. $\qquad \square$

8.3.10. LEMMA. *Let G be a finitely generated residually finite group and let A be an abelian normal subgroup of G. Suppose that the following conditions hold:*

(i) *There exists $L \lhd G$ such that $A \leq L$, L/A is nilpotent minimax and G/L is abelian-by-finite;*

(ii) *$\mathbf{Tor}(L/A)$ is finite;*

(iii) *A is G-hyperfinite.*

Then A is finite.

PROOF. Let C be the upper L-hypercentre of A, so that C is G-invariant. By Corollary 8.2.16, $L/C = A/C \rtimes H/C$, for some subgroup H. Lemma 8.2.9 shows that $C_{A/C}(H/C) = 1$ and hence Proposition 8.2.12 shows that $G/C = A/C \rtimes K/C$ for some subgroup K. In particular $K \cap A = C$. By Lemma 8.3.7 A/C is finite and Corollary 2.4.5 shows that there is a normal subgroup H of G such that G/H is finite. By Proposition 1.2.13 H is finitely generated, so Lemma 8.3.9 shows that C is finite. Hence A is finite. □

8.3.11. COROLLARY. *Let G be a finitely generated residually finite group and let A be a soluble normal subgroup of G. Suppose that the following conditions hold:*

(i) *There exists $L \lhd G$ such that $A \leq L$, L/A is nilpotent minimax and G/L is abelian-by-finite;*

(ii) *$\mathbf{Tor}(L/A)$ is finite;*

(iii) *A is G-hyperfinite.*

Then A is finite.

PROOF. The hypotheses imply that G/A is a soluble-by-finite minimax group. Let $S/A = \mathbf{Tor}(G/A)$. Since G/L satisfies the maximal condition, $SL/L \cong S/(S \cap L)$ is finite. Also $(S \cap L)/A$ is finite, by hypothesis, so S/A is finite. Hence, by Proposition 8.3.6, G/A is residually finite. Then clearly A has a finite series of G-invariant subgroups

$$1 = A_0 \leq A_1 \leq \cdots \leq A_{n-1} \leq A_n = A$$

such that the factors A_{j+1}/A_j are abelian and G/A_j is residually finite, for $0 \leq j \leq n-1$.

We use induction on n. If $n = 1$, then A is abelian and the result follows from Lemma 8.3.10. Then we may suppose that $n > 1$ and that we have already proved that A/A_1 is finite. Then $B/A_1 = \mathbf{Tor}(G/A_1)$ is finite. Corollary 2.4.5 shows that there is a normal subgroup K such that G/K is finite and $K \cap B = A_1$. Proposition 1.2.13 implies that K is finitely generated and Lemma 8.3.10 may be applied to K to deduce that A_1 is finite. Since A/A_1 is finite we see that A is finite, as required.   □

8.3.12. THEOREM. *Let G be a finitely generated soluble residually finite group with finite 0-rank. If $\mathbf{Tor}\,(G)$ is G-hyperfinite, then G is minimax.*

PROOF. By Theorem 2.4.11 there is a normal subgroup L of G such that $T = \mathbf{Tor}\,(G) \leq L$, L/T is torsion-free nilpotent and G/L is finitely generated abelian-by-finite. By Proposition 8.3.1, L/T is minimax and Corollary 8.3.11 shows that T is finite. Since G/T is minimax, it follows that G is minimax.   □

8.3.13. COROLLARY. *Let G be a finitely generated soluble-by-finite residually finite group, having finite 0-rank. If $\mathbf{Tor}\,(G)$ is G-hyperfinite, then G is minimax.*

PROOF. Let K be a soluble normal subgroup of G such that $|G : K|$ is finite. Proposition 1.2.13 shows that K is finitely generated and Theorem 8.3.12 shows that K is minimax. Hence G is also minimax.   □

8.3.14. COROLLARY. *Let G be a finitely generated soluble group having finite section rank. If the Sylow p-subgroups of G are finite for each $p \in \Pi(G)$, then G is minimax and residually finite.*

PROOF. By Theorem 4.2.1, G contains normal subgroups $T \leq L$ such that T is periodic, L/T is a torsion-free nilpotent group of finite 0-rank and G/L is abelian-by-finite. Furthermore, G has finite 0-rank. By Proposition 8.3.1 L/T, and hence G/T, is minimax. Since the Sylow p-subgroups of T are finite, Corollary 3.3.12 implies that $T/\mathbf{O}_{p'}(T)$ is a finite soluble group for each prime p and Proposition 8.3.6 implies that $G/\mathbf{O}_{p'}(T)$ is residually finite. Since $\cap_{p \in \Pi(G)}\mathbf{O}_{p'}(T) = 1$, Remak's Theorem implies that G is also residually finite. Since G is a soluble group with finite Sylow subgroups, T is G-hyperfinite. Hence G satisfies the hypotheses of Theorem 8.3.12 and it follows that G is minimax.   □

8.3.15. COROLLARY. *Let G be a finitely generated soluble-by-finite group having finite section rank. If the Sylow p-subgroups of G are finite for each $p \in \Pi(G)$, then G is minimax and residually finite.*

8.3.16. COROLLARY. *Let G be a finitely generated soluble-by-finite group having finite section rank. If G is residually finite, then G is minimax.*

PROOF. Using Theorem 4.2.1 we see that the Sylow p-subgroups of G are finite for all primes p. Then Corollary 8.3.15 gives the result. □

8.3.17. COROLLARY. *Let G be a finitely generated soluble-by-finite group having finite special rank. If G is residually finite, then G is minimax.*

8.4. Supplements to Divisible Abelian Normal Subgroups

Let G be a finitely generated soluble group with finite abelian section rank and let $T = \mathbf{Tor}\,(G)$. If T_p is a Sylow p-subgroup of G, then T_p is Chernikov, for all primes p. Let D_p denote the divisible part of T_p and let $D = \langle D_p | p \in \Pi(G) \rangle$, a characteristic divisible subgroup such that T/D has finite Sylow p-subgroups for all $p \in \Pi(G)$, by Theorem 4.1.6. From Corollary 8.3.14 it follows that G/D is a residually finite minimax group and hence T/D is finite. Therefore the next phase of our consideration of finitely generated soluble groups of finite section rank is reduced to the case when G is an extension of a divisible abelian layer finite subgroup D by a minimax group. The current section is dedicated to the consideration of this problem, and the main question which arises here is when does the subgroup D have a proper supplement?

For the group G, an infinite periodic abelian normal subgroup is called G-*quasifinite* if every proper G-invariant subgroup of G is finite. We observe that every such subgroup A is a p-group for some prime p. Indeed, if $\Pi(A)$ contains more than one prime p, then $A = P \times Q$, where P is the Sylow p-subgroup of A and Q is the Sylow p'-subgroup of A. Since P, Q are proper G-invariant subgroups of A, then they are finite and hence A must be finite, contrary to the definition of A. Next let B be the lower layer of A, so that $B = \Omega_1(A)$. If $B \neq A$, then B is finite since it is G-invariant. Then A is a Chernikov group, with divisible part D, say and D is an infinite G-invariant subgroup of A, so $A = D$. Consequently if A is G-quasifinite, then either A is a divisible Chernikov p-group, or A is an elementary abelian p-group, for some prime p.

If G is a periodic abelian group, then we say that G is *thin layer finite* if the Sylow p-subgroups of G are finite, for all primes $p \in \Pi(G)$.

8.4.1. PROPOSITION. *Let G be a group and let A be an abelian normal subgroup of G. Suppose that there exists $g \in G$ such that $C_G(A) \neq gC_G(A) \in \zeta(G/C_G(A))$ and that the following conditions hold:*

(i) *G/A is nilpotent;*

(ii) $A = \underset{p \in \Pi(A)}{Dr} C_p$, where $C_p = \underset{1 \le j \le n(p)}{Dr} P_{j,p}$ and $P_{j,p}$ is a divisible $\langle g \rangle$-invariant Chernikov p-subgroup, for $1 \le j \le n(p)$, $p \in \Pi(G)$, $n(p) \in \mathbb{N}$;

(iii) $P_{j,p}$ is $\langle g \rangle$-quasifinite and $P_{j,p} = [P_{j,p}, g] \ne 1$, for $1 \le j \le n(p)$, $p \in \Pi(G)$.

Then there is a subgroup R such that $G = AR$ and $A \cap R = B$ is a characterstic thin layer finite subgroup. If Y is another subgroup satisfying $G = AY$ and $A \cap Y = B$, then there is a characteristic thin layer finite subgroup E containing B such that YE/E and RE/E are conjugate.

PROOF. By Proposition 8.2.6 there is a subgroup L such that $G = AL$ and $L = \langle \mathbf{E}_{G,m}(g) \rangle$, for some natural number m and $L \cap A$ is a proper G-invariant subgroup of A. It follows that $L \cap P_{j,p}$ is a G-invariant subgroup of $P_{j,p}$ and, since every subgroup of $P_{j,p}$ is G-quasifinite, either $L \cap P_{j,p}$ is finite or $L \cap P_{j,p} = P_{j,p}$. Lemma 8.2.5 shows that there is a natural number k such that $L \subseteq \mathbf{E}_{G,k}(g)$. Thus, if $L \cap P_{j,p} = P_{j,p}$, then $P_{j,p} \subseteq \mathbf{E}_{G,k}(g)$ and $\langle P_{j,p}, g \rangle$ is nilpotent, contrary to (iii). Therefore, $L \cap P_{j,p}$ is finite and there is a natural number $t(p)$ such that

$$\langle L \cap P_{j,p} | 1 \le j \le n(p) \rangle \le B_p = \Omega_{t(p)}(C_p).$$

Let $B = \underset{p \in \Pi(A)}{Dr} B_p$, a characteristic subgroup of A whose Sylow p-subgroups are finite for all primes p and let $R = BL$. Then $G = RA$ and $R \cap A = BL \cap A = B(A \cap L) = B$.

Let Y be another supplement to A in G such that $Y \cap A = B$. Then $g \in AY$, so that $g = yu$, for certain elements $u \in A, y \in Y$ and $y = gv$, where $v = u^{-1}$. Since $A = [g, A]$, there exists $w \in A$ such that $v = [g, w]$, so $y = gv = w^{-1}gw$. We have

$$Y/B = Y/(Y \cap A) \cong YA/A = G/A = LA/A.$$

Let $\phi : Y/B \longrightarrow LA/A$ denote this isomorphism. Then $\phi(yB) = gA$. By definition of L we have

$$LA/A = \langle \mathbf{E}_{G,m}(g) \rangle A/A \le \langle \mathbf{E}_{G/A,m}(gA) \rangle$$

and it follows that $Y/B \le \langle \mathbf{E}_{G/B,m}(yB) \rangle$. Then

$$\langle \mathbf{E}_{G/B,m}(yB) \rangle = (Y/B)((A/B) \cap \langle \mathbf{E}_{G/B,m}(yB) \rangle)$$

and similarly,

$$R/B = LB/B = \langle \mathbf{E}_{G,m}(g) \rangle B/B \le \langle \mathbf{E}_{G/B,m}(gB) \rangle,$$

so that

$$\langle \mathbf{E}_{G/B,m}(gB) \rangle = (R/B)((A/B) \cap \langle \mathbf{E}_{G/B,m}(gB) \rangle).$$

Since $g = yu$ and $u \in A$ it is then easy to see that

$$(A/B) \cap \langle \mathbf{E}_{G/B,m}(gB) \rangle = (A/B) \cap \langle \mathbf{E}_{G/B,m}(yB) \rangle.$$

We know already that $\langle \mathbf{E}_{G,m}(g) \rangle \subseteq \mathbf{E}_{G,k}(g)$ and hence $\langle \mathbf{E}_{G/B,m}(gB) \rangle \subseteq \mathbf{E}_{G/B,k}(gB)$. Using this inclusion and repeating the arguments above we deduce that $\langle \mathbf{E}_{G/B,m}(gB) \rangle \cap C_p B / B$ is finite and hence there is a characteristic subgroup E_p such that

$$\langle \mathbf{E}_{G/B,m}(gB) \rangle \cap C_p B / B \leq E_p / B.$$

We note that E_p / B depends only on gB. We let $E/B = \underset{p \in \Pi(A)}{\mathrm{Dr}}\, E_p B / B$, a characteristic subgroup of A with finite Sylow p-subgroups for all primes p. We now have

$$\langle \mathbf{E}_{G/B,m}(gB) \rangle E / E = RE/E \text{ and } \langle \mathbf{E}_{G/B,m}(yB) \rangle = YE/E.$$

Since $g^w = y$, $(gB)^{wB} = yB$ and

$$\langle \mathbf{E}_{G/B,m}(gB) \rangle^{wB} = \langle \mathbf{E}_{G/B,m}(gB^{wB}) \rangle = \langle \mathbf{E}_{G/B,m}(yB) \rangle,$$

we have

$$YE/E = \langle \mathbf{E}_{G/B,m}(yB) \rangle E / E = \langle \mathbf{E}_{G/B,m}(gB) \rangle^{wB} E / E$$
$$= (\langle \mathbf{E}_{G/B,m}(gB) \rangle E / E)^{wE} = (RE/E)^{wE},$$

as required.   □

8.4.2. LEMMA. *Let G be a group and let A be an abelian normal subgroup of G. Suppose that the following conditions hold:*

(i) *G/A is a finite nilpotent group;*

(ii) *$A = \underset{p \in \Pi(A)}{\mathrm{Dr}}\, C_p$, where $C_p = \underset{1 \leq j \leq n(p)}{\mathrm{Dr}}\, P_{j,p}$ and $P_{j,p}$ is a divisible G-invariant Chernikov p-subgroup, for $1 \leq j \leq n(p)$, $p \in \Pi(G), n(p) \in \mathbb{N}$;*

(iii) *$P_{j,p}$ is G-quasifinite and $P_{j,p} = [P_{j,p}, G] \neq 1$, for $1 \leq j \leq n(p)$, $p \in \Pi(G)$.*

Then there is a subgroup R such that $G = AR$ and $A \cap R = B$ is a characterstic thin layer finite subgroup. If Y is another subgroup satisfying $G = AY$ and $A \cap Y = B$, then there is a characteristic thin layer finite subgroup E containing B such that YE/E and RE/E are conjugate.

PROOF. We let $Z = C_G(A)$ and use induction on $|G/Z|$, noting that (iii) implies that $Z \neq G$. Since G/Z is nilpotent there is a nontrivial element $gZ \in \zeta(G/Z)$ and this choice of g implies that $C_A(g)$ is G-invariant. Since $P_{j,p}$ is G-quasifinite, either $C_A(g) \cap P_{j,p} = P_{j,p}$ or $C_A(g) \cap P_{j,p}$ is finite. In the latter case we have $[P_{j,p}, g] \cong P_{j,p}/(C_A(g) \cap$

$P_{j,p}$ and since $[P_{j,p}, g]$ is also G-invariant we obtain that $[P_{j,p}, g] = P_{j,p}$. Thus either $[P_{j,p}, g] = 1$ or $[P_{j,p}, g] = P_{j,p}$. Let

$$\sigma(p) = \{j | 1 \leq j \leq n(p) \text{ and } [P_{j,p}, g] = P_{j,p}\}, C_{p,1} = \langle P_{j,p} | j \in \sigma(p) \rangle \text{ and,}$$

$$\nu(p) = \{j | 1 \leq j \leq n(p) \text{ and } [P_{j,p}, g] = 1\}, C_{p,2} = \langle P_{j,p} | j \in \nu(p) \rangle.$$

Then $[C_{p,1}, g] = C_{p,1}$ and $[C_{p,2}, g] = 1$, for each prime p. Let

$$A_1 = \operatorname*{Dr}_{p \in \Pi(A)} C_{p,1} \text{ and } A_2 = \operatorname*{Dr}_{p\Pi(A)} C_{p,2}.$$

Then $[A_1, g] = A_1$, $A_2 \leq C_A(g)$ and $A = A_1 \times A_2$, so that $A = A_1 C_A(g)$. It also follows that $A_1 = [A, g]$. We also note that the subgroups A_1, A_2 are G-invariant.

The factor group G/A_2 satisfies the hypotheses of Proposition 8.4.1 so that $A_1 \cong A/A_2$ contains thin layer finite characteristic subgroups B_1, E_1 and there is a subgroup R_1/A_2 such that $B_1 \leq E_1$, $G/A_2 = (A/A_2)(R_1/A_2)$ and $(A/A_2) \cap (R_1/A_2) = B_1 A_2/A_2$. Furthermore, if V/A_2 is a subgroup with the properties $G/A_2 = (A/A_2)(V/A_2)$ and $A/A_2 \cap V/A_2 = B_1 A_2/A_2$, then $R_1 E_1$ and $V E_1$ are conjugate. Without loss of generality we may suppose that $B_1 = 1$. We also observe that, from the construction of R_1, we have $g \in R_1$.

We now consider R_1. We have $A_2 = A \cap R_1$ and $R_1/A_2 = R_1/(A \cap R_1) \cong R_1 A/A = G/A$. Thus every R_1-invariant subgroup of A_2 is also G-invariant. If $j \in \nu(p)$, for $p \in \Pi(A_2)$, then $P_{j,p}$ is R_1-invariant and $[P_{j,p}, R_1] = P_{j,p}$. Since $g \in C_G(A_2)$ we have $|G/C_G(A)| > |R_1/C_{R_1}(A_2)|$. By the induction hypothesis, A_2 contains thin characteristic subgroups B_2, E_2 and there is a subgroup R such that $B_2 \leq E_2$, $R_1 = A_2 R$ and $B_2 = A_2 \cap R$. Furthermore, if W is a subgroup with the properties $R_1 = A_2 W, B_2 = A_2 \cap W$, then $R E_2/E_2$ and $W E_2/E_2$ are conjugate.

Again without loss of generality, we may assume that $B_2 = 1$. We now have $G = R_1 A$ and $R_1 = A_2 R$ so that $G = AR$. Since $A_2 = A \cap R_1$ and $A_2 \cap R = 1$, we have $A \cap R = 1$.

Suppose now that Y is a subgroup of G such that $G = AY$ and $A \cap Y = 1$. For the factor group G/A_2 we have

$$G/A_2 = (A/A_2)(Y A_2/A_2) \text{ and } Y A_2 \cap A = A_2,$$

so that $Y A_2/A_2$ is a complement to A/A_2 in G/A_2. By our proof above, A_1 contains a characteristic thin layer finite subgroup E_1 such that $(Y E_1 A_2)^x = R_1 E_1$, for some element x. Since $E_1 \leq A_1$ we have $E_1 \cap R_1 = 1$. It follows that $R_1 E_1/E_1 \cong R_1$. We set $U = Y^x$. Then we have

$$(U E_1/E_1)(A_2 E_1/E_1) = R_1 E_1/E_1$$

and

$$UE_1 \cap A_2 E_1 = E_1(UE_1 \cap A_2) \leq E_1(UE_1 \cap A) = E_1(U \cap A) = E_1.$$

This means that UE_1/E_1 is a complement to $R_1 E_1/E_1$. Our proof above shows that A_2 contains a characteristic thin layer finite group E_2 such that $(UE_1 E_2)^z = RE_1 E_2$, for some element z. It follows that $(YE_1 E_2)^{xz} = RE_1 E_2$. The result follows on setting $E = E_1 E_2$.   □

8.4.3. LEMMA. *Let G be a group and let A be an abelian normal subgroup of G. Suppose that the following conditions hold:*

(i) *G/A is a nilpotent minimax group;*

(ii) *$A = \underset{p \in \Pi(A)}{Dr}\, C_p$, where $C_p = \underset{1 \leq j \leq n(p)}{Dr}\, P_{j,p}$ and $P_{j,p}$ is a divisible G-invariant Chernikov p-subgroup, for $1 \leq j \leq n(p)$, $p \in \Pi(G), n(p) \in \mathbb{N}$;*

(iii) *$P_{j,p}$ is G-quasifinite and $P_{j,p} = [P_{j,p}, G] \neq 1$, for $1 \leq j \leq n(p)$, $p \in \Pi(G)$.*

Then there is a subgroup R such that $G = AR$ and $A \cap R = B$ is a characterstic thin layer finite subgroup. If Y is another subgroup satisfying $G = AY$ and $A \cap Y = B$, then there is a characteristic thin layer finite subgroup E containing B such that YE/E and RE/E are conjugate.

PROOF. We let $Z = C_G(A)$ and use induction on $\mathbf{r}_0(G/Z)$, noting that Lemma 8.4.2 proves the result in the case when $\mathbf{r}_0(G/Z) = 0$. We suppose that $\mathbf{r}_0(G/Z) > 0$ and let $T/Z = \mathbf{Tor}\,(G/Z)$, a Chernikov group, since it is minimax. Let D/Z be the divisible part of T/Z. Since C_p is Chernikov, $\Omega_n(C_p)$ is finite for all $n \in \mathbb{N}$ and since $\Omega_n(C_p) \lhd G$ it follows that $G/C_G(\Omega_n(C_p))$ is finite. Hence $D \leq C_G(\Omega_n(C_p))$, for all $n \in \mathbb{N}$, and so $D \leq C_G(C_p)$. Since this is true for all $p \in \Pi(G)$ we have $D \leq C_G(A)$. Hence T/Z is finite and, by Corollary 2.4.5, G/Z has a torsion-free normal subgroup U/Z such that G/U is finite. Clearly U/Z is nontrivial and, since it is a normal subgroup of G/Z, it follows that $U/Z \cap \zeta(G/Z) \neq 1$. Hence $\zeta(G/Z)$ contains an element gZ of infinite order and the subgroup $C_A(g)$ is G-invariant.

Since $P_{j,p}$ is G-quasifinite, either $C_A(g) \cap P_{j,p} = P_{j,p}$ or $C_A(g) \cap P_{j,p}$ is finite. In the latter case we have $[P_{j,p}, g] \cong P_{j,p}/(C_A(g) \cap P_{j,p})$ and since $[P_{j,p}, g]$ is also G-invariant we obtain that $[P_{j,p}, g] = P_{j,p}$. Thus either $[P_{j,p}, g] = 1$ or $[P_{j,p}, g] = P_{j,p}$. Let

$$\sigma(p) = \{j | 1 \leq j \leq n(p) \text{ and } [P_{j,p}, g] = P_{j,p}\}, C_{p,1} = \langle P_{j,p} | j \in \sigma(p) \rangle \text{ and,}$$

$$\nu(p) = \{j | 1 \leq j \leq n(p) \text{ and } [P_{j,p}, g] = 1\}, C_{p,2} = \langle P_{j,p} | j \in \nu(p) \rangle.$$

Then $[C_{p,1}, g] = C_{p,1}$ and $[C_{p,2}, g] = 1$, for all primes p. Let

$$A_1 = \operatorname*{Dr}_{p \in \Pi(A)} C_{p,1} \text{ and } A_2 = \operatorname*{Dr}_{p \Pi(A)} C_{p,2}.$$

Then $[A_1, g] = A_1, A_2 \leq C_A(g)$ and $A = A_1 \times A_2$, so that $A = A_1 C_A(g)$. It also follows that $A_1 = [A, g]$ and we note that the subgroups A_1, A_2 are G-invariant.

The factor group G/A_2 satisfies the hypotheses of Proposition 8.4.1 so that $A_1 \cong A/A_2$ contains thin layer finite characteristic subgroups B_1, E_1 and there is a subgroup R_1/A_2 such that $B_1 \leq E_1$, $G/A_2 = (A/A_2)(R_1/A_2)$ and $(A/A_2) \cap (R_1/A_2) = B_1 A_2/A_2$. Furthermore, if V/A_2 is a subgroup with the properties $G/A_2 = (A/A_2)(V/A_2)$, $A/A_2 \cap V/A_2 = B_1 A_2/A_2$, then $R_1 E_1$ and $V E_1$ are conjugate. Without loss of generality we may suppose that $B_1 = 1$. We also observe that, from the construction of R_1 we have $g \in R_1$.

We now consider R_1. We have $A_2 = A \cap R_1$ and $R_1/A_2 = R_1/(A \cap R_1) \cong R_1 A/A = G/A$. Thus every R_1-invariant subgroup of A_2 is also G-invariant. If $j \in \nu(p)$, for $p \in \Pi(A_2)$, then $P_{j,p}$ is R_1-invariant and $[P_{j,p}, R_1] = P_{j,p}$, so since $g \in C_G(A_2)$, we have $\mathbf{r}_0(G/C_G(A)) > \mathbf{r}_0(R_1/C_{R_1}(A_2))$. By the induction hypothesis, A_2 contains thin charactersitic subgroups B_2, E_2 and there is a subgroup R such that $B_2 \leq E_2$, $R_1 = A_2 R$ and $B_2 = A_2 \cap R$. Furthermore, if W is a subgroup with the properties $R_1 = A_2 W, B_2 = A_2 \cap W$, then $R E_2/E_2$ and $W E_2/E_2$ are conjugate.

The remainder of the proof now follows as in the last part of the proof of Lemma 8.4.2. $\square$

8.4.4. THEOREM. *Let G be a group containing normal subgroups A and L such that $A \leq L$. Suppose that the following conditions hold:*

(i) *L/A is a nilpotent minimax group;*

(ii) *$A = \operatorname*{Dr}_{p \in \Pi(A)} C_p$, where $C_p = \operatorname*{Dr}_{1 \leq j \leq n(p)} P_{j,p}$ and $P_{j,p}$ is a divisible G-invariant Chernikov p-subgroup, for $1 \leq j \leq n(p)$, $p \in \Pi(G), n(p) \in \mathbb{N}$;*

(iii) *$P_{j,p}$ is G-quasifinite and $P_{j,p} = [P_{j,p}, g] \neq 1$, for $1 \leq j \leq n(p)$, $p \in \Pi(G)$.*

Then there is a subgroup U such that $G = AU$ and $A \cap U = D$ is a characteristic thin layer finite subgroup.

PROOF. Since $P_{j,p}$ is Chernikov, it contains a minimal infinite L-invariant subgroup Q, every proper L-invariant subgroup of which must be finite. Hence Q is L-quasifinite and if Q is G-invariant, then $Q = P_{j,p}$.

Suppose that there is an element $x \in G$ such that $Q_1 = Q^x \neq Q$. Then $Q_1 \cap Q$ is a proper L-invariant subgroup of Q and hence is finite. If $Q_1 Q \neq P_{j,p}$, then there is an element y such that $Q_1 Q$ does not contain Q^y and hence $Q^y \cap Q_1 Q$ is finite. Using similar arguments we obtain, after finitely many steps, L-invariant L-quasifinite subgroups $Q_0 = Q, Q_1, Q_2, \ldots, Q_k$ such that $P_{j,p} = Q_0 Q_1 \ldots Q_k$ and all intersections $Q_j \cap Q_0 Q_1 \ldots Q_{j-1} Q_{j+1} \ldots Q_k$ are finite. Then there is a natural number $m(j)$ such that $\Omega_{m(j)}(P_{j,p})$ contains all these intersections and it follows that

$$P_{j,p}/\Omega_{m(j)}(P_{j,p}) = Q_0 \Omega_{m(j)}(P_{j,p})/\Omega_{m(j)}(P_{j,p}) \times \cdots$$
$$\times Q_k \Omega_{m(j)}(P_{j,p})/\Omega_{m(j)}(P_{j,p}).$$

Suppose, for a contradiction, that $[Q, L] \neq Q$, so that the L-invariant subgroup $[Q, L]$ is finite. Then, since Q_j is conjugate to Q, $[Q_j, L]$ is also finite, for $1 \leq j \leq k$ and hence $[P_{j,p}, L] = [Q_0, L] \ldots [Q_k, L]$ is finite, which contradicts (iii). Hence $[Q, L]$ is infinite, so $[Q, L] = Q$, since Q is L-quasifinite. Then $[Q_j, L] = Q_j$, for $1 \leq j \leq k$. Let $s(p) = \max\{m(0), \ldots, m(k)\}$. Then $C_p/\Omega_{s(p)}(C_p)$ is the direct product of finitely many L-invariant, L-quasifinite subgroups. The subgroup $\underset{p \in \Pi(A)}{\text{Dr}} \Omega_{s(p)}(C_p)$ is characteristic and thin layer finite. Factoring by $\underset{p \in \Pi(A)}{\text{Dr}} \Omega_{s(p)}(C_p)$, we may suppose that $\underset{p \in \Pi(A)}{\text{Dr}} \Omega_{s(p)}(C_p) = 1$. Thus we may suppose that $C_p = \underset{1 \leq j \leq s(p)}{\text{Dr}} Q_{j,p}$, where $Q_{j,p}$ is a divisible L-invariant Chernikov p-group, which is L-quasifinite and $Q_{j,p} = [Q_{j,p}, L]$, for $1 \leq j \leq s(p)$, $p \in \Pi(A)$.

By Lemma 8.4.3 there is a subgroup R such that $L = AR$ and $A \cap L = B$ is a characteristic thin layer finite subgroup. Furthermore, if Y is another such subgroup satisfying $L = AY$ and $A \cap Y = B$, then there is a characteristic thin layer finite subgroup E containing B such that YE/E and RE/E are conjugate. Again without loss of generality, we may suppose that $B = 1$.

Let $x \in G$. Since $L = RA$ and $R \cap A = 1$ we have $L = L^x = (RA)^x = R^x A$ and $R^x \cap A = (R \cap A)^x = 1$. From what we have shown above, there is an element $w \in L$ such that $(RE/E)^{xE} = (RE/E)^{wE}$. Hence $xw^{-1}E \in N_{G/E}(RE/E)$ so $xE \in (L/E)N_{G/E}(RE/E)$. Therefore $G/E = (L/E)N_{G/E}(RE/E)$. Since $L/E = (RE/E)(A/E)$ we have $G/E = (A/E)N_{G/E}(RE/E)$.

Now suppose that $aE \in (A/E) \cap N_{G/E}(RE/E)$ and let $xE \in RE/E, yE \in L/E$. Then $yE = bEx_1 E$, where $x_1 \in R$ and $b \in A$. Let $x_2 = x_1 x x_1^{-1} \in R$. We have

$$[(aE)^{yE}, xE] = [(aE)^{x_1 E}, xE] = [aE, x_2 E]^{x_1 E} \in RE/E.$$

Hence $A/E \cap N_{G/E}(RE/E)$ is L-invariant. For $1 \le j \le s(p)$ and for all $p \in \Pi(A)$ each subgroup $Q_{j,p}$ is R-quasifinite, since $L = RA$. Since E is thin layer finite, $E \cap Q_{j,p}$ is finite. It follows that $Q_{j,p}E/E$ is R-quasifinite and $[Q_{j,p}E/E, RE/E] = Q_{j,p}E/E$. By the definition of R we have

$$RE \cap A = E(R \cap A) = E \text{ so } RE/E \cap A/E = 1.$$

Let $zE \in A/E \cap N_{G/E}(RE/E)$. Then $[zE, RE/E] \le RE/E$. On the other hand, A is normal in G, so $[zE, RE/E] \le A/E$. Hence $[zE, RE/E] \le A/E \cap RE/E = 1$, which shows that

$$(A/E) \cap N_{G/E}(RE/E) = C_{G/E}(RE/E).$$

Repeating the argument above, we deduce that $C_{A/E}(RE/E)$ lies in a characteristic thin layer finite subgroup D/E. Finally, set $U/E = (D/E)N_{G/E}(RE/E)$. Then, by construction, $G = UA$ and $U \cap A = D$. $\qquad\square$

8.4.5. COROLLARY. *Let G be a group containing normal subgroups A and L such that $A \le L$. Suppose that the following conditions hold:*

(i) *L/A is a nilpotent minimax group and G/L is abelian-by-finite;*

(ii) *$A = \underset{p \in \Pi(A)}{Dr}\, C_p$, where $C_p = \underset{1 \le j \le n(p)}{Dr}\, P_{j,p}$ and $P_{j,p}$ is a divisible G-invariant Chernikov p-subgroup, for $1 \le j \le n(p)$, $p \in \Pi(G), n(p) \in \mathbb{N}$;*

(iii) *$P_{j,p}$ is G-quasifinite and $P_{j,p} = [P_{j,p}, g] \ne 1$, for $1 \le j \le n(p)$, $p \in \Pi(G)$.*

Then G is not finitely generated.

PROOF. By Theorem 8.4.4, G contains a subgroup U such that $G = AU$ and $A \cap U = D$ is a characteristic thin layer finite subgroup. It follows that $\Pi(A/D) = \Pi(A)$. Suppose, for a contradiction, that $G = \langle g_1, \ldots, g_m \rangle$ is finitely generated. Then $g_j = u_j a_j$, where $u_j \in U, a_j \in A$, for $1 \le j \le m$. Then

$$G/D = \langle g_1 D, \ldots, g_m D \rangle = (U/D)\langle a_1 D, \ldots, a_m D \rangle^{U/D}.$$

It follows that $A/D = \langle a_1 D, \ldots, a_m D \rangle^{U/D}$. Since A/D is abelian and layer finite, $\langle a_1 D, \ldots, a_m D \rangle^{U/D}$ must be finite, so that A/D is finite, which is the desired contradiction. $\qquad\square$

8.4.6. COROLLARY. *Let G be a finitely generated group containing abelian normal subgroups A and L such that $A \le L$. Suppose that the following conditions hold:*

(i) *L/A is a nilpotent minimax group and G/L is abelian-by-finite;*

(ii) $A = \underset{p\in\Pi(A)}{\mathrm{Dr}}\, C_p$, where $C_p = \underset{1\leq j\leq n(p)}{\mathrm{Dr}}\, P_{j,p}$ and $P_{j,p}$ is a divisible G-invariant Chernikov p-subgroup, for $1 \leq j \leq n(p)$, $p \in \Pi(G), n(p) \in \mathbb{N}$;

(iii) $P_{j,p}$ is G-quasifinite, for $1 \leq j \leq n(p)$, $p \in \Pi(G)$.

Then $\Pi(A)$ is finite, so A is a Chernikov group.

PROOF. Since $L \lhd G$, it follows that $[P_{j,p}, L] \lhd G$ and hence is either finite or $P_{j,p}$. Let

$$\sigma(p) = \{j | 1 \leq j \leq n(p) \text{ and } [P_{j,p}, L] = K_{j,p} \text{ is finite}\} \text{ and}$$

$$\nu(p) = \{1, \ldots, n(p)\} \setminus \sigma(p).$$

Then $R_p = \underset{j\in\sigma(p)}{\mathrm{Dr}}\, K_{j,p}$ is a finite G-invariant subgroup and $R = \underset{p\in\Pi(A)}{\mathrm{Dr}}\, R_p$ is a thin layer finite G-invariant subgroup. It follows that $\Pi(A/R) = \Pi(A)$, so without loss of generality, we may suppose that $R = 1$. Hence we may suppose that either $[P_{j,p}, L] = P_{j,p}$ or $[P_{j,p}, L] = 1$, for $1 \leq j \leq n(p)$ and $p \in \Pi(A)$.

Let $S_p = \underset{j\in\nu(p)}{\mathrm{Dr}}\, P_{j,p}$, $S = \underset{p\in\Pi(A)}{\mathrm{Dr}}\, S_p$, $Z_p = \underset{j\in\sigma(p)}{\mathrm{Dr}}\, P_{j,p}$ and $Z = \underset{p\in\Pi(A)}{\mathrm{Dr}}\, Z_p$. The subgroups S, Z are G-invariant and $A/S \leq \zeta(L/S)$, so L/S is nilpotent. By Proposition 8.3.3, G/S is minimax and hence A/S is Chernikov. Since $A = S \times Z$, it follows that Z is a G-invariant Chernikov subgroup and $Z \leq C_A(L)$. Consider G/Z. Then $P_{j,p}Z/Z$ is G/Z-quasifinite and $[P_{j,p}Z/Z, G/Z] = P_{j,p}Z/Z$, for all j and all primes p. Corollary 8.4.5 shows that G/Z is not finitely generated, contrary to our hypothesis concerning G. Hence $S = 1$ and $A = Z$ is Chernikov. $\square$

8.4.7. COROLLARY. Let G be a finitely generated group containing abelian normal subgroups A and L such that $A \leq L$. Suppose that the following conditions hold:

(i) L/A is a nilpotent minimax group and G/L is abelian-by-finite;

(ii) $A = \underset{p\in\Pi(A)}{\mathrm{Dr}}\, C_p$, where C_p is a divisible Chernikov p-subgroup, for $p \in \Pi(G)$;

Then $\Pi(A)$ is finite, so A is a Chernikov group.

PROOF. Since C_p is Chernikov it contains a minimal G-invariant subgroup Q_1, every proper G-invariant subgroup of which is finite, so Q_1 is G-quasifinite. If $C_p \neq Q_1$, then C_p/Q_1 is Chernikov and so contains a G-invariant, G-quasifinite subgroup Q_2/Q_1, a process which can be continued. By Corollary 5.1.9, C_p has finite Zaitsev rank so this process terminates in finitely many steps. In particular, we may choose a G-invariant subgroup R_p of C_p such that C_p/R_p is G-quasifinite. Let $R = \operatorname*{Dr}_{p \in \Pi(A)} R_p$ and note that $\Pi(A/R) = \Pi(A)$. We apply Corollary 8.4.6 to G/R to deduce that $\Pi(A/R)$ is finite. Hence $\Pi(A)$ is finite and A is a Chernikov group. $\qquad\square$

8.4.8. COROLLARY. *Let G be a finitely generated soluble group. If G has finite section rank, then G is minimax.*

PROOF. By Theorem 4.2.1, G contains normal subgroups $T \leq L$ such that T is periodic, the Sylow p-subgroups of T are Chernikov, for all primes p, L/T is a torsion-free nilpotent group of finite 0-rank and G/L is abelian-by-finite. Furthermore, T contains a G-invariant divisible subgroup D such that the Sylow p-subgroups of T/D are finite. It follows from Corollary 8.3.14 that G/D is minimax and T/D is finite. Corollary 2.4.5 implies that there is a normal subgroup H such that $T \cap H = D$ and G/H is finite. Proposition 1.2.13 implies that H is finitely generated. Clearly D is a direct product of divisible Chernikov p-subgroups, so we may apply Corollary 8.4.7 to H, according to which D is Chernikov and since G/D is minimax, it follows that G is minimax. $\qquad\square$

8.4.9. COROLLARY. *Let G be a finitely generated soluble-by-finite group. If G has finite section rank, then G is minimax.*

We next observe the result first proved by D. J. S. Robinson [**226**].

8.4.10. COROLLARY. *Let G be a finitely generated soluble group. If every elementary abelian section of G is finite, then G is minimax.*

PROOF. Theorem 4.2.10 shows that G has finite section rank and the result follows by Corollary 8.4.8. $\qquad\square$

8.4.11. COROLLARY. *Let G be a finitely generated soluble-by-finite group. If G has finite bounded section rank, then G is minimax.*

We also note an earlier result of Robinson [**224**].

8.4.12. COROLLARY. *Let G be a finitely generated soluble-by-finite group. If G has finite special rank, then G is minimax.*

Our aim now is to generalize these results to generalized radical groups. To this end we must consider the transfinite derived series of a group G, which we term the *lower derived series* consisting of characteristic subgroups,

$$G = \delta_0(G) \geq \delta_1(G) \geq \dots \delta_\alpha(G) \geq \delta_{\alpha+1}(G) \geq \dots \delta_\tau(G),$$

defined recursively by $\delta_1(G) = G', \delta_{\alpha+1}(G) = [\delta_\alpha(G), \delta_\alpha(G)]$, for ordinals α and $\delta_\lambda(G) = \bigcap_{\mu<\lambda} \delta_\mu(G)$, for all limit ordinals λ. We remark that $\delta_\tau(G) = [\delta_\tau(G), \delta_\tau(G)]$. The least ordinal τ with this property is called the hypo-derived length of G. If the series terminates at 1, then we recall that G is called a *hypoabelian group*.

8.4.13. LEMMA. *Let G be an infinite periodic hyperabelian group and suppose that the Sylow p-subgroups of G are finite for all primes p. Then there is a natural number k such that $G/\delta_k(G)$ is finite.*

PROOF. If G is actually soluble, then the result is clear. We note also that a group of the type in question is infinite if and only if $\pi(G)$ is infinite. The group G is easily seen to be hyperfinite, so there is actually an ascending normal series

$$1 = H_0 \le H_1 \le \dots H_\alpha \le H_{\alpha+1} \le \dots H_\gamma = G$$

with finite abelian factors.

Suppose first that $\gamma = \omega$, the first infinite ordinal, and suppose, for a contradiction, that G/G' is finite. Let $\pi = \pi(G/G')$, a finite set of primes and let P be a Sylow π-subgroup of G. Then P is finite, so $P \le H_n$ for some natural number n. Since the Sylow π-subgroups of G are conjugate, G/H_n is a π'-group and clearly there is a prime $q \in \pi(G/H_n)$. If $R/H_n = O_{q'}(G/H_n)$ then $G \ne R$ and Corollary 3.3.12 implies that G/R is a finite soluble group. Certainly G/R is a π'-group which yields the desired contradiction, since G/G' is a π-group. The result therefore follows in this case.

Next, suppose that γ is a limit ordinal and $\gamma > \omega$. Then there is a limit ordinal β such that $\beta + \omega = \gamma$. Then G/H_β is infinite so, by the first case above applied to G/H_β, it follows easily that G/G' is infinite.

If γ is not a limit ordinal then there is a limit ordinal β and a natural number m such that $\beta + m = \gamma$. Then G/H_β is a finite soluble group and clearly H_β is infinite. Since β is a limit ordinal there is a limit ordinal ρ (note that 0 is a limit ordinal) such that $\rho + \omega = \beta$ and, as in the case when γ is a limit ordinal, H_β/H'_β must be infinite. Hence G/H'_β is an infinite soluble group and the result now follows. $\square$

8.4.14. THEOREM. *Let G be a finitely generated radical group. If G has finite section rank, then G is minimax.*

PROOF. By Theorem 4.2.1, G contains normal subgroups $T \le L$ such that T is periodic, the Sylow p-subgroups of T are Chernikov, for all primes p, L/T is a torsion-free nilpotent group of finite 0-rank and

G/L is abelian-by-finite. Furthermore, T contains a G-invariant divisible subgroup D such that the Sylow p-subgroups of T/D are finite. If T/D is infinite, then Lemma 8.4.13 implies that there is a G-invariant subgroup R such that $D \le R$ and T/R is an infinite soluble group. Since G/T is also soluble, G/R is soluble and finitely generated. Corollary 8.4.8 then implies that G/R is minimax and hence T/R is finite, a contradiction. Hence T/D is finite, so G is soluble and we may now apply Corollary 8.4.8 again to deduce the result. $\qquad\square$

8.4.15. COROLLARY. *Let G be a finitely generated generalized radical group. If G has finite bounded section rank, then G is minimax and soluble-by-finite.*

PROOF. By Corollary 4.4.4, G contains a normal hyperabelian subgroup H of finite index. Proposition 1.2.13 shows that H is finitely generated and Theorem 8.4.14 gives the result. $\qquad\square$

One further consequence of all this work is the following result which appears in [67].

8.4.16. COROLLARY. *Let G be a finitely generated generalized radical group. If G has finite special rank, then G is minimax and soluble-by-finite.*

CHAPTER 9

The Influence of Important Families of Subgroups of Finite Rank

In this chapter, we shall discuss the influence of different families of subgroups on the structure of a group. More specifically, we shall investigate the question

> Let S be a family of subgroups of a group G and suppose that all the subgroups in S satisfy some condition on ranks. When does G satisfy the same condition on ranks?

We have already briefly touched upon this topic in Theorems 2.3.3, 4.2.11, 4.4.6 and Corollary 6.2.19. In this chapter, obtaining answers to this question is our sole purpose.

Probably the most important family of subgroups in a group is the family of all abelian subgroups. The influence of the structure of the abelian subgroups of a group on the entire group is highly significant for many classes of groups, especially locally generalized radical groups. So, for example, if every abelian subgroup of a locally generalized radical group is finite, then the group itself is finite, a fact which follows from a theorem due to M. I. Kargapolov [**128**] and P. Hall and C. R. Kulatilaka [**111**]. Similarly, if all abelian subgroups of a locally generalized radical group satisfy the minimal condition, then the group is Chernikov, which follows from a theorem due to V. P. Shunkov and O. H. Kegel and B. A. F. Wehrfritz (see [**134**, Theorem 5.8], for example). Here we shall not consider the influence of the abelian subgroups on the properties of the entire group to its fullest extent, since the book of S. N. Chernikov [**41**] and the surveys [**35**] and [**275**] cover this topic in some detail. In this chapter, we shall consider the structure of groups whose abelian subgroups satisfy a finiteness condition on ranks.

We begin by considering groups whose abelian subgroups have finite 0-rank. We shall require a very useful result concerned with the structure of groups whose Hirsch–Plotkin radical has finite 0-rank, a result which has been obtained (in more general form) by D. I. Zaitsev [**271**] and J. C. Lennox and D. J. S. Robinson [**173**, Corollary, Theorem H].

Ranks of Groups: The Tools, Characteristics, and Restrictions
By Martyn R. Dixon, Leonid A. Kurdachenko and Igor Ya Subbotin
Copyright © 2017 John Wiley & Sons, Inc.

The proof given in the latter paper uses homological methods, so we use here the group theoretical approach of Zaitsev.

9.1. The Existence of Supplements to the Hirsch–Plotkin Radical

In this section we generalize some of the results from Section 8.2 and begin with some new terminology. To this end, let n be a natural number. We say that the group G *splits n-conjugately over its normal subgroup H* if H has a complement in G and there are exactly n conjugacy classes of complements to H in G. In particular, if $n = 1$, then G *splits conjugately over H*. We start with the following very useful lemma of D. I. Zaitsev [271].

9.1.1. LEMMA. *Let G be a group and let A, B be normal subgroups of G such that $A \leq B$. Suppose that*
(i) *A is abelian and $C_A(B) = 1$;*
(ii) *B splits n-conjugately over A, for some $n \in \mathbb{N}$.*
Then G contains a subgroup L such that $B \leq L$, $|G : L| \leq n$ and L splits k-conjugately over A for some $k \leq n$. Furthermore, if A has a complement in G, then we may let $L = G$.

PROOF. Let $C_1, \ldots, C_n$ be subgroups of G such that C_j, C_m are not conjugate in B, whenever $j \neq m$ and every complement to A in B is conjugate to some C_j, for $1 \leq j \leq n$. Let $\mathcal{L}_j = \{C_j^x | x \in B\}$, for $1 \leq j \leq n$. Since $B \triangleleft G$ we have $\mathcal{L}_j^g = \{C_j^{xg} | x \in B\} = \mathcal{L}_m$, for some m. This implies that there is a homomorphism from G to the permutation group over the set $\{\mathcal{L}_j | 1 \leq j \leq n\}$. Let $L_j = \{g \in G | \mathcal{L}_j^g = \mathcal{L}_j\}$, a subgroup of G and clearly $|G : L_j| \leq n$, for $1 \leq j \leq n$. Furthermore,

$$A \cap N_G(C_j) = A \cap C_G(C_j) = 1.$$

If $g \in L_j$, then $\mathcal{L}_j^g = \mathcal{L}_j$, so $C_j^g \in \mathcal{L}_j$ and $C_j^g = C_j^b$, for some $b \in B$. Hence $gb^{-1} \in N_G(C_j)$ so that $g \in BN_G(C_j)$ and hence $L_j = BN_G(C_j)$. On the other hand, $B = AC_j$ so that $L_j = AC_j N_G(C_j) = AN_G(C_j)$. Hence

$$L_j = A \rtimes N_G(C_j).$$

Choose d such that $|G : L_d| \leq |G : L_j|$, for all j such that $1 \leq j \leq n$ and let $L = L_d$. Then $|G : L| \leq n$. Let S be a complement to A in L. Since $L = AS$ and $A \leq B \leq L$ we have $B = B \cap AS = A(B \cap S)$. Then there is a natural number m such that $1 \leq m \leq n$ and $(B \cap S)^u = C_m$, for some $u \in B$. Then $u = va$, where $a \in A, v \in B \cap S$, so $(B \cap S)^u = (B \cap S)^a$. Clearly, $B \cap S \triangleleft S$ so $S \leq N_G(B \cap S)$ and hence

$$S^a \leq N_G(B \cap S)^a = N_G((B \cap S)^a) = N_G(C_m).$$

Therefore, since $A \leq L$, we have

$$L = L^a = (A \rtimes S)^a = A \rtimes S^a \leq A \rtimes N_G(C_m) = L_m.$$

Since $|G : L| \leq |G : L_m|$ we deduce that $L = L_m$. Since

$$L = A \rtimes S = (A \rtimes S)^a = A \rtimes S^a = A \rtimes N_G(C_m)$$

we obtain

$$N_G(C_m) = N_G(C_m) \cap L = (N_G(C_m) \cap A) \rtimes S^a = S^a$$

and hence every complement to A in L is conjugate to a subgroup of the form $N_G(C_m)$. It follows that L splits k-conjugately over A for some $k \leq n$.

Finally, suppose that $G = A \rtimes K$, for some subgroup K. Then $B = A \rtimes (B \cap K)$ and we may set $B \cap K = C_j$, for some j. Then $C_j \lhd K$ and hence

$$G = A \rtimes K \leq A \rtimes N_G(C_j) = L_j$$

and we may set $L = G$.	□

The next proposition will be useful at this juncture and later.

9.1.2. PROPOSITION (Heineken and Kurdachenko [113]). *Let G be a nilpotent group and suppose that $\boldsymbol{Tor}\,(G)$ is bounded. Then G contains a torsion-free normal subgroup A such that G/A is bounded.*

PROOF. Let $T = \boldsymbol{Tor}\,(G)$, let b be a natural number such that $T^b = 1$ and note that Proposition 1.2.11 implies that G/T is torsion-free. Corollary 1.2.9 shows that the terms of the upper central series of G/T are also torsion-free so there is a series of normal subgroups

$$T = Z_0 \leq Z_1 \leq \cdots \leq Z_k = G$$

such that $Z_{j+1}/Z_j \leq \zeta(G/Z_j)$ is torsion-free, for $0 \leq j \leq k - 1$. We use induction on k and first discuss the case $k = 1$, so G/T is abelian. Then G has a central series

$$1 = C_0 \leq C_1 \leq \cdots \leq C_t = T \leq G,$$

where $C_{j+1}/C_j = (T/C_j) \cap \zeta(G/C_j)$, for $0 \leq j \leq t - 1$ and we use induction on t to prove that G contains a subgroup of the type required. If $t = 1$, then $T \leq \zeta(G)$, $G' \leq T$ and, for $g \in G \setminus T$, we consider $\sigma_g : G \longrightarrow G$ defined by $\sigma_g(x) = [x, g]$. By Lemma 3.2.5, σ_g is an endomorphism of G with kernel $C_G(g)$ and image $[G, g] \leq T$. Then $G/C_G(g) \cong [G, g] \leq T$ and it follows that $(G/C_G(g))^b = 1$. However, $\bigcap_{g \in G \setminus T} C_G(g) = \zeta(G)$ and Remak's Theorem gives us the embedding

$$G/\zeta(G) \longrightarrow \underset{g \in G \setminus T}{\mathrm{Cr}}\, G/C_G(g).$$

Hence $G/\zeta(G)$ is bounded, $B = \zeta(G)^b$ is torsion-free and G/B is also bounded.

Suppose now that $t > 1$ and that G/C_1 contains a torsion-free normal subgroup B_1/C_1 with bounded factor group G/B_1. Now B_1/C_1 is abelian and $\mathbf{Tor}\,(B_1) = C_1 \leq \zeta(B_1)$. This is the situation of the preceding argument, so B_1 contains a torsion-free normal subgroup E with $(B_1/E)^d = 1$, for some $d \in \mathbb{N}$. Then $B_1^d \leq E$ and B_1^d is a G-invariant torsion-free subgroup. Furthermore, G/B_1^d is bounded.

Finally assume that $k > 1$ and that Z_{k-1} contains a torsion-free normal subgroup B_2 such that $(Z_{k-1}/B_2)^w = 1$, for some $w \in \mathbb{N}$. Then $B_3 = Z_{k-1}^w \leq B_2$ and B_3 is a G-invariant torsion-free subgroup. Also $\mathbf{Tor}\,(G/B_3)$ is bounded and $(G/B_3)/(Z_{k-1}/B_3)$ is abelian and torsion-free. By the induction hypothesis, it follows that there is a torsion-free normal subgroup A/B_3 of G/B_3 such that $(G/B_3)/(A/B_3) \cong G/A$ is bounded and it is clear that A is torsion-free. This completes the proof. $\square$

It is a well-known fact (which we ask the reader to prove below) that a Chernikov group contains no proper subgroup isomorphic to itself. This is a property enjoyed by groups in certain other group theoretical classes and we give some examples below.

9.1.3. LEMMA. *Let G be a Chernikov group and let B be a subgroup of G. Then $B \cong G$ if and only if $G = B$.*

EXERCISE 9.1. *Prove Lemma 9.1.3.*

9.1.4. COROLLARY. *Let G be a periodic locally nilpotent group of finite section rank and let B be a subgroup of G. Then $B \cong G$ if and only if $G = B$.*

PROOF. Assume that $B \cong G$. Then $\Pi(B) = \Pi(G)$ and we let $p \in \Pi(G)$. Let P be the Sylow p-subgroup of G and let V be the Sylow p-subgroup of B. Then P and V are isomorphic and moreover $V \leq P$, since P is uniquely determined by p. Theorem 4.1.3 shows that P is Chernikov and Lemma 9.1.3 implies that $P = V$. Since this is true for each prime p, we have $G = B$. $\square$

We next prove a result due to V. S. Charin [**33**].

9.1.5. PROPOSITION. *Let A be a torsion-free abelian group of finite 0-rank and let B be a subgroup of A. If A and B are isomorphic, then A/B is finite.*

PROOF. Let $\phi : A \longrightarrow B$ be an isomorphism. Since A has finite 0-rank it contains a finitely generated subgroup L such that A/L is periodic. Let $Y = \phi(L)$ and note that $A/L \cong B/Y$. Since $\mathbf{r}_0(L) = \mathbf{r}_0(Y)$ it follows that A/Y is periodic. Furthermore, $L/(L \cap Y) \cong LY/Y$ and $Y/(Y \cap L) \cong YL/L$ are finite. Let W/L and W_1/Y be the divisible parts of A/L and B/Y, respectively. Since LY/Y is finite we have

$$W_1/Y \cong (W_1/Y)(LY/Y)/(LY/Y) \cong W_1 L/LY$$

and since LY/L is finite we have, similarly,

$$W/L \cong WY/LY.$$

Since $W_1/Y \cong W/L$ it follows that $W_1 L/LY \cong WY/LY$ and it follows that W_1/Y is the divisible part of B/Y.

Now set $\rho = \Pi(A/Y) \cup \Pi(LY/Y) \cup \Pi(LY/L)$ and let $q \in \rho$. Let Q/L be the Sylow q-subgroup of A/L and let Q_1/Y be the Sylow q-subgroup of B/Y. Then

$$Q_1/Y \cong (Q_1/Y)(LY/Y)/(LY/Y) \cong Q_1 L/LY$$

and similarly we have

$$Q/L \cong QY/LY.$$

Since $Q_1/Y \cong Q/L$ we have $Q_1 L/LY = QY/LY$. However, Q_1/Y and Q/L are Chernikov, so Q_1/Y is the Sylow q-subgroup of A/Y. Hence the Sylow ρ-subgroup of B/Y is the Sylow ρ-subgroup of A/Y. Consequently, $\Pi((A/Y)/(B/Y))$ is finite and $(A/Y)/(B/Y)$ is Chernikov, since B/Y has finite section rank. However, the divisible part of A/Y lies in B/Y, so $(A/Y)/(B/Y) \cong A/B$ is finite. This completes the proof. $\qquad \square$

We note that this proof is somewhat simpler than that given in [33].

9.1.6. COROLLARY. *Let A be an abelian group of finite section rank and let B be a subgroup of A. If $A \cong B$ then A/B is finite.*

PROOF. Let $T = \mathbf{Tor}\,(A)$ and $U = \mathbf{Tor}\,(B)$, so that T, U are isomorphic. Corollary 9.1.4 shows that $T = U$. Since $A \cong B$, the torsion-free groups A/T and B/T are isomorphic and $B/T \leq A/T$. By Proposition 3.2.3, A/T has finite 0-rank and an application of Proposition 9.1.5 shows that B/T has finite index in A/T. Since $A/B \cong (A/T)/(B/T)$ the result follows. $\qquad \square$

Next we obtain a further result of D. I. Zaitsev [271].

9.1.7. LEMMA. *Let G be a group and let A be a torsion-free normal abelian subgroup. Suppose that*

(i) A *is rationally G-irreducible of finite 0-rank;*
(ii) G/A *is nilpotent.*

Then G is nilpotent-by-finite or G contains a subgroup L of finite index such that L splits n-conjugately over A for some natural number n.

PROOF. Let $C = C_G(A)$ so that $A \leq \zeta(C)$ and clearly C is nilpotent. If G/C is finite, then G is nilpotent-by-finite so we may suppose that G/C is infinite. Let $T/C = \mathbf{Tor}\,(G/C)$. Since $\mathbf{r}_0(A)$ is finite, A is isomorphic to a subgroup of $\mathbb{Q} \times \cdots \times \mathbb{Q}$ and T/C can be thought of as a subgroup of $GL_n(\mathbb{Q})$, for some $n \in \mathbb{N}$. Hence T/C is finite, by Proposition 1.4.2, so G/C is not periodic. Because G/C is nilpotent, Proposition 9.1.2 implies that it contains a (nontrivial) torsion-free normal subgroup D/C such that G/D is bounded. Then there is a nontrivial element $gC \in D/C \cap \zeta(G/C) \neq 1$. The mapping $\sigma_g : A \longrightarrow A$ defined by $\sigma_g(a) = [a, g]$ is an endomorphism with kernel $C_A(g)$ and image $[A, g]$, so $A/C_A(g) \cong [A, g]$. Since $g \notin C$, it follows that $A/C_A(g)$ is nontrivial and torsion-free. Let $c \in C_A(g)$ and $x \in G$. By the choice of g we have $g^x = gu$, where $u \in C$. Then

$$1 = [c, g] = [c, g]^x = [c^x, g^x] = [c^x, gu] = [c^x, u][c^x, g]^u = [c^x, g]^u.$$

Hence $c^x \in C_A(g)$, so $C_A(g)$ is a G-invariant subgroup of A. If $C_A(g) \neq 1$, then $A/C_A(g)$ is periodic, as A is G-rationally irreducible, which contradicts our earlier assertion. Hence $C_A(g) = 1$ and $A \cong [A, g]$. By Proposition 9.1.5, $|A : [A, g]|$ is finite. Let $\{a_1, \ldots, a_m\}$ be a transversal to $[A, g]$ in A.

Let $H = A \rtimes \langle g \rangle$. If $y \in H$ and $H = A \rtimes \langle y \rangle$, then $y = ga$ for some element $a \in A$. Then $a = ba_j$ for some $b \in [A, g]$ and j such that $1 \leq j \leq m$, so $y = gba_j$. Since $b \in [A, g]$, there exists $b_1 \in A$ such that $b = [g, b_1]$ and we have

$$y = gba_j = g[g, b_1]a_j = b_1^{-1}gb_1a_j = b_1^{-1}(ga_j)b_1.$$

Therefore, $\langle y \rangle$ and $\langle ga_j \rangle$ are conjugate and it follows that H splits m_0-conjugately over A for some $m_0 \leq m$.

Since G/A is nilpotent, $H = A \rtimes \langle g \rangle$ is subnormal in G and we let

$$H = H_0 \triangleleft H_1 \triangleleft \ldots \triangleleft H_k = G$$

be a subnormal series, where each of the factors H_{j+1}/H_j is nontrivial. Using Lemma 9.1.1 we see that H_1 contains a subgroup U such that $H \leq U$, $|H_1 : U|$ is finite and U splits t-conjugately over A for some natural number $t \leq m_0$. Since $|H_1 : U|$ is finite, U contains an H_1-invariant subgroup U_1 of finite index in H_1 and, as $H_1 \triangleleft H_2$, we have

$U_1^x \lhd H_1$, for all $x \in H_2$. Let $|H_1 : U_1| = r$ and let $U_2 = \bigcap_{x \in H_2} U_1^x$. Then $U_2 \lhd H_2$ and Remak's Theorem gives the embedding

$$H_1/U_2 \longrightarrow \underset{x \in H_2}{\mathrm{Cr}} \; H_1/U_1^x.$$

Since $|H_1 : U_1^x| = r$, for each $x \in H_2$ we have $H_1^r \leq U_2$. Hence $\langle g_1 \rangle = \langle g \rangle \cap U_2$ has index at most r in $\langle g \rangle$. We have $U = A \rtimes E$, for some subgroup E. Since $A \leq U_2 \leq U$ it follows that $U_2 = A \rtimes E_1$, where $E_1 = E \cap U_2$ and $A \rtimes \langle g_1 \rangle = A(\langle g \rangle \cap U_2) = A\langle g \rangle \cap U_2$. Since $H = A\langle g \rangle \lhd H_1$, it follows that $A\langle g_1 \rangle \lhd U_2$ and, since gC has infinite order, so does $g_1 C$. Repeating the previous argument for the element g_1 instead of g, we deduce that $A \times \langle g_1 \rangle$ splits m_1-conjugately over A, for some $m_1 \in \mathbb{N}$. Lemma 9.1.1 shows that U_2 splits s-conjugately over A for some $s \leq m_1$.

Repeating this argument a finite number of times we find that G contains a subgroup L of finite index such that L splits n-conjugately over A, for some $n \in \mathbb{N}$, as required.   □

We need only the following particular case of a theorem due to D. I. Zaitsev [**271**] and J. C. Lennox and D. J. S. Robinson [**173**, Corollary to Theorem H].

9.1.8. THEOREM. *Let G be a group and let A be a torsion-free nilpotent normal subgroup of G. Suppose that A has finite 0-rank and G/A is abelian-by-finite. Then either G is almost nilpotent or G contains a nilpotent subgroup S such that $|G : AS|$ is finite.*

PROOF. Let H be a subgroup of G such that $|G : H|$ is finite. Since A is torsion-free nilpotent of finite 0-rank there is a finite series of H-invariant subgroups

$$1 = B_0 \leq B_1 \leq \cdots \leq B_n = A$$

such that B_{j+1}/B_j is torsion-free abelian of finite 0-rank, for $0 \leq j \leq n - 1$. We may refine this series to one in which the torsion-free factors are rationally H-irreducible, so without loss of generality we may assume that our series has these properties.

If U is a subgroup of H such that $|H : U|$ is finite, then we can construct a refinement

$$1 = C_0 \leq C_1 \leq \cdots \leq C_k = A$$

of this series whose factors are torsion-free and rationally U-irreducible. Since A has finite 0-rank, this process must terminate. Hence G contains a subgroup L of finite index with the property that A has a finite series of L-invariant subgroups

$$1 = D_0 \leq D_1 \leq \cdots \leq D_m = A$$

such that D_{j+1}/D_j is torsion-free abelian of finite 0-rank and rationally V-irreducible for each subgroup V of L having finite index in L, for $0 \leq j \leq m - 1$. We now proceed by induction on m. If $m = 1$, then Lemma 9.1.7 proves that L contains a subgroup E of finite index such that either E is nilpotent or $E = A \rtimes K$ for some subgroup K. Since G/A is nilpotent, K is nilpotent and the result holds for $m = 1$.

Suppose now that $m > 1$ and that we have already proved that L/D_1 contains a subgroup E/D_1 of finite index such that either E/D_1 is nilpotent or contains a nilpotent subgroup K/D_1 such that $E/D_1 = (A/D_1)(K/D_1)$. In the latter case, it follows that every K-invariant subgroup of D_1 is also E-invariant, since A is nilpotent. The definition of L implies that D_1 is rationally E-irreducible. Lemma 9.1.7 shows that either E (and hence G) is nilpotent-by-finite or K contains a nilpotent subgroup S such that SD_1 has finite index in K. Then AS has finite index in E and hence in G. This completes the proof. $\square$

9.2. Groups Whose Locally Minimax Subgroups Have Finite Rank

As we already saw in Chapter 8, a finitely generated soluble-by-finite group with finite section rank is minimax. This shows that locally minimax subgroups play a special role in groups satisfying finite rank conditions. Clearly every abelian group is locally minimax and we have already noted that the family of all abelian subgroups plays an influential role on the structure of a group. In this section we discuss groups whose abelian subgroups satisfy some finiteness condition on their rank. We start with groups whose abelian subgroups have finite 0-rank and begin with an interesting result showing that, for certain generalized radical groups, if the abelian subgroups have finite 0-rank, then the 0-rank of all abelian subgroups is bounded, which in turn forces the entire group to have finite 0-rank. Our first result appeared in [67].

9.2.1. THEOREM. *Let G be a generalized radical group and suppose that $\mathbf{Tor}\,(G) = 1$.*

(i) *If every abelian subgroup of G has finite 0-rank, then there is a natural number r such that every abelian subgroup has finite 0-rank at most r;*

(ii) *If r is a natural number such that every abelian subgroup of G has finite 0-rank at most r, then G has finite 0-rank and there is a function f_7 such that $\mathbf{r}_0(G) \leq f_7(r) = r(r+1)$.*

PROOF. (i) Let L be the Hirsch–Plotkin radical of G. By Proposition 1.2.11 the set of elements of finite order in a locally nilpotent group is a characteristic subgroup and the hypotheses imply that $\mathbf{Tor}\,(L) = 1$, so L is torsion-free. By Theorem 2.3.3, L is nilpotent and if B is a maximal normal abelian subgroup of L, then $\mathbf{r}_0(B)$ is finite, say $\mathbf{r}_0(B) = b$. In addition, $\mathbf{r}_0(L) \leq t = b(b+1)/2$. It follows that L has a finite series of G-invariant subgroups

$$1 = A_0 \leq A_1 \leq \cdots \leq A_n = L,$$

every factor of which is torsion-free, central in L and G-rationally irreducible. Furthermore $n \leq t$. Let $C_j = C_G(A_{j+1}/A_j)$, for $0 \leq j \leq n-1$. We can view $G_j = G/C_j$ as an irreducible subgroup of $GL_{k(j)}(\mathbb{Q})$, where $k(j) = \mathbf{r}_0(A_{j+1}/A_j) \leq t$. By Corollary 1.4.12 G_j is abelian-by-finite, for $1 \leq j \leq n$.

Let $C = \bigcap_{j=0}^{n-1} C_j$. By Remak's Theorem there is an embedding

$$G/C \longrightarrow G/C_1 \times \cdots \times G/C_{n-1}$$

so that G/C is abelian-by-finite. By Proposition 4.2.8 $C = L$ and hence G/L contains a normal abelian subgroup V/L such that G/V is finite.

In V/L we choose a free abelian subgroup S/L such that V/S is periodic. By Theorem 9.1.8 either S contains a nilpotent subgroup U of finite index or S contains a nilpotent subgroup W such that LW has finite index in S.

In the former case,

$$\mathbf{r}_0(V) = \mathbf{r}_0(S) = \mathbf{r}_0(U).$$

Let A be a maximal normal abelian subgroup of U and let $\mathbf{r}_0(A) = c$. Using Theorem 2.3.3 we see that U has finite 0-rank and $\mathbf{r}_0(U) \leq c(c+1)/2$. It follows that V has finite 0-rank and $\mathbf{r}_0(V) \leq c(c+1)/2$. Since G/V is finite, $\mathbf{r}_0(V) = \mathbf{r}_0(G) \leq c(c+1)/2$.

In the latter case, we have

$$\mathbf{r}_0(V/L) = \mathbf{r}_0(S/L) = \mathbf{r}_0(WL/L) = \mathbf{r}_0(W/(W \cap L)) \leq \mathbf{r}_0(W).$$

Let E be a maximal normal abelian subgroup of W and suppose that $\mathbf{r}_0(E) = e$. By Theorem 2.3.3 we see that W has finite 0-rank and $\mathbf{r}_0(W) \leq e(e+1)/2$. Hence V has finite 0-rank at most $b(b+1)/2 + e(e+1)/2$ and since G/V is finite, we have $\mathbf{r}_0(V) = \mathbf{r}_0(G) \leq b(b+1)/2 + e(e+1)/2$.

(ii) We observe that $b, c, e \leq r$ and hence

$$\mathbf{r}_0(G) \leq r(r+1)/2 + r(r+1)/2 = r(r+1) = f_7(r).$$

This completes the proof of the theorem. $\square$

For the consideration of groups G with $\mathbf{Tor}\,(G) \neq 1$ we need some further restrictions since we need to transfer information concerning G to its factor group $G/\mathbf{Tor}\,(G)$.

9.2.2. LEMMA. *Let G be a group and suppose that every finitely generated subgroup of G is a soluble-by-finite minimax group. If T is a normal periodic subgroup of G and U/T is a torsion-free abelian group of finite 0-rank, then U contains a torsion-free abelian subgroup V such that $\mathbf{r}_0(V) = \mathbf{r}_0(U/T)$.*

EXERCISE 9.2. *Prove Lemma 9.2.2.*

9.2.3. COROLLARY. *Let G be a group and suppose that every finitely generated subgroup of G is a soluble-by-finite minimax group. If T is a normal periodic subgroup of G and A/T is a free abelian group of infinite 0-rank, then A contains a free abelian subgroup C of infinite 0-rank.*

PROOF. Without loss of generality we may suppose that A/T has countable 0-rank and we let $A/T = \underset{n \in \mathbb{N}}{\mathrm{Dr}} \langle a_n T \rangle$, where $\langle a_n T \rangle$ is infinite cyclic, for $n \in \mathbb{N}$. Let $U_n = \langle a_j | 1 \leq j \leq n \rangle$. By Lemma 9.2.2 we see that U_n contains a free abelian subgroup V_n such that $\mathbf{r}_0(V_n) = n$. The proof of Lemma 9.2.2 shows that U_{n+1} is centre-by-finite and hence there is a natural number s such that $b_{n+1} = a_{n+1}^s \in C_{U_n}(V_n)$. Then $V_{n+1} = V_n \rtimes \langle b_{n+1} \rangle$ is a free abelian subgroup such that $\mathbf{r}_0(V_{n+1}) = n+1$. In this way we may construct an infinite ascending chain of free abelian subgroups,

$$V_1 \leq V_2 \leq \cdots \leq V_n \leq V_{n+1} \leq \cdots,$$

such that $\mathbf{r}_0(V_n) = n$, for all $n \in \mathbb{N}$. Then $V = \bigcup_{n \in \mathbb{N}} V_n$ is free abelian of infinite 0-rank. $\square$

We may now deduce the following result from [**67**].

9.2.4. THEOREM. *Let G be a generalized radical group and suppose that every finitely generated subgroup of G is minimax. If every abelian subgroup of G has finite 0-rank, then G has finite 0-rank.*

PROOF. Let $T = \mathbf{Tor}\,(G)$ and let U/T be a torsion-free abelian subgroup. From Corollary 9.2.3 it follows that U/T has finite 0-rank, otherwise there is a torsion-free abelian subgroup V of G of infinite

0-rank. Hence each abelian subgroup of G/T has finite 0-rank and we may apply Theorem 9.2.1 to deduce that G has finite 0-rank.   □

This theorem allows us to obtain further results using different rank restrictions, as is evident from the following corollaries.

9.2.5. COROLLARY. *Let G be a generalized radical group and suppose that every finitely generated subgroup of G is minimax. If every abelian subgroup of G has finite section p-rank for some prime p, then G has finite section p-rank.*

PROOF. By Proposition 3.2.3 every abelian subgroup of G has finite 0-rank. Theorem 9.2.4 shows that G has finite 0-rank and Corollary 6.2.3 shows that $G/\mathbf{Tor}\,(G)$ has finite special rank, so has finite section p-rank. By Theorem 3.3.5 $\mathbf{Tor}\,(G)$ has finite section p-rank. Since $\mathbf{sr}_p(G/\mathbf{Tor}\,(G))$ and $\mathbf{sr}_p(\mathbf{Tor}\,(G))$ are finite, we see that G has finite section p-rank.   □

9.2.6. COROLLARY. *Let G be a generalized radical group and suppose that every finitely generated subgroup of G is minimax. If every abelian subgroup of G has finite section rank, then G has finite section rank.*

In particular, we obtain the following classical result.

9.2.7. COROLLARY. *Let G be a locally finite group. If every abelian subgroup of G has finite section rank, then G has finite section rank.*

A group G is called *nearly radical* if G has an ascending series whose factors are locally nilpotent or finite. In the paper [**17**], the term "generalized radical group" was used for such groups. However we use this term for a much wider class of groups.

9.2.8. LEMMA. *Let G be a nearly radical group. If G has finite section rank, then G is radical-by-finite.*

EXERCISE 9.3. *Prove Lemma 9.2.8.*

9.2.9. COROLLARY. *Let G be a finitely generated nearly radical group. If G has finite section rank, then G is a soluble-by-finite minimax group.*

PROOF. Lemma 9.2.8 implies that G contains a normal radical subgroup R and Proposition 1.2.13 implies that R is also finitely generated. Theorem 8.4.14 shows that R is a soluble minimax group.   □

9.2.10. COROLLARY. *Let G be a nearly radical group. If every abelian subgroup of G has finite section rank, then G is radical-by-finite and has finite section rank.*

PROOF. By Corollary 9.2.9 every finitely generated subgroup of G is minimax and Corollary 9.2.6 gives the result. □

We next deduce a result of R. Baer and H. Heineken [**17**].

9.2.11. COROLLARY. *Let G be a radical group. If every abelian subgroup of G has finite 0-rank and finite p-rank, for all primes p, then G has finite section rank.*

PROOF. By Proposition 3.2.3 and Lemma 3.1.3, every abelian subgroup of G has finite section rank and we can apply Corollary 9.2.10 to deduce the result. □

In the case of groups with bounded section rank, we can obtain a more general result, whose proof we now discuss.

9.2.12. LEMMA. *Let G be a periodic locally nilpotent group. If every abelian subgroup of G has finite bounded section rank, then G has finite bounded section rank.*

EXERCISE 9.4. *Prove Lemma 9.2.12.*

The following lemma was proved in [**127**] in the case of the special rank. The proof is the same in the case of bounded section rank.

9.2.13. LEMMA. *Let G be a periodic group and let A be a locally nilpotent normal subgroup of G such that G/A is abelian. If every abelian subgroup of G has finite bounded section rank, then G/A has finite bounded section rank.*

PROOF. By Corollary 9.2.7, G has finite section rank, so $G/A = \underset{p \in \Pi(G/A)}{\mathrm{Dr}} G_p/A$, where G_p/A is the Chernikov Sylow p-subgroup of G/A. Let $S/A = \underset{p \in \Pi(G/A)}{\mathrm{Dr}} S_p/A$, where $S_p/A = \Omega_1(G_p/A)$. By Lemma 3.1.3, $\mathrm{sr}_p(G_p/A) = \log_p(|S_p/A|)$ and it is therefore enough to prove that S has finite bounded section rank. Every Sylow p-subgroup of S lies in the normal subgroup S_p, which is almost locally nilpotent. It follows that the Sylow p-subgroups of S are conjugate (see [**49**, Proposition 2.2.4], for example).

Let $\Pi(S/A) = \{p_n | n \in \mathbb{N}\}$ and, relabelling, let S_n/A now denote the Sylow p_n-subgroup of S/A. We shall define a sequence of subgroups

$$S = H_0 \geq H_1 \geq \cdots \geq H_n \geq H_{n+1} \geq \cdots$$

such that $S = H_n A$, for each $n \in \mathbb{N}$. Suppose that H_n has already been chosen and let P_{n+1} be a Sylow p_{n+1}-subgroup of H_n. Let

$$H_{n+1} = H_n \cap N_S(P_{n+1}).$$

Then P_{n+1} is a Sylow p_{n+1}-subgroup of $H_n \cap S_{n+1}$ and $H_n \cap S_{n+1}$ is a normal subgroup of H_n. If $x \in H_n$, then P_{n+1}^x is a Sylow p_{n+1}-subgroup of $H_n \cap S_{n+1}$. Since the Sylow p-subgroups of S are conjugate, for each prime p, there exists $y \in H_n \cap S_{n+1}$ such that $P_{n+1}^x = P_{n+1}^y$ and hence $xy^{-1} \in H_n \cap N_S(P_{n+1})$. Thus $x \in (H_n \cap N_S(P_{n+1}))(H_n \cap S_{n+1})$, so

$$H_n = H_{n+1}(H_n \cap S_{n+1}).$$

Since $(H_n \cap S_{n+1})/(H_n \cap A)$ is a finite p_{n+1}-subgroup, $H_n \cap S_{n+1} = P_{n+1}(H_n \cap A)$. Hence $H_n = H_{n+1}(H_n \cap A)$ and $S = H_n A = H_{n+1} A$, as required.

We observed above that G has finite section rank and Corollary 3.2.21 shows that P_n is Chernikov. Hence the chain

$$H_1 \cap P_n \geq H_2 \cap P_n \geq \cdots \geq H_k \cap P_n \geq \ldots$$

terminates in finitely many steps and there is a natural number $j(n) \geq n$ such that

$$H_{j(n)} \cap P_n = H_{j(n)+k} \cap P_n, \text{ for all } k \in \mathbb{N}.$$

If $k \geq j(n)$ and P is an arbitrary p_n-subgroup of H_k, then P normalizes P_n since $H_k \leq H_{j(n)} \leq H_n$. Hence $P \leq H_k \cap P_n = T_n$ and T_n is the unique Sylow p_n-subgroup of H_k. Consequently, $[T_n, T_m] = 1$, whenever $n \neq m$ and we deduce that

$$T = \langle T_n | n \in \mathbb{N} \rangle = \underset{n \in \mathbb{N}}{\mathrm{Dr}}\, T_n.$$

Finally, $S_n = H_k A \cap S_n = (H_k \cap S_n)A$ and if $k \geq j(n)$, then we also have $H_k \cap S_n = T_n(H_k \cap A)$. Therefore $S_n = T_n A$ and since $S = \langle S_n | n \in \mathbb{N} \rangle A$ we have $S = TA$. Clearly T is locally nilpotent and Lemma 9.2.12 shows that T has finite bounded section rank. Hence $S/A = TA/A \cong T/(T \cap A)$ has finite bounded section rank, as required. $\quad\square$

9.2.14. THEOREM. *Let G be a locally finite group. If every abelian subgroup of G has finite bounded section rank, then G has finite bounded section rank.*

PROOF. By Corollary 4.4.4, G contains normal subgroups $L \leq A$ such that L is hypercentral, A/L is abelian and G/A is finite. Lemmas 9.2.12 and 9.2.13 show that A has finite bounded section rank and hence the same is true of G, since $|G : A|$ is finite. $\quad\square$

We next obtain a classic result originally due to V. P. Shunkov [242].

9.2.15. COROLLARY. *Let G be a locally finite group. If every abelian subgroup of G has finite special rank, then G has finite special rank.*

PROOF. Let A be an arbitrary abelian subgroup of G. By Proposition 6.1.6 A has finite bounded section rank. Using Theorem 9.2.14 we see that G has finite bounded section rank and Theorem 6.2.18 implies that G has finite special rank. $\square$

The result above generalized the following classical result of Yu. M. Gorchakov [**94**].

9.2.16. COROLLARY. *Let G be a periodic locally soluble group. If every abelian subgroup of G has finite special rank, then G has finite special rank.*

We turn our attention to radical groups next.

9.2.17. COROLLARY. *Let G be a radical group. If every abelian subgroup of G has finite bounded section rank, then G has finite bounded section rank.*

PROOF. By Corollary 9.2.10, G has finite section rank. Theorem 4.2.1 implies that G contains normal subgroups $T \leq L \leq K$ such that T is locally finite, L/T is torsion-free nilpotent of finite 0-rank, K/L is finitely generated torsion-free abelian and G/K is finite. Theorem 6.2.2 shows that K/T has finite bounded section rank and since G/K is finite, this means that G/T has finite bounded section rank. Since T is a periodic radical group, it is locally finite and Theorem 9.2.14 implies that T has finite bounded section rank. Thus T and G/T have bounded section rank, as does G. $\square$

This now allows us to obtain the desired result for generalized radical groups.

9.2.18. THEOREM. *Let G be a generalized radical group. If every abelian subgroup of G has finite bounded section rank, then G has finite bounded section rank. Furthermore, G is radical-by-finite.*

PROOF. Let R be the maximal normal radical subgroup of G and let L/R be the maximal normal locally finite subgroup of G/R. If A/R is an abelian subgroup of L/R, then A is radical and Corollary 9.2.17 shows that A has finite bounded section rank. Hence A/R has finite bounded section rank and Theorem 9.2.14 implies that L/R has finite bounded section rank. By Corollary 4.4.4, L/R must be radical-by-finite and since R is the maximal normal radical subgroup of G, it follows that L/R is finite. Then Lemma 3.4.1 implies that G/R is finite, so G is radical-by-finite. Corollary 9.2.17 shows that R has finite bounded section rank and hence so does G. □

A specific case of this result appears in [158], from which other results are given later.

9.2.19. COROLLARY. *Let G be a generalized radical group. If every abelian subgroup of G has finite special rank, then G has finite special rank.*

PROOF. Let A be an arbitrary abelian subgroup of G. By Proposition 6.1.6, A has finite bounded section rank and Theorem 9.2.18 implies that G has finite bounded section rank. Theorem 6.2.18 shows that G has finite special rank. □

We also have the following fundamental special case, which is a result of M. I. Kargapolov [127].

9.2.20. COROLLARY. *Let G be a soluble group. If every abelian subgroup of G has finite special rank, then G has finite special rank.*

In Theorem 4.4.6 we considered the influence of locally radical subgroups of finite bounded section rank on the structure of locally generalized radical groups. Now we obtain a considerable strengthening of this result.

A group G is called *locally (soluble and minimax)* if every finitely generated subgroup of G is soluble and minimax.

9.2.21. THEOREM. *Let G be a locally generalized radical group. If every locally (soluble and minimax) subgroup of G has finite bounded section rank, then G has finite bounded section rank.*

PROOF. Theorem 4.4.6 implies that we must prove that every locally radical subgroup of G has finite bounded section rank. Thus without loss of generality we may assume that G is a locally radical group. Let F be a finitely generated subgroup of G, so that F is radical. If A is an abelian subgroup of F, then A is locally minimax and hence has finite special rank. As Corollary 9.2.17 shows this means F has finite special rank. Then Theorem 8.4.14 implies that F is minimax, so that G is locally minimax and hence of finite bounded section rank. □

9.2.22. COROLLARY (Kurdachenko, Otal and Subbotin [158]). *Let G be a locally generalized radical group. If every locally (soluble and minimax) subgroup of G has finite special rank, then G has finite special rank.*

PROOF. Let M be a locally (soluble and minimax) subgroup of G. Theorem 6.2.18 shows that M has finite bounded section rank and hence G has finite bounded section rank, by Theorem 9.2.21. Theorem 6.2.18 can be applied again to deduce that G has finite special rank, as required. $\square$

9.2.23. COROLLARY (Kurdachenko, Otal and Subbotin [158]). *Let G be a locally generalized radical group. If every locally soluble subgroup of G has finite special rank, then G has finite special rank.*

In Theorem 4.2.1 we considered the influence of the locally soluble subgroups with finite section rank on the structure of locally (soluble-by-finite) groups. We obtain a generalization of this result also.

9.2.24. THEOREM (Kurdachenko, Otal and Subbotin [158]). *Let G be a locally nearly radical group. If every locally (soluble and minimax) subgroup of G has finite section rank, then G has finite section rank.*

PROOF. Let F be a finitely generated subgroup of G, so that F is nearly radical and let A be an abelian subgroup of F. Then A is locally minimax and hence has finite section rank, by hypothesis. Then F has finite section rank, by Corollary 9.2.10 and it contains a normal radical subgroup R of finite index. That R is finitely generated follows from Proposition 1.2.13, so Theorem 8.4.14 shows that R is minimax. This argument shows that every finitely generated subgroup of G is minimax, so has finite 0-rank and is soluble-by-finite.

Suppose, for a contradiction, that the 0-ranks of the finitely generated subgroups of G are unbounded. Then there is a family of finitely generated subgroups, $\{F_n | n \in \mathbb{N}\}$, such that

$$\mathbf{r}_0(F_1) < \mathbf{r}_0(F_2) \cdots < \mathbf{r}_0(F_n) < \ldots.$$

Let $K_n = \langle F_1, \ldots, F_n \rangle$, for $n \in \mathbb{N}$. Then K_n is finitely generated and hence contains a normal soluble minimax subgroup R_n of finite index and we have $\mathbf{r}_0(R_n) = \mathbf{r}_0(K_n) \geq \mathbf{r}_0(F_n)$. It follows that

$$\mathbf{r}_0(R_1) < \mathbf{r}_0(R_2) < \cdots < \mathbf{r}_0(R_n) < \ldots.$$

Since $R_n \lhd K_n$ we may form each of the products $R_1 R_2 \ldots R_n$ and, in particular, $R_1 \ldots R_n$ has a finite series of normal subgroups whose factors are soluble minimax. Hence $R_1 \ldots R_n$ is a soluble minimax group, for all $n \in \mathbb{N}$ and $E = \bigcup_{n \in \mathbb{N}} (R_1 \ldots R_n)$ is locally (soluble and minimax). By hypothesis, E has finite section rank and Theorem 4.2.1 shows that E has finite 0-rank. Hence there is a natural number m such that $\mathbf{r}_0(E) < \mathbf{r}_0(R_m)$, but since $R_m \leq E$, this is the desired contradiction. Consequently, there is a natural number k such that $\mathbf{r}_0(F) \leq k$

for every finitely generated subgroup F of G and now Proposition 2.5.2 shows that G has finite 0-rank.

It follows from Theorem 2.4.13 that G contains normal subgroups $T \leq L \leq B \leq G$ such that T is locally finite, L/T is torsion-free nilpotent, B/L is finitely generated torsion-free abelian and G/B is finite. Each Sylow p-subgroup S of T is locally nilpotent and hence locally (soluble and minimax), so has finite section rank and this is true for each prime p. By Corollary 3.2.21 S is Chernikov and Theorem 4.1.3 shows that T has finite section rank. It is clear that $\mathbf{r}_0(G) = \mathbf{r}_0(B) = \mathbf{r}_0(B/T) \geq \mathbf{r}_0(L/T)$ and Corollary 3.2.24 implies that L/T has finite section rank. Also B/L has finite section rank, as does B/T by Lemma 3.2.2. Then G/T, and hence G itself, has finite section rank, again by Lemma 3.2.2, as required.   □

There are a couple of very easy consequences of this result which we state. They are analogous to Corollary 6.2.20.

9.2.25. COROLLARY (Kurdachenko, Otal and Subbotin [**158**]). *Let G be a locally (soluble-by-finite) group. If every locally (soluble and minimax) subgroup of G has finite section rank, then G has finite section rank.*

9.2.26. COROLLARY (Kurdachenko, Otal and Subbotin [**158**]). *Let G be a locally (soluble-by-finite) group. If every locally soluble subgroup of G has finite section rank, then G has finite section rank.*

9.3. Groups Whose Abelian Subgroups Have Finite Rank

In earlier sections we have discussed groups G whose abelian subgroups satisfy some finite rank condition and we saw that for generalized radical groups this condition ensures that the rank of the group G is also finite. However these results cannot be extended to other classes of groups and in particular cannot be extended to locally soluble groups. The corresponding example was constructed by Yu. I. Merzlyakov in the papers [**194, 195**]. (The construction was first expounded in [**194**] and some gaps were later repaired in [**195**].) The example given is of a locally polycyclic group of infinite special rank whose abelian subgroups have finite special rank. Moreover, every elementary abelian section of this group is finite. However, for each prime p and every natural number n this group has a section of order p^n, so it has infinite section p-rank for each prime p. The construction of this example and its properties was described in detail in the book [**130**, 25.3], so we shall not repeat it here.

Thus, on the one hand, we have classes of groups (locally nilpotent groups, locally supersoluble groups, locally finite groups, radical groups, generalized radical groups and others) in which the fact that all abelian subgroups have finite bounded section rank (or finite special rank) implies that the group has finite bounded section rank (or finite special rank). On the other hand, we have the class of locally soluble (and even locally polycyclic) groups and all sorts of extensions of these classes where the abelian subgroups of finite rank exert a much weaker influence on the entire group. Of course this raises the question of the exact boundary and in the final section of this chapter we shall try to approach this border more closely. Our results here are based on the observation that in the group constructed by Merzlyakov the ranks of the chief factors are not bounded.

Let k be a natural number. A group G is called k-*generalized radical* if G has an ascending series of normal subgroups whose factors are locally nilpotent or locally finite of special rank at most k. The group G will be called k-*radical* if G has an ascending series of normal subgroups whose factor groups are locally nilpotent of special rank at most k. Finally a group G is called k-*hyperabelian* if G has an ascending series of normal subgroups whose factors are abelian of special rank at most k.

9.3.1. PROPOSITION. *A group G is k-radical if and only if it is k-hyperabelian.*

EXERCISE 9.5. *Prove Proposition 9.3.1.*

9.3.2. PROPOSITION. *Let k be a natural number and let G be a k-radical group. Then G has normal subgroups $Z \leq K$ such that Z is hypercentral, K/Z is abelian and G/K is isomorphic to a subgroup of a Cartesian product of finite groups of order at most $\mu(k)$. Furthermore, G/K is soluble and $\mathbf{dl}(G/K) \leq \zeta(k)$.*

Here μ is the Maltsev function and ζ is the Zassenhaus function and we refer the reader to Section 1.4 for their definitions.

EXERCISE 9.6. *Prove Proposition 9.3.2.*

9.3.3. COROLLARY. *Let k be a natural number and let G be a locally k-radical group. Then there is a natural number t such that $\delta_t(G)$ is locally nilpotent.*

PROOF. Let $t = \zeta(k) + 1$ and let M be a finite subset of $D = \delta_t(G)$. Then there is a finitely generated subgroup F such that $M \subseteq \delta_t(F)$. By Proposition 9.3.2, $\delta_t(F)$ is hypercentral and a hypercentral group is locally nilpotent, by Corollary 1.2.6. It follows that $\langle M \rangle$ is nilpotent and hence D is locally nilpotent. This completes the proof. □

9.3.4. COROLLARY. *Let k be a natural number and let G be a locally k-radical group. If every abelian subgroup of G has finite special rank, then G has finite special rank.*

PROOF. By Corollary 9.3.3, G is a radical group, so the result follows by Corollary 9.2.19. □

We are now ready to give some more of the results occurring in [**158**].

9.3.5. THEOREM. *Let k be a natural number and let G be a locally k-generalized radical group. If every abelian subgroup of G has finite special rank, then G has finite special rank.*

PROOF. Let L be a locally radical subgroup of G. Then L is locally k-radical and Corollary 9.3.4 implies that L has finite special rank. Theorem 6.2.18 shows that L has finite bounded section rank and Theorem 4.4.6 shows that G too has finite bounded section rank. A further application of Theorem 6.2.18 then shows that G has finite special rank. □

9.3.6. COROLLARY (Kurdachenko, Otal and Subbotin [**158**]). *Let k be a natural number and let G be a group. Suppose that G has an ascending series whose factors are locally k-generalized radical groups. If every abelian subgroup of G has finite special rank, then G has finite special rank.*

PROOF. Let

$$1 = K_0 \leq K_1 \leq \ldots K_\alpha \leq K_{\alpha+1} \leq \ldots K_\gamma = G$$

be an ascending series whose factors are locally k-generalized radical groups. We prove using transfinite induction that G is generalized radical. Since K_1 is locally k-generalized radical, Theorem 9.3.5 implies that K_1 has finite special rank. By Theorem 6.2.18 and Corollary 4.4.4 K_1 is hyperabelian-by-finite and hence K_1 contains a normal subgroup which is either abelian or finite. Furthermore K_1 is generalized radical. Suppose we know that K_α is generalized radical. We next show that for all α, $K_{\alpha+1}/K_\alpha$ is also generalized radical. Let A/K_α be an abelian subgroup of $K_{\alpha+1}/K_\alpha$. Then A is generalized radical and every abelian subgroup of A has finite special rank, so A has finite special rank, by Corollary 9.2.19. Thus every abelian subgroup of $K_{\alpha+1}/K_\alpha$ has finite special rank and Theorem 9.3.5 implies that $K_{\alpha+1}/K_\alpha$ has finite special rank. As with K_1, this means that $K_{\alpha+1}/K_\alpha$ is hyperabelian-by-finite, so $K_{\alpha+1}$ is generalized radical. It follows that G is generalized radical and now Corollary 9.2.19 gives the result.   □

As we saw in Corollary 9.2.15 the finiteness of the special rank of the abelian subgroups in a locally finite group implies the finiteness of the special rank of the entire group. On the other hand, the example of Merzlyakov, mentioned above, shows that in torsion-free groups the situation is quite different. We remark that in this example the 0-ranks of the abelian subgroups are unbounded. This suggests that the reason for the growth of the special rank occurs because of the growth of the 0-rank of the abelian subgroups. The last main result of this chapter concerns this and again appears in [**158**].

9.3.7. THEOREM. *Let G be a locally nearly radical group and suppose that there is a natural number k such that $r_0(A) \leq k$, for all abelian subgroups A of G. If all the abelian subgroups of G have finite section rank, then G has finite section rank and $r_0(G) \leq k(k+1)$.*

PROOF. Let F be a finitely generated subgroup of G, so that F has finite section rank and there is a normal radical subgroup R of finite index, by Corollary 9.2.10. Then R is finitely generated, by Proposition 1.2.13. Theorem 8.4.14 shows that R is minimax, so that F is minimax and soluble-by-finite. Let $T = \mathbf{Tor}\,(F)$ and let U/T be an arbitrary finitely generated torsion-free abelian subgroup of F/T. Lemma 9.2.2 shows that F contains a torsion-free abelian subgroup V such that $r_0(V) = r_0(U/T)$ and it follows from the hypotheses that $r_0(U/T) \leq k$, for each such subgroup U/T. Theorem 9.2.1 implies that $r_0(F) \leq k(k+1)$ and Proposition 2.5.2 shows that G is a generalized radical group such that $r_0(G) \leq k(k+1)$. As we proved above, every finitely generated subgroup of G is minimax, so Corollary 9.2.6 proves that G has finite section rank.   □

We highlight the following special case of this result.

9.3.8. COROLLARY. *Let G be a locally (soluble-by-finite) group and suppose that there is a natural number k such that $r_0(A) \leq k$, for all abelian subgroups A of G. If all the abelian subgroups of G have finite section rank, then G has finite section rank and $r_0(G) \leq k(k+1)$.*

A variation of this result also holds for locally generalized radical groups.

9.3.9. COROLLARY. *Let G be a locally generalized radical group and suppose that there is a natural number k such that $r_0(A) \leq k$, for* *all abelian subgroups A of G. If all the abelian subgroups of G have finite bounded section rank, then G has finite bounded section rank and $r_0(G) \leq k(k+1)$.*

PROOF. Let F be a finitely generated subgroup of G, so that F has finite section rank and contains a normal radical subgroup R of finite index, by Theorem 9.2.18. Then R is finitely generated, by Proposition 1.2.13 and Theorem 8.4.14 shows that R is minimax. Hence F is minimax and soluble-by-finite. Thus G is a locally nearly radical group and Theorem 9.3.7 shows that G has finite section rank and that $r_0(G) \leq k(k+1)$. By Theorem 4.2.1, G is generalized radical and a further application of Theorem 9.2.18 shows that G has finite bounded section rank. □

9.3.10. COROLLARY. *Let G be a locally generalized group and suppose that there is a natural number k such that $r_0(A) \leq k$, for all abelian subgroups A of G. If all the abelian subgroups of G have finite special rank, then G has finite special rank and $r_0(G) \leq k(k+1)$.*

PROOF. Let A be an arbitrary abelian subgroup of G. By Proposition 6.1.6 A has finite bounded section rank and Corollary 9.3.9 shows that G has finite bounded section rank. Theorem 6.2.18 then implies that G has finite special rank. □

We now immediately obtain the following result which appears in [51].

9.3.11. COROLLARY. *Let G be a locally (soluble-by-finite) group and suppose that there is a natural number k such that $r_0(A) \leq k$, for all abelian subgroups A of G. If all the abelian subgroups of G have finite special rank, then G has finite special rank and $r_0(G) \leq k(k+1)$.*

CHAPTER 10

A Brief Discussion of Other Interesting Results

There are numerous problems in group theory which are related to the theory of groups of finite rank. In the previous chapters we have, in a sense, given a "mandatory program" of results which we and others believe are the foundation and first floor of the building called "the theory of groups of finite rank". This building looks very surreal. Its architects do not have a plan, not all floors are complete in the different directions leaving the side branches and they can be connected with other floors, both upper and lower. We now wish to provide a glimpse of the rest of the picture, but do so schematically. More specifically, we now give some results that we believe are important and that are associated with the theme of this book, but do so without providing proofs, whose inclusion would significantly lengthen our exposition. We also give some unsolved problems which might provide some additional impetus in the theory of groups of finite rank.

10.1. Recent Work

In previous chapters, we saw that the class of locally generalized radical groups is a class of groups that is particularly receptive to the methods which prove useful in the theory of groups of finite rank. However there are other broad classes of groups in which it is possible to obtain a structure theory for groups of finite rank. One class of groups where some progress has been made concerns the class of residually finite groups in which the following result, due to A. Lubotzky and A. Mann [**180**], has been obtained. This theorem was proved using the theory of analytic pro-p groups, at the time a relatively new branch of group theory.

10.1.1. THEOREM. *Let G be a residually finite group of finite special rank. Then G is almost locally soluble.*

Using this result it is not difficult to obtain the following interesting corollary.

10.1.2. COROLLARY. *Let G be a periodic locally graded group of finite special rank. Then G is locally finite.*

261
Ranks of Groups: The Tools, Characteristics, and Restrictions
By Martyn R. Dixon, Leonid A. Kurdachenko and Igor Ya Subbotin
Copyright © 2017 John Wiley & Sons, Inc.

Theorem 10.1.1 was generalized by N. S. Chernikov [36], shortly thereafter. He introduced the minimal class of groups $\mathfrak{X}$, containing the class of all periodic locally graded groups and satisfying the additional properties:

(a) If G is a group whose finitely generated subgroups belong to $\mathfrak{X}$, then $G \in \mathfrak{X}$. Thus $\mathrm{L}\mathfrak{X} = \mathfrak{X}$.

(b) If G has a family $\mathcal{S}$ of normal subgroups such that $G/H \in \mathfrak{X}$, for all $H \in \mathcal{S}$, and $\cap_{H \in \mathcal{S}} H = 1$, then $G \in \mathfrak{X}$. Thus $\mathrm{R}\mathfrak{X} = \mathfrak{X}$.

(c) If G is a group having an ascending series of normal subgroups whose factors belong to $\mathfrak{X}$, then $G \in \mathfrak{X}$.

(d) If G is a group having a descending series of normal subgroups whose factors belong to $\mathfrak{X}$, then $G \in \mathfrak{X}$.

Groups in the class $\mathfrak{X}$ will be called *strongly locally graded*. It is quite easy to see that all strongly locally graded groups are locally graded. We note that this use of the terminology is different from that used in [79].

N. S. Chernikov [36] obtained the following result concerning strongly locally graded groups.

10.1.3. THEOREM. *Let G be a strongly locally graded group of finite special rank. Then G is almost locally soluble.*

A quite broad generalization of Theorem 10.1.1 has been obtained in [53] and [56], where certain subclasses of $\mathfrak{X}$ were studied. In these papers the class of groups which are residually of special rank at most r, for some fixed natural number r, is studied. Thus a group G is residually (locally soluble of finite special rank r) if G has a family of normal subgroups $\mathcal{L}$ such that $\cap_{H \in \mathcal{L}} H = 1$ and G/H is locally soluble of finite special rank at most r for all $H \in \mathcal{L}$. The following two results were proved, among others.

10.1.4. THEOREM. *Let G be a group that is residually (locally soluble of special rank r), for some natural number r. Then there are subgroups M, N of G such that $M \lhd N \lhd G$ satisfying*

(i) M *is hyperabelian and locally nilpotent,*

(ii) N/M *is residually (linear of r-bounded degree) and*

(iii) G/N *is soluble of r-bounded derived length.*

For the next theorem we say that the group is residually (locally finite of special rank r) if G has a family $\mathcal{L}$ of normal subgroups such that G/H is locally finite of special rank at most r, for all $H \in \mathcal{L}$ and $\cap_{H \in \mathcal{L}} H = 1$, for some fixed r.

10.1.5. THEOREM. *Let G be a periodic group and suppose that G is residually (locally finite of special rank r) for some natural number r. Then G is locally finite.*

These results are similar to ones obtained by D. Segal [**235**]. We also note that, in general, p-groups which are residually finite need not be locally finite, there being several constructions of groups of this kind. Among these examples we mention the elegant construction due to R. I. Grigorchuk [**98**], which, although simple, is rather sophisticated, as evidenced by the many questions concerning its structure which remain unsolved.

Theorem 6.2.12 implies that a locally finite group having finite special rank is hyperfinite. For locally finite groups that are residually (locally finite of special rank r) this conclusion is no longer valid, as shown by the following result in [**53**].

10.1.6. THEOREM. *Let r be a natural number, let p, q be distinct primes and let $G = P \wr Q$, where P is of order p and Q is an infinite elementary abelian q-group. Then G is residually (soluble of special rank r), but has no non-trivial finite normal subgroups.*

We note also that groups which satisfy the conditions of Theorem 10.1.4 need not be locally soluble, as the following well-known result stated in [**53**] shows.

10.1.7. THEOREM. *Let p be a fixed prime and let G be a free group of countable free rank. Then G has a descending normal series of subgroups N_i such that G/N_i is a finite p-group of special rank 9 for all i and $\cap_{i=1}^{\infty} N_i = 1$.*

However, under some natural additional restrictions, the situation changes radically. We shall say that G is *strongly residually (finite-p of special rank r)* if G has a descending chain of normal subgroups N_i, for $i \geq 1$, such that G/N_i is a finite p-group of special rank at most r and $\cap_{i \geq 1} N_i = 1$. The next result appeared in [**53**].

10.1.8. THEOREM. *Let p be a prime and suppose that the group G is strongly residually (finite-p of special rank r), for the fixed natural number r.*

(a) *If G is locally soluble, then G is soluble and there exists a natural number k, depending only on r, such that $\delta_k(G)$ is a finite p-group of special rank r;*

(b) *If G is locally nilpotent, then G is nilpotent and there exists a natural number c, depending only on r such that $\gamma_c(G)$ is a finite p-group of special rank r.*

Theorem 6.2.12 shows that every locally nilpotent group of finite special rank is hypercentral. This result cannot be extended to locally nilpotent groups that are residually (special rank r) as the following theorem from [53] shows.

10.1.9. THEOREM. *There exists a locally nilpotent group G that is strongly residually (special rank r), but G is not hypercentral (and not even hypercentral-by-soluble).*

Furthermore, Proposition 6.2.4 shows that every torsion-free locally nilpotent group of finite special rank is nilpotent. This is no longer true for the kinds of group we are currently discussing.

10.1.10. THEOREM. *There exists a metabelian, torsion-free locally nilpotent group that is residually (special rank 2), but which is neither nilpotent nor of finite special rank.*

Theorem 10.1.4 provides us with a general picture describing groups which are residually (finite special rank r), where r is fixed. If G is a group satisfying the hypotheses of Theorem 10.1.4, the structure of each of the groups M and G/N is reasonably transparent, but less can be said concerning the structure of N/M. Indeed, according to Theorem 10.1.7, this group could be a free group. So the question naturally arises concerning the conditions under which G will be locally (soluble-by-finite). Some answers to this question were obtained in [56], where results analogous to the Tits alternative for linear groups were given (see Theorem 1.4.3 and also [22]).

10.1.11. THEOREM. *Let G be a group that is residually (locally (soluble-by-finite) of special rank r). Then either G is locally (soluble-by-finite) or G contains a non-abelian free subgroup.*

10.1.12. THEOREM. *Let G be a finitely generated residually (soluble of finite rank r) group. Then either G is nilpotent-by-abelian-by-finite or G contains a non-abelian free subgroup.*

As is so often the case concerning the theory of generalized soluble groups, an essential step is to clarify the structure and properties of certain modules over group rings. This is true in many of the aforementioned results concerning groups of finite special rank, where one is often lead to consider group rings of groups with finite special rank. The study of such group rings is a separate, specialized topic which we do not pursue here. We note only that some of the ideas have been discussed in the books [152] and [153]. Certain modules arise naturally in some questions concerning groups of finite special rank as we now discuss.

If G is a group and M is a G-invariant subset of G, then $C_G(M)$ is a normal subgroup of G. The factor group $G/C_G(M)$ is called the *cocentralizer* of the set M in G, a definition which was introduced in the paper [145]. Clearly, $G/C_G(M)$ is isomorphic to some subgroup of $\mathrm{Aut}\,(\langle M \rangle)$.

The influence of the cocentralizers of objects related to a given group is relevant in many branches of group theory. For example, Formation Theory has played a significant role in the Theory of Finite Groups. In this theory local formations are defined via the cocentralizers of the chief factors of the groups involved.

In the theory of infinite groups, much interesting work has concerned the structure of groups and the cocentralizers of their conjugacy classes. In particular if $\mathfrak{H}$ is a class of groups, then we say that an element $g \in G$ has an $\mathfrak{H}$-conjugacy class if $G/C_G(x^G) \in \mathfrak{H}$. If $\mathfrak{H} = \mathfrak{I}$, the class of trivial groups, then it follows that $g \in \zeta(G)$ and if $\mathfrak{H} = \mathfrak{F}$, the class of finite groups, then g has only finitely many conjugacy classes.

A question which usually arises early in the discussion of this type of problem concerns the structure of the normal closure of an element g. If the conjugacy class g^G is finite, then the normal closure $\langle g \rangle^G$ is an extension of a finite group by a free abelian group of finite 0-rank, by a result in [137]. It is always the case that we have $C_G(g^G) = C_G(\langle g \rangle^G)$, so that $C_G(g^G) \cap \langle g \rangle^G \le \zeta(\langle g \rangle^G)$. It follows, using Corollary 7.2.35, that if $\mathfrak{H}$ is a class of soluble-by-finite groups of finite special rank, then $D = (\langle g \rangle^G)'$ is soluble-by-finite of finite special rank. The factor group $\langle g \rangle^G/D$ is abelian and naturally it can be regarded as a cyclic $\mathbb{Z}H$-module, where $H = G/C_G(g^G)$ is a soluble-by-finite group of finite special rank.

A group G is said to have $\mathfrak{H}$-conjugacy classes (or G is an $\mathfrak{H}C$-group) if $G/C_G(g^G) \in \mathfrak{H}$ for each element $g \in G$. Thus if $\mathfrak{H} = \mathfrak{I}$, then the class of $\mathfrak{H}C$-groups is the class of all abelian groups and if $\mathfrak{H} = \mathfrak{F}$, then we obtain the class of FC-groups, or groups with finite conjugacy classes. As we have seen, this class is a natural extension of both the class of finite groups and the class of abelian groups, so that it inherits many of the properties of these classes. The theory of FC-groups is one of the most well-studied branches of the theory of infinite groups and many authors have made significant contributions to this theory (see, for example, the books of Yu. M. Gorchakov [95], M. J. Tomkinson [253] and S. N. Chernikov [41], and the surveys [254, 214]).

The natural case when the conjugacy classes of elements are finite of bounded order was considered by B. H. Neumann [203]. He showed in this paper that this class of groups is quite easily and simply described and indeed is rather restricted.

10.1.13. THEOREM (B. H. Neumann [**203**]). *Let G be a group and suppose that there is a natural number d such that $|x^G| = |G : C_G(x)| \le d$, for each element $x \in G$. Then G' is finite.*

A further question arises from Theorem 10.1.13 as follows.

Let G be a group and suppose that there is a natural number d such that $|x^G| \le d$ for each element $x \in G$. Is there a function ν such that $|G'| \le \nu(d)$?

A positive answer to this question was obtained by J. Wiegold [**260**], as seen in the following theorem.

10.1.14. THEOREM. *Let G be a group and suppose that there is a natural number d such that $|x^G| = |G : C_G(x)| \le d$, for each element $x \in G$. Then there exists a function ν, depending on d only, such that $|G'| \le \nu(d)$. Furthermore, $\nu(d) = d^m$, where*

$$m = \frac{1}{2}d(d-1)\log_2(dk) \text{ and } k = d(d-1)(\log_2 d)^2 + 1.$$

Also, if d is prime, then $\nu(d) = d$.

This rather complex function obtained by Wiegold has been the subject of much refinement over time, involving many algebraists, and we here digress to list some of the papers that contain these refinements. In all these cases $\nu(d) \le d^m$, where

(a) $m = 6d(\log_2 d)^3$ (due to I. D. MacDonald [**184**]);
(b) $m = k^2(1+k)(1+k+k^2)/2$, where $k = \log_2 d$ (J. A. Shepperd and J. Wiegold [**240**]);
(c) $m = (3+5\log_2 d)/2$ and, for soluble groups, $m = (5+\log_2 d)/2$ (P. M. Neumann and M. R. Vaughan-Lee [**205**]);
(d) $m = (41 + \log_2 d)/2$ (M. Cartwright [**29**]);
(e) $m = (13 + \log_2 d)/2$ (D. Segal and A. Shalev [**237**]);
(f) $m = (7 + \log_2 d)/2$ (R. Guralnick and A. Maróti [**104**]).

Generally in the determination of m, the reduction of the problem to the finite case is almost immediate. It should be noted that Wiegold conjectured that $m = (1 + \log_2 d)/2$.

Theorem 10.1.13 was extended to groups of finite section rank in the paper [**87**] of S. Franciosi, F. de Giovanni and L. A. Kurdachenko.

10.1.15. THEOREM. *Let G be a group and suppose that $G/C_G(x^G)$ is a soluble group of finite section rank for each element $x \in G$. Suppose that there is a natural number d and, for each prime p, a natural number r_p such that $\boldsymbol{dl}(G/C_G(x^G)) \le d$ and $\boldsymbol{sr}_p(G/C_G(x^G)) \le r_p$, for each $x \in G$. Then G' has finite section rank and there exist functions $\mathfrak{s}_p$ such that $\boldsymbol{sr}_p(G') \le \mathfrak{s}_p(r_p, d)$, for each prime p.*

A special case of this result is as follows, also recorded in [**87**].

10.1.16. THEOREM. *Let G be a group and suppose that $G/C_G(x^G)$ is a soluble group of finite special rank for each element $x \in G$. Suppose that there are natural numbers d, r such that $dl(G/C_G(x^G)) \leq d$ and $r(G/C_G(x^G)) \leq r$, for all $x \in G$. Then G' has finite special rank and there is a function $\mathfrak{k}$ such that $r(G') \leq \mathfrak{k}(r, d)$.*

It should be noted that in [**87**] these theorems use stricter conditions. Indeed it is assumed that $G/C_G(x^G)$ is isomorphic to a fixed section of a soluble group U of finite section (respectively finite special) rank. However, this fact is not really used in the proof and is really only used so that the derived length and section p-rank (respectively special rank) of all cocentralizers $G/C_G(x^G)$ are bounded.

As a consequence of this result it can be deduced that in such groups, the normal closure $\langle g \rangle^G$ of each element g of G has finite section (respectively finite special) rank and, furthermore, we have $\mathbf{sr}_p(\langle g \rangle^G) \leq b_p$ for every prime p (respectively $r(\langle g \rangle) \leq b$).

In [**250**], H. Smith began work concerning groups in which the special rank of the normal closure $\langle g \rangle^G$ is at most d, for every element $g \in G$ and a fixed natural number d. The main result of this paper is as follows.

10.1.17. THEOREM. *Let d be a natural number. Let G be a locally (soluble-by-finite) group and suppose that $r(\langle g \rangle^G) \leq d$, for each $x \in G$. Then G' has finite special rank and there is a function $\mathfrak{e}$ such that $r(G') \leq \mathfrak{e}(d)$.*

This result was generalized in [**178**] to all groups. We remark that if G is a group and if G' has special rank at most k, then $\langle x \rangle^G$ has special rank at most $k + 1$, for all $x \in G$.

10.1.18. THEOREM. *Let d be a natural number. Let G be a group and suppose that $r(\langle g \rangle^G) \leq d$, for each $x \in G$. Then G' has finite special rank, bounded in terms of d only.*

Smith's result was extended further in the paper [**148**] of L. A. Kurdachenko, J. M. Muñoz-Escolano and J. Otal in which it was assumed that the section p-rank of $\langle g \rangle^G$ is bounded for each element $g \in G$ and for each prime p. The main result of this paper is as follows.

10.1.19. THEOREM. *Let G be a locally generalized radical group in which the normal closure of every cyclic subgroup has section p-rank at most $\kappa(p)$ for some function $\kappa : \mathbb{P} \longrightarrow \mathbb{N}$. Then G' has section p-rank at most $\lambda(\kappa(p))$ for some function λ.*

We now discuss a different topic which is relevant in this chapter concerned with how the structure of a group of infinite special rank is influenced by the structure of the proper subgroups of infinite special rank. Many recent papers have discussed groups in which the proper subgroups of infinite special rank satisfy some property such as nilpotency or normality and their generalizations. Here we give a brief summary of these papers. Generally, let $\mathcal{P}$ be a group theoretical property or class of groups and, as usual, say that G is a $\mathcal{P}$-group or that $G \in \mathcal{P}$ if G has the property $\mathcal{P}$ or belongs to the class $\mathcal{P}$. We let $\mathcal{P}^*$ denote the class of groups G in which every proper subgroup of G is a $\mathcal{P}$-group or is of finite special rank. Clearly, if $\mathcal{P}$ is closed under taking subgroups, then every $\mathcal{P}$-group is also in the class $\mathcal{P}^*$ and of course groups of finite special rank are also $\mathcal{P}^*$-groups. Thus we are interested in the questions:

Let $G \in \mathcal{P}^*$. What can be said concerning G? Is G either of finite special rank or a $\mathcal{P}$-group? Are there other possibilities for G and if so what?

Usually we will need to place restrictions on G at the start in order to exclude examples such as those due to A. Yu. Ol'shanskii mentioned earlier in the book. We remind the reader that $\mathfrak{N}_c$ denotes the class of groups which are nilpotent of class at most c and $\mathfrak{S}_d$ denotes the class of groups which are soluble of derived length at most d. Suppose that $G \in \mathfrak{N}_1^*$, so all proper subgroups of G are abelian or of finite special rank. Clearly groups which are themselves abelian or are of finite special rank fall into this class. It was proved in [55] that if c is a natural number and if $G \in \mathfrak{X} \cap \mathfrak{N}_c^*$, then either $G \in \mathfrak{N}_c$ or $G \in \mathfrak{R}$, where, as earlier, $\mathfrak{R}$ is the class of groups of finite special rank. This paper contains other results of this nature. For example, if $G \in \mathfrak{X} \cap (\mathfrak{L}\mathfrak{N})^*$, then either G is locally nilpotent or G has finite rank and in either case G is almost locally soluble.

Of course there is no corresponding result if we replace $\mathfrak{N}_c$ by $\mathfrak{N}$, the class of all nilpotent groups, since the Heineken-Mohamed [114] groups are locally nilpotent, non-nilpotent and of infinite special rank, but have all proper subgroups nilpotent. However the existence of such groups is essentially the only reason why we cannot replace $\mathfrak{N}_c$ by $\mathfrak{N}$ in the results mentioned above.

At this point it is natural to ask if one can obtain similar results for groups in the class $\mathfrak{S}_d^*$. One potential obstacle to such a classification is the existence of simple locally finite groups of infinite special rank in which every proper subgroup is finite or metabelian. For example, the group $PSL(2, \mathbb{F})$ has this property if $\mathbb{F}$ is a locally finite field with no proper infinite subfields (see [**212**]). In order to settle this question, information concerning soluble groups with all proper subgroups soluble of derived length d is required. The results mentioned above require a well-known result of Schmidt [**232**], which states that a finite group with all proper subgroups nilpotent is soluble. For groups with all proper subgroups soluble of bounded derived length, a slight extension of a theorem of D. I. Zaitsev [**266**], that an infinite soluble group of derived length exactly d has a proper subgroup of derived length precisely d, is required.

Using this result the following theorem can be obtained (see [**52**]).

10.1.20. THEOREM. *Let $G \in \mathfrak{X} \cap (\mathbf{L}\mathfrak{S})^*$. Then either*

(i) $G \in \mathbf{L}\mathfrak{S}$ *or*

(ii) *G has finite special rank or*

(iii) *G is isomorphic to one of $SL(2, \mathbb{F}), PSL(2, \mathbb{F})$ or $Sz(\mathbb{F})$ for some infinite locally finite field $\mathbb{F}$ in which every proper subfield is finite.*

Here $Sz(\mathbb{F})$ denotes the Suzuki group over the field $\mathbb{F}$. We note that each of the groups in (iii) actually is in the class under consideration.

A number of very recent interesting important papers concerning groups of infinite special rank whose proper subgroups of infinite special rank satisfy some further condition have further extended our knowledge in this area (see [**43, 44, 45, 71, 72, 73, 74, 75, 76, 78, 79, 80, 81, 82**], for example). In these papers the authors begin with a group of infinite special rank and assume that the proper subgroups of infinite special rank satisfy some additional property. Thus in [**45**] the authors assume that the proper subgroups of infinite rank have polycyclic conjugacy classes; in [**79**] M. De Falco, F. de Giovanni, C. Musella and Ya. P. Sysak prove the lovely result that if G is a locally (soluble-by-finite) group in which every non-abelian subgroup of infinite special rank is normal, then G is either of finite special rank or is metahamiltonian. Here a group G is *metahamiltonian* if every non-abelian subgroup of G is normal in G.

As is well known a group in which every subgroup is normal is called a *Dedekind* group, and such groups were classified by Dedekind in [**46**]. What then happens if every subgroup of infinite rank is a normal subgroup? In [**85**] a number of results were obtained in this direction. First the authors showed in [**85**, Theorem 2.3] that a locally graded group with all subgroups Chernikov or normal is itself either Chernikov or a Dedekind group, a result having a number of other consequences. They show in [**85**, Corollary 2.5] that if G is a locally graded group with all proper subgroups soluble of finite section rank, then G is also soluble with finite section rank. In [**85**, Theorem 2.7] the authors obtain the following theorem.

10.1.21. THEOREM. *Let G be a locally graded group. Then every non-normal subgroup of G is soluble with finite section rank if and only if G satisfies one of the following conditions:*

(i) *G is a Dedekind group;*

(ii) *G is soluble with finite section rank;*

(iii) *G has finite section rank and contains a finite normal minimal non-soluble subgroup N such that G/N is a Dedekind group. In particular, G' is finite.*

This result is used to obtain precise conditions for the structure of a locally graded group with all non-normal subgroups polycyclic.

In general the problem of deciding the structure of a locally graded group whose subgroups are normal or of finite special rank seems more complicated, partly because our knowledge of locally graded groups with all proper subgroups of finite special rank is so limited.

The papers [**167, 165**] and [**70**] start to rectify this. A well-known theorem of Roseblade [**230**] asserts that if G is a group in which every subgroup is subnormal of defect at most d, then G is a nilpotent group (with class dependent on d). In [**70**] the authors study the class of $\mathfrak{X}$-groups all of whose proper subgroups are of finite special rank or subnormal of defect at most d. It is proved there that if G is an $\mathfrak{X}$-group of infinite special rank in which every proper subgroup of infinite special rank is subnormal of defect at most d, then G is nilpotent of nilpotency class bounded by a function of d. In particular, if every subgroup of infinite special rank is normal then G is a Dedekind group, so is nilpotent of class at most 2.

Earlier in [**167**] it was proved that if G is a soluble group of infinite special rank whose proper subgroups of infinite special rank are subnormal, then G is a Baer group, which is to say that every finitely generated subgroup of G is subnormal in G. This result was taken a step further in [**165**] where it is shown that a locally (soluble-by-finite) group of infinite rank, all of whose proper subgroups of infinite rank are subnormal, is in any case soluble therefore generalizing a famous result of Möhres [**198**], who showed that a group in which all subgroups are subnormal is soluble. Indeed, for torsion-free locally (soluble-by-finite) groups G of infinite special rank whose proper subgroups of infinite special rank are subnormal, the conclusion is that G is nilpotent [**165**]. In this case, however, an example is exhibited of a metabelian locally nilpotent group G of infinite special rank, whose torsion subgroup T has finite special rank and which contains a non-subnormal subgroup.

It turns out though that every subgroup of infinite special rank is subnormal.

There are other natural generalizations of normality such as permutability. It is well known that O. Ore [210] proved that in a finite group G all permutable subgroups of G are subnormal in G. Permutable subgroups have also been called quasinormal subgroups by some authors, and during his investigations of permutable subgroups, S. E. Stonehewer [251] proved that all permutable subgroups of a group G are ascendant in G.

Groups with all subgroups permutable, so called quasihamiltonian groups, were classified quite precisely by Iwasawa [120]. The following theorem summarizes some of the main results and proofs can be found in [233], for example.

10.1.22. THEOREM. *Let G be a non-abelian quasihamiltonian group. Then*

(i) *G is locally nilpotent;*

(ii) *G is metabelian;*

(iii) *If G is not periodic then $\boldsymbol{Tor}(G)$ is abelian and $G/\boldsymbol{Tor}(G)$ is a torsion-free abelian group of 0-rank 1.*

Groups all of whose subgroups of infinite special rank are permutable were studied in [65] where it was shown that, for the class $\mathfrak{X}$, if $G \in \mathfrak{X}$ has infinite special rank and every subgroup of infinite special rank is permutable then G is quasihamiltonian. This result was generalized in a certain subclass of $\mathfrak{X}$ in [76], where it was also proved that if G is a periodic locally graded group of infinite rank and if all subgroups of infinite special rank are modular, then G has modular subgroup lattice.

The study of groups with all proper subgroups of a certain type has also taken other directions. Finite groups with all subgroups nilpotent were studied in [232] and shown to be soluble, but again the examples of A. Yu. Ol'shanskii [208] show that this is not true in general. In a series of papers B. Bruno [25, 24, 26] and B. Bruno and R. E. Phillips [28] showed that a locally graded, non-perfect group, all of whose proper subgroups are nilpotent-by-finite is itself nilpotent-by-finite or periodic. This continued a line of research concerned with the notion of minimal non-$\mathfrak{Y}$-groups, where $\mathfrak{Y}$ is some class of groups. A *minimal non-$\mathfrak{Y}$-group* is a group not in the class $\mathfrak{Y}$ all of whose proper subgroups are in $\mathfrak{Y}$. In the paper of J. Otal and J. M. Peña [211], locally graded groups all of whose proper subgroups are Chernikov-by-nilpotent and locally graded groups with all proper subgroups nilpotent-by-Chernikov were considered and, in a follow-up paper, Napolitani and

Pegoraro [201] proved that a locally graded group with all proper subgroups nilpotent-by-Chernikov is either nilpotent-by-Chernikov or is a perfect locally finite p-group in which all proper subgroups are nilpotent. In a tour de force A. O. Asar proved in [3] that a locally nilpotent p-group in which every proper subgroup is nilpotent-by-Chernikov is itself nilpotent-by-Chernikov. In particular this meant that the following theorem is true.

10.1.23. THEOREM. *Let G be a locally graded group all of whose proper subgroups is nilpotent-by-Chernikov. Then G is nilpotent-by-Chernikov and hence is soluble.*

This result has inspired many generalizations. For example, in [59] and [58], groups with all proper subgroups finite rank-by-nilpotent were considered. The following result was obtained.

10.1.24. THEOREM. *Let G be a locally (soluble-by-finite) group and suppose that every proper subgroup of G belongs to $\mathfrak{RN}$*

(a) *If G is a p-group for some prime p then either $G \in \mathfrak{RN}$ or $G/G' \cong C_{p^\infty}$ and every proper subgroup of G is nilpotent.*
(b) *If G is not a p-group then $G \in \mathfrak{RN}$.*

When a bound is placed on the nilpotency classes things are more straightforward (see [54]). In this case, if G is a locally (soluble-by-finite) group and every proper subgroup of G belongs to the class $\mathfrak{RN}_c$, then $G \in \mathfrak{RN}_c$.

The opposite case, so to speak, of groups with all proper subgroups soluble-by-finite special rank was discussed in [62].

10.1.25. THEOREM. *Let G be a locally (soluble-by-finite) group with all proper subgroups soluble-by-finite special rank. Then either*

(i) *G is locally soluble, or*
(ii) *G is soluble-by-finite special rank and almost locally soluble, or*
(iii) *G is soluble-by-$PSL(2, F)$, or*
(iv) *G is soluble-by-$Sz(F)$,*

where F is an infinite locally finite field with no infinite proper subfields.

This generalized the work of [57] and depended in an essential way on the paper [201]. There are a number of other results of the types discussed here, including [5] and [2], for example.

10.2. Questions

It is only fitting in this final section to mention some of the apparently open problems which remain unsolved concerning groups satisfying certain rank conditions. Some of these will no doubt have occurred to the reader already, some are well known and some are not really germaine to our subject but seem to be important stepping stones whose solution would be of tremendous interest to the group theoretic community. As we have seen in this book locally generalized radical groups are reasonably well behaved when a finiteness condition is imposed on its special rank (or other ranks that have been considered here). There are however a couple of questions that we should append here concerning this class of groups. The first of these questions relates to the work in Section 4.2, where a very strong structure theorem was obtained for locally generalized radical groups of finite section rank. Still the following problem is elusive, although we remark that for nearly radical groups, a positive result is given in Corollary 9.2.9.

QUESTION 10.1. *Let G be a generalized radical group of finite section rank. Is G locally (soluble-by-finite)?*

We also left a loose end in the work of Chapter 7 concerning generalized radical groups, so the following question is still to be decided.

QUESTION 10.2. *Let G be a group such that $G/\zeta_\infty(G)$ is a generalized radical group of finite 0-rank. Is there a normal subgroup H of G such that H has finite 0-rank and G/H is hypercentral?*

For those readers interested in general series of subgroups we offer the following question.

QUESTION 10.3. *Let G be a group of finite special rank. Suppose that G has a Kurosh-Chernikov system whose factors are finite. Is G locally (soluble-by-finite)?*

We now turn attention to the murkier situation where we consider residually finite groups and, more generally, locally graded groups.

Since in much of the last section we were concerned with the class of locally graded groups and since rather a lot is known concerning the class $\mathfrak{X}$ introduced by N. S. Chernikov in [**36**] we start with:

QUESTION 10.4. *How close is the class of strongly locally graded groups to the class of all locally graded groups?*

Theorem 10.1.3 asserts that if G is a strongly locally graded group of finite special rank, then G is almost locally soluble. It seems to be a very difficult problem to determine what happens for locally graded groups of finite special rank in general. But it seems natural to hope that they will be almost locally soluble also.

QUESTION 10.5. *If G is a locally graded group of finite special rank, then is G almost locally soluble?*

A further natural question that arises in this direction concerns groups all of whose proper subgroups have finite special rank. If G is strongly locally graded, then every subgroup of G is also strongly locally graded and it is rather easy to see, using the argument of [**76**, Lemma 1], that in this case G has finite special rank and hence is almost locally soluble, using Theorem 10.1.3. In light of this we ask the following question.

QUESTION 10.6. *If G is locally graded and every proper subgroup of G has finite special rank, then does G have finite special rank and is G then almost locally soluble?*

For our next question we return to the class of periodic locally graded groups. We saw in Theorem 9.3.5 that if G is a locally k-generalized radical group and every abelian subgroup of G has finite special rank, then G has finite special rank.

QUESTION 10.7. *Let G be a periodic locally graded group with all locally soluble (respectively abelian) subgroups of finite special rank. Is G of finite special rank? Is G locally finite?*

A particular case of this question can be given for residually finite groups.

QUESTION 10.8. *Is a periodic residually finite group with all abelian subgroups of finite special rank also of finite special rank?*

Of course a free group has all its abelian subgroups of (bounded!) finite special rank and does not have finite special rank, hence the restriction to periodic groups here.

A question that is related to this and which arises in other contexts is the following specific case.

QUESTION 10.9. *Are there infinite residually finite groups with all subgroups either finite or of finite index?*

There are some very natural questions that arose from much of the work mentioned in Section 10.1. These questions are probably less well known than those that have gone before and some of them may have easy solutions. First we consider the class $\mathfrak{N}_c^*$.

QUESTION 10.10. *If G is locally graded and $G \in \mathfrak{N}_c^*$, then is $G \in \mathfrak{N}_c \cup \mathfrak{R}$? (with similar questions for $\mathfrak{S}_d^*$ (and so on)).*

Next we have:

QUESTION 10.11. *Let G be a periodic strongly residually (rank r) group for some fixed r. Is G almost locally soluble? Is G locally finite?*

Finally, there are many natural questions concerning the class $\mathfrak{N}_c\mathfrak{R}$. We remark that B. Bruno and F. Napolitani [27] have studied groups all of whose proper subgroups are nilpotent of class at most c-by-Chernikov.

QUESTION 10.12.

(a) *If G is locally graded and all proper subgroups of G are $\mathfrak{N}_c\mathfrak{R}$ what can be said concerning G?*

(b) *If G is a locally finite p-group and every proper subgroup of G is in the class $\mathfrak{N}_c\mathfrak{R}$, then is $G \in \mathfrak{N}_c\mathfrak{R}$?*

(c) *If G is locally graded and every proper subgroup of G is in the class $\mathfrak{R}\mathfrak{N}_c$, then is G in the class $\mathfrak{R}\mathfrak{N}_c$?*

Our hope is that readers of this book will pursue these and other questions to add further to the fascinating theory of groups satisfying conditions on their ranks.

Bibliography

S. I. Adyan

[1] *Certain torsion-free groups*, Izv. Akad. Nauk SSSR Ser. Mat. **35** (1971), 459–468.

A. Arikan and N. Trabelsi

[2] *On groups whose proper subgroups are Chernikov-by-Baer or (periodic divisible abelian)-by-Baer*, J. Algebra Appl. **12** (2013), no. 6, 1350015, 15.

A. O. Asar

[3] *Locally nilpotent p-groups whose proper subgroups are hypercentral or nilpotent-by-Chernikov*, J. London Math. Soc. **61** (2000), 412–422.

M. Aschbacher and R. Guralnick

[4] *Some applications of the first cohomology group*, J. Algebra **90** (1984), 446–460.

A. Badis and N. Trabelsi

[5] *Groups whose proper subgroups are Baer-by-Chernikov or Baer-by-(finite rank)*, Cent. Eur. J. Math. **9** (2011), no. 6, 1344–1348.

R. Baer

[6] *Situation der Untergruppen und Struktur der Gruppe*, Sitz. Ber. Heidelberg Akad **2** (1933), 12–17.

[7] *Representations of groups as quotient groups. II. Minimal central chains of a group*, Trans. Amer. Math. Soc. **58** (1945), 348–389.

[8] *Endlichkeitskriterien für Kommutatorgruppen*, Math. Ann. **124** (1952), 161–177.

[9] *Das Hyperzentrum einer Gruppe. III*, Math. Z. **59** (1953), 299–338.

[10] *Finite extensions of Abelian groups with minimum condition*, Trans. Amer. Math. Soc. **79** (1955), 521–540.

[11] *Noethersche Gruppen*, Math. Z. **66** (1956), 269–288.

[12] *Abzählbar erkennbare gruppentheoretische Eigenschaften*, Math. Z. **79** (1962), 344–363.

[13] *Local and global hypercentrality and supersolubility. I, II*, Nederl. Akad. Wetensch. Proc. Ser. A 69, Indag. Math. **28** (1966), 93–110; 111–126.

Ranks of Groups: The Tools, Characteristics, and Restrictions
By Martyn R. Dixon, Leonid A. Kurdachenko and Igor Ya Subbotin
Copyright © 2017 John Wiley & Sons, Inc.

[14] *Polyminimaxgruppen*, Math. Ann. **175** (1968), 1–43.

[15] *Lokal endlich-auflösbare Gruppen mit enlichen Sylowuntergruppen*, J. Reine Angew. Math. **239/240** (1970), 109–144.

[16] *Durch Formationen bestimmte Zerlegungen von Normalteilern endlicher Gruppen*, J. Algebra **20** (1972), 38–56.

R. Baer and H. Heineken

[17] *Radical groups of finite abelian subgroup rank*, Illinois J. Math. **16** (1972), 533–580.

A. Ballester-Bolinches, S. Camp-Mora, L. A. Kurdachenko, and J. Otal

[18] *Extension of a Schur theorem to groups with a central factor with a bounded section rank*, J. Algebra **393** (2013), 1–15.

G. Baumslag and N. Blackburn

[19] *Groups with cyclic upper central factors*, Proc. London Math. Soc. (3) **10** (1960), 531–544.

G. Baumslag

[20] *Lecture Notes on Nilpotent Groups*, American Mathematical Society, Providence, R.I., 1971.

V. V. Belyaev

[21] *Locally finite groups with Černikov Sylow p-subgroups*, Algebra i Logika **20** (1981), 605–619 (Russian), English transl. in Algebra and Logic, **20** (1981) 393-402.

S. Black

[22] *A finitary Tits' alternative*, Arch. Math. (Basel) **72** (1999), no. 2, 86–91.

C. J. B. Brookes

[23] *Ideals in group rings of soluble groups of finite rank*, Math. Proc. Cambridge Philos. Soc. **97** (1985), no. 1, 27–49.

B. Bruno

[24] *Groups whose proper subgroups contain a nilpotent subgroup of finite index*, Boll. Un. Mat. Ital. D (6) **3** (1984), no. 1, 179–188 (1985).

[25] *On groups with "abelian by finite" proper subgroups*, Boll. Un. Mat. Ital. B (6) **3** (1984), no. 3, 797–807.

[26] *On p-groups with nilpotent-by-finite proper subgroups*, Bollettino U. M. I. (Series A) **3** (1989), 45–51.

B. Bruno and F. Napolitani

[27] *A note on nilpotent-by-Chernikov groups*, Glasg. Math. J. **46** (2004), no. 2, 211–215.

B. Bruno and R. E. Phillips

[28] *A note on groups with nilpotent-by-finite proper subgroups*, Arch. Math. **65** (1995), 369–374.

M. Cartwright

[29] *The order of the derived group of a BFC-group*, J. London Math. Soc. (2) **30** (1984), no. 2, 227–243.

V. S. Charin

[30] *On the theory of locally nilpotent groups*, Mat. Sbornik N.S. **29(71)** (1951), 433–454.

[31] *On groups of automorphisms of nilpotent groups*, Ukrain. Mat. Ž. **6** (1954), 295–304.

[32] *On locally solvable groups of finite rank*, Mat. Sb. N.S. **41(83)** (1957), 37–48.

[33] *Solvable groups of type A_4*, Mat. Sb. (N.S.) **52 (94)** (1960), 895–914.

[34] *Solvable groups of type A_3*, Mat. Sb. (N.S.) **54 (96)** (1961), 489–499.

V. S. Charin and D. I. Zaitsev

[35] *Groups with finiteness conditions and other restrictions for subgroups*, Ukrain. Mat. Zh. **40** (1988), no. 3, 277–287, 405.

N. S. Chernikov

[36] *A theorem on groups of finite special rank*, Ukrain. Mat. Zh. **42** (1990), 962–970 (Russian), English transl. in Ukrainian Math. J. **42**, (1990), 855-861.

S. N. Chernikov

[37] *Complete groups possessing ascending central series*, Rec. Math. [Mat. Sbornik] N.S. **18(60)** (1946), 397–422.

[38] *On complete groups with an ascending central series*, Doklady Akad. Nauk SSSR (N.S.) **70** (1950), 965–968.

[39] *On special p-groups*, Mat. Sbornik N.S. **27(69)** (1950), 185–200.

[40] *Infinite locally finite groups with finite Sylow subgroups*, Mat. Sb. (N.S.) **52 (94)** (1960), 647–652.

[41] *Gruppy s zadannymi svoistvami sistemy podgrupp (groups with Prescribed Properties of Systems of Subgroups*, "Nauka", Moscow, 1980, Sovremennaya Algebra. [Modern Algebra].

F. de Giovanni

[42] *Infinite groups with rank restrictions on subgroups*, Zap. Nauchn. Sem. S.-Peterburg. Otdel. Mat. Inst. Steklov. (POMI) **414** (2013), no. Voprosy Teorii Predstavlenii Algebr i Grupp. 25, 31–39.

F. de Giovanni and F. Saccomanno

[43] *A note on groups of infinite rank whose proper subgroups are abelian-by-finite*, Colloq. Math. **137** (2014), no. 2, 165–170.

[44] *A note on infinite groups whose subgroups are close to be normal-by-finite*, Turkish J. Math. **39** (2015), no. 1, 49–53.

F. de Giovanni and M. Trombetti

[45] *Groups whose proper subgroups of infinite rank have polycyclic conjugacy classes*, Algebra Colloq. **22** (2015), no. 2, 181–188.

R. Dedekind

[46] *Über Gruppen deren sämtliche Teiler Normalteiler sind*, Math. Ann. **48** (1897), 548–561.

A. P. Dietzmann

[47] *On p-groups*, Doklady Akad. Nauk. SSSR **15** (1937), 71–76.

J. D. Dixon and B. Mortimer

[48] *Permutation Groups*, Graduate Texts in Mathematics, vol. 163, Springer-Verlag, New York, 1996.

M. R. Dixon

[49] *Sylow Theory, Formations and Fitting Classes in Locally Finite Groups*, Series in Algebra, World Scientific, Singapore, 1994, Volume 2.

[50] *Certain rank conditions on groups*, Note Mat. **28** (2008), no. suppl. 2, 155–180 (2009).

M. R. Dixon, M. J. Evans, and H. Smith

[51] *Locally (soluble-by-finite) groups of finite rank*, J. Algebra **182** (1996), 756–769.

[52] *Locally (soluble-by-finite) groups with all proper insoluble subgroups of finite rank*, Arch. Math. (Basel) **68** (1997), 100–109.

[53] *On groups that are residually of finite rank*, Israel J. Math. **107** (1998), 1–16.

[54] *Groups with all proper subgroups (finite rank)-by-nilpotent*, Arch. Math. **72** (1999), 321–327.

[55] *Locally (soluble-by-finite) groups with all proper non-nilpotent subgroups of finite rank*, J. Pure Appl. Algebra **135** (1999), 33–43.

[56] *A Tits alternative for groups that are residually of bounded rank*, Israel J. Math. **109** (1999), 53–59.

[57] *Groups with all proper subgroups nilpotent-by-finite rank*, Arch. Math **75** (2000), 81–91.

[58] *Groups with all proper subgroups (finite rank)-by-nilpotent. II*, Comm. Algebra **29** (2001), no. 3, 1183–1190.

[59] *Groups with some minimal conditions on non-nilpotent subgroups*, J. Group Theory **4** (2001), 207–215.

[60] *Locally soluble-by-finite groups with the weak minimal condition on non-nilpotent subgroups*, J. Algebra **249** (2002), 226–246.

[61] *A finiteness condition on subgroups of large derived length*, J. Algebra **280** (2004), 762–771.

[62] *Groups with all proper subgroups soluble-by-finite rank*, J. Algebra **289** (2005), 135–147.

[63] *Locally (soluble-by-finite) groups with various rank restrictions on subgroups of infinite rank*, Glasgow Math. J. **47** (2005), 309–317.

[64] *Omissible extensions of $SL_2(k)$ where k is a field of positive characteristic*, Int. J. Group Theory **2** (2013), no. 1, 145–155.

M. R. Dixon and Y. Karatas

[65] *Groups with all subgroups permutable or of finite rank*, Cent. Eur. J. Math. **10** (2012), no. 3, 950–957.

M. R. Dixon, L. A. Kurdachenko, and J. Otal

[66] *On groups whose factor-group modulo the hypercentre has finite section p-rank*, J. Algebra **440** (2015), 489–503.

M. R. Dixon, L. A. Kurdachenko, and N. V. Polyakov

[67] *Locally generalized radical groups satisfying certain rank conditions*, Ricerche di Matematica **56** (2007), 43–59.

M. R. Dixon, L. A. Kurdachenko, and I. Ya. Subbotin

[68] *On various rank conditions in infinite groups*, Algebra Discrete Math. (2007), no. 4, 23–44.

M. R. Dixon and I. Ya. Subbotin

[69] *Groups with finiteness conditions on some subgroup systems: a contemporary stage*, Algebra Discrete Math. (2009), no. 4, 29–54.

M. J. Evans and Y. Kim

[70] *On groups in which every subgroup of infinite rank is subnormal of bounded defect*, Comm. Algebra **32** (2004), no. 7, 2547–2557.

M. De Falco, F. de Giovanni, and C. Musella

[71] *Groups whose proper subgroups of infinite rank have a transitive normality relation*, Mediterr. J. Math. **10** (2013), no. 4, 1999–2006.

[72] *Groups with finitely many conjugacy classes of non-normal subgroups of infinite rank*, Colloq. Math. **131** (2013), no. 2, 233–239.

[73] *Groups with normality conditions for subgroups of infinite rank*, Publ. Mat. **58** (2014), no. 2, 331–340.

[74] *A note on soluble groups with the minimal condition on normal subgroups*, J. Algebra Appl. **13** (2014), no. 4, 1350134, 5.

[75] *Groups in which every normal subgroup of infinite rank has finite index*, Southeast Asian Bull. Math. **39** (2015), no. 2, 195–201.

[76] *A note on groups of infinite rank with modular subgroup lattice*, Monatsh. Math. **176** (2015), no. 1, 81–86.

M. De Falco, F. de Giovanni, C. Musella, and Y. P. Sysak

[77] *On the upper central series of infinite groups*, Proc. Amer. Math. Soc. **139** (2011), no. 2, 385–389.

[78] *Groups of infinite rank in which normality is a transitive relation*, Glasg. Math. J. **56** (2014), no. 2, 387–393.

[79] *On metahamiltonian groups of infinite rank*, J. Algebra **407** (2014), 135–148.

M. De Falco, F. de Giovanni, C. Musella, and N. Trabelsi

[80] *Groups whose proper subgroups of infinite rank have finite conjugacy classes*, Bull. Aust. Math. Soc. **89** (2014), no. 1, 41–48.

[81] *Groups with restrictions on subgroups of infinite rank*, Rev. Mat. Iberoam. **30** (2014), no. 2, 537–550.

[82] *Groups of infinite rank with finite conjugacy classes of subnormal subgroups*, J. Algebra **431** (2015), 24–37.

Yu G. Federov

[83] *On infinite groups whose non-trivial subgroups have finite index*, Uspehi Mat. Nauk. **6** (1958), 187–189.

W. Feit and J. G. Thompson

[84] *Solvability of groups of odd order*, Pacific J. Math. **13** (1963), 775–1029.

S. Franciosi, F. de Giovanni, and M. L. Newell

[85] *Groups with polycyclic non-normal subgroups*, Algebra Colloq. **7** (2000), no. 1, 33–42.

S. Franciosi, F. de Giovanni, and L. A. Kurdachenko

[86] *On groups with many almost normal subgroups*, Ann. Mat. Pura Appl. **169** (1995), 35–65.

[87] *The Schur property and groups with uniform conjugacy classes*, J. Algebra **174** (1995), 823–847.

L. Fuchs

[88] *Infinite Abelian Groups. Vol. I*, Pure and Applied Mathematics, Vol. 36, Academic Press, New York, 1970.

[89] *Infinite Abelian Groups. Vol. II*, Pure and Applied Mathematics. Vol. 36-II, Academic Press, New York, 1973.

W. Gaschütz

[90] *Zu einem von B. H. und H. Neumann gestellten Problem*, Math. Nachr. **14** (1955), 249–252.

282   BIBLIOGRAPHY

V. M. Glushkov

[91] *On the theory of ZA-groups*, Doklady Akad. Nauk SSSR (N.S.) **74** (1950), 885–888.

[92] *On locally nilpotent groups without torsion*, Doklady Akad. Nauk SSSR (N.S.) **80** (1951), 157–160.

[93] *On some questions of the theory of nilpotent and locally nilpotent groups without torsion*, Mat. Sbornik N.S. **30(72)** (1952), 79–104.

Yu. M. Gorchakov

[94] *On the existence of Abelian subgroups of infinite rank in locally solvable groups*, Dokl. Akad. Nauk SSSR **156** (1964), 17–20 (Russian), English transl. in Soviet Math. Dokl., **5** (1964), 591-594.

[95] *Gruppy s konechnymi klassami sopryazhennykh elementov[groups with finite classes of conjugate elements]*, "Nauka", Moscow, 1978, Sovremennaya Algebra. [Modern Algebra Series].

D. Gorenstein

[96] *Finite Groups*, Harper & Row Publishers, New York, 1968.

[97] *Finite Simple groups*, University Series in Mathematics, Plenum Publishing Corp., New York, 1982, An Introduction to Their Classification.

R. I. Grigorchuk

[98] *On Burnside's problem on periodic groups*, Funktsional. Anal. i Prilozhen. **14** (1980), no. 1, 53–54.

K. W. Gruenberg

[99] *Cohomological Topics in Group Theory*, Lecture Notes in Mathematics, Vol. 143, Springer-Verlag, Berlin-New York, 1970.

O. Grün

[100] *Beiträge zur Gruppentheorie. I*, J. Reine Angew. Math. **174** (1936), 1–14.

R. M. Guralnick

[101] *On a result of Schur*, J. Algebra **59** (1979), no. 2, 302–310.

[102] *Generation of simple groups*, J. Algebra **103** (1986), no. 1, 381–401.

[103] *On the number of generators of a finite group*, Arch. Math. (Basel) **53** (1989), no. 6, 521–523.

R. M. Guralnick and A. Maróti

[104] *Average dimension of fixed point spaces with applications*, Adv. Math. **226** (2011), no. 1, 298–308.

P. Hall

[105] *Finiteness conditions for soluble groups*, Proc. London Math. Soc. (3) **4** (1954), 419–436.
[106] *Some sufficient conditions for a group to be nilpotent*, Illinois J. Math. **2** (1958), 787–801.
[107] *On the finiteness of certain soluble groups*, Proc. London Math. Soc. (3) **9** (1959), 595–622.
[108] *The Frattini subgroups of finitely generated groups*, Proc. London Math. Soc. (3) **11** (1961), 327–352.
[109] *The Edmonton Notes on Nilpotent Groups*, Queen Mary College Mathematics Notes, London, 1969.

P. Hall and G. Higman

[110] *On the p-length of p-soluble groups and reduction theorems for Burnside's problem*, Proc. London Math. Soc. (3) **6** (1956), 1–42.

P. Hall and C. R. Kulatilaka

[111] *A property of locally finite groups*, J. London Math. Soc. **39** (1964), 235–239.

B. Hartley and M. J. Tomkinson

[112] *Splitting over nilpotent and hypercentral residuals*, Math. Proc. Cambridge Philos. Soc. **78** (1975), no. 2, 215–226.

H. Heineken and L. A. Kurdachenko

[113] *Groups with subnormality for all subgroups that are not finitely generated*, Ann. Mat. Pura Appl. **169** (1995), 203–232.

H. Heineken and I. J. Mohamed

[114] *A group with trivial centre satisfying the normalizer condition*, J. Algebra **10** (1968), 368–376.

K. A. Hirsch

[115] *On infinite soluble groups I*, Proc. London Math. Soc. (2) **44** (1938), 53–60.
[116] *Über lokal-nilpotente Gruppen*, Math. Z. **63** (1955), 290–294.

B. Huppert

[117] *Lineare auflösbare Gruppen*, Math. Z. **67** (1957), 479–518.
[118] *Endliche Gruppen. I*, Die Grundlehren der Mathematischen Wissenschaften, Band 134, Springer-Verlag, Berlin, 1967.

K. Iwasawa

[119] *Einege sätze über freie gruppen*, Proc. Imp. Acad.Tokyo **19** (1943), 272–274.
[120] *On the structure of infinite M-groups*, Jap. J. Math. **18** (1943), 709–728.

N. V. Kalashnikova and L. A. Kurdachenko

[121] *Groups which are dual to layer-finite*, Infinite Groups 1994 (Berlin) (F de Giovanni and M. L. Newell, eds.), Walter de Gruyter, 1996, pp. 103–109.

L. Kaloujnine

[122] *Über gewisse Beziehungen zwischen einer Gruppe und ihren Automorphismen*, Bericht über die Mathematiker-Tagung in Berlin, Januar, 1953, Deutscher Verlag der Wissenschaften, Berlin, 1953, pp. 164–172.

M. J. Karbe

[123] *Unendliche Gruppen mit schwachen Kettenbedingungen für endlich erzeugte Untergruppen*, Arch. Math. (Basel) **45** (1985), no. 2, 97–110.

M. J. Karbe and L. A. Kurdachenko

[124] *Just infinite modules over locally soluble groups*, Arch. Math. (Basel) **51** (1988), no. 5, 401–411.

M. I. Kargapolov

[125] *Some problems in the theory of nilpotent and solvable groups*, Dokl. Akad. Nauk SSSR **127** (1959), 1164–1166.
[126] *Locally finite groups having normal systems with finite factors*, Sibirsk. Mat. Ž **2** (1961), 853–873 (Russian).
[127] *On solvable groups of finite rank*, Algebra i Logika Sem. 1 (1962), no. 5, 37–44.
[128] *On a problem of O. Yu. Schmidt*, Sibirsk. Mat. Ž. **4** (1963), 232–235.

M. I. Kargapolov and Yu. I. Merzlyakov

[129] *Fundamentals of the Theory of Groups*, Graduate Texts in Mathematics, vol. 62, Springer Verlag, Berlin, Heidelberg, New York, 1979.
[130] *Osnovy Teorii Grupp (Foundations of the Theory of Groups*, third ed., "Nauka", Moscow, 1982.

L. S. Kazarin and L. A. Kurdachenko

[131] *Finiteness conditions and factorizations in infinite groups*, Uspekhi Mat. Nauk **47:3** (1992), 75–114 (Russian), English transl. in Math. Surveys, **47:3** (1992), 81-126.

L. S. Kazarin, L. A. Kurdachenko, and I. Ya. Subbotin

[132] *On groups saturated with abelian subgroups*, Internat. J. Algebra Comput. **8** (1998), no. 4, 443–466.

O. H. Kegel

[133] *Uber einfache, lokal endliche Gruppen*, Math. Zeit. **95** (1967), 169–195.

O. H. Kegel and B. A. F. Wehrfritz

[134] *Locally Finite Groups*, North-Holland Mathematical Library, North-Holland, Amsterdam, London, 1973, Volume 3.

E. I. Khukhro and H. Smith

[135] *Locally finite groups with all subgroups normal-by-(finite rank)*, J. Algebra **200** (1998), no. 2, 701–717.

L. G. Kovács

[136] *On finite soluble groups*, Math. Z. **103** (1968), 37–39.

L. A. Kurdachenko

[137] *FC-groups whose periodic part is imbedded in a direct product of finite groups*, Mat. Zametki **21** (1977), no. 1, 9–20.

[138] *Groups that satisfy weak minimality and maximality conditions for normal subgroups*, Sibirsk. Mat. Zh. **20** (1979), no. 5, 1068–1076, 1167.

[139] *Groups satisfying weak minimality and maximality conditions for subnormal subgroups*, Mat. Zametki **29** (1981), no. 1, 19–30, 154.

[140] *Some conditions for embeddability of an FC-group in a direct product of finite groups and a torsion-free abelian group*, Mat. Sb. **114** (1981), 566–582 (Russian), English transl in Math. Sb. **42** (1982), 499–514.

[141] *Locally nilpotent groups with the weak minimal condition for normal subgroups*, Sibirsk. Mat. Zh. **25** (1984), no. 4, 99–106.

[142] *Locally nilpotent groups with weak minimality and maximality conditions for normal subgroups*, Dokl. Akad. Nauk Ukrain. SSR Ser. A (1985), no. 8, 9–12, 86.

[143] *Locally nilpotent groups with the condition* min-∞-*n*, Ukrain. Mat. Zh. **42** (1990), no. 3, 340–346.

[144] *Some classes of groups with weak minimality and maximality conditions for normal subgroups*, Ukrain. Mat. Zh. **42** (1990), no. 8, 1050–1056.

[145] *Groups with minimax classes of conjugate elements*, Infinite groups and related algebraic structures (Russian), Akad. Nauk Ukrainy, Inst. Mat., Kiev, 1993, pp. 160–177.

[146] *Artinian modules over groups of finite rank and the weak minimal condition for normal subgroups*, Ricerche Mat. **44** (1995), no. 2, 303–335 (1996).

L. A. Kurdachenko and V. E. Goretsky

[147] *Groups with weak minimality and maximality conditions for subgroups that are not normal*, Ukrain. Mat. Zh. **41** (1989), no. 12, 1705–1709, 1728.

L. A. Kurdachenko, J. M. Muñoz-Escolano, and J. Otal

[148] *Groups in which the normal closures of cyclic subgroups have finite section rank*, Internat. J. Algebra Comput. **22** (2012), no. 4, 1250032, 20.

L. A. Kurdachenko and J. Otal

[149] *Frattini properties of groups with minimax conjugacy classes*, Topics in infinite groups, Quad. Mat., vol. 8, Dept. Math., Seconda Univ. Napoli, Caserta, 2001, pp. 221–237.

[150] *The rank of the factor-group modulo the hypercenter and the rank of the some hypocenter of a group*, Cent. Eur. J. Math. **11** (2013), no. 10, 1732–1741.

[151] *Groups with Chernikov factor-group by hypercentral*, Rev. R. Acad. Cienc. Exactas Fís. Nat. Ser. A Math. RACSAM **109** (2015), no. 2, 569–579.

L. Kurdachenko, J. Otal, and I. Ya. Subbotin

[152] *Groups with Prescribed Quotient Groups and Associated Module Theory*, Series in Algebra, World Scientific, Singapore, 2002, Volume 8.

[153] *Artinian Modules over Group Rings*, Frontiers in Mathematics, Birkhäuser Verlag, Basel, 2007.

[154] *On influence of contranormal subgroups on the structure of infinite groups*, Comm. Algebra **37** (2009), no. 12, 4542–4557.

[155] *Criteria of nilpotency and influence of contranormal subgroups on the structure of infinite groups*, Turkish J. Math. **33** (2009), no. 3, 227–237.

[156] *Some criteria for existence of supplements to normal subgroups and their applications*, Internat. J. Algebra Comput. **20** (2010), no. 5, 689–719.

[157] *On a generalization of Baer theorem*, Proc. Amer. Math. Soc. **141** (2013), no. 8, 2597–2602.

[158] *Some remarks about groups of finite special rank*, Advances in Group Theory and Applications **1** (2016), 55–76.

L. A. Kurdachenko and N. N. Semko

[159] *Groups with a weak maximality condition for nonnilpotent subgroups*, Ukraïn. Mat. Zh. **58** (2006), no. 8, 1068–1083.

L. A. Kurdachenko, N. N. Semko, and A. A. Pypka

[160] *On some relationships between the upper and lower central series in finite groups*, Francisk Scorina Gomel State University Proceedings **3** (2014), 66–71.

L. A. Kurdachenko and P. Shumyatsky

[161] *The ranks of central factor and commutator groups*, Math. Proc. Cambridge Philos. Soc. **154** (2013), no. 1, 63–69.

L. A. Kurdachenko, P. Shumyatsky, and I. Ya. Subbotin

[162] *Groups with many nilpotent subgroups*, Algebra Colloq. **8** (2001), 3.

L. A. Kurdachenko and H. Smith

[163] *Groups with the weak minimal condition for non-subnormal subgroups*, Ann. Mat. Pura Appl. **173** (1997), 299–312.

[164] *Groups with the weak maximal condition for non-subnormal subgroups*, Ricerche Mat. **47** (1998), no. 1, 29–49.

[165] *Groups in which all subgroups of infinite rank are subnormal*, Glasg. Math. J. **46** (2004), no. 1, 83–89.

[166] *Groups with the weak minimal condition for non-subnormal subgroups. II*, Comment. Math. Univ. Carolin. **46** (2005), no. 4, 601–605.

L. A. Kurdachenko and P. Soules

[167] *Groups with all non-subnormal subgroups of finite rank*, Groups St. Andrews 2001 in Oxford. Vol. II, London Math. Soc. Lecture Note Ser., vol. 305, Cambridge Univ. Press, Cambridge, 2003, pp. 366–376.

L. A. Kurdachenko and I. Ya. Subbotin

[168] *On some properties of the upper and lower central series*, Southeast Asian Bull. Math. **37** (2013), no. 4, 547–554.

L. A. Kurdachenko, A. V. Tushev, and D. I. Zaitsev

[169] *Noetherian modules over nilpotent groups of finite rank*, Arch. Math. (Basel) **56** (1991), no. 5, 433–436.

A. G. Kurosh

[170] *Zur zerlegung unendlicher gruppen*, Math. Annalen **106** (1932), 107–113.

[171] *Zur Theorie der teilweise geordneten Systeme von endlichen Mengen*, Mat. Sbornik N.S. **5** (1939), 343-346.

[172] *The Theory of Groups*, Chelsea Publishing Co., New York, 1960, Translated from the Russian and edited by K. A. Hirsch. 2nd English ed. 2 volumes.

J. C. Lennox and D. J. S. Robinson

[173] *Soluble products of nilpotent groups*, Rend. Sem. Mat. Univ. Padova **62** (1980), 261–280.

[174] *The Theory of Infinite Soluble Groups*, Oxford Mathematical Monographs, The Clarendon Press Oxford University Press, Oxford, 2004.

P. Longobardi and M. Maj

[175] *On the number of generators of a finite group*, Arch. Math. (Basel) **50** (1988), no. 2, 110–112.

P. Longobardi, M. Maj, and H. Smith

[176] *A finiteness condition on non-nilpotent subgroups*, Comm. Algebra **24** (1996), 3567–3588.

[177] *Locally nilpotent groups with all subgroups normal-by-(finite rank)*, J. Group Theory **1** (1998), no. 3, 291–299.

[178] *Groups in which normal closures of elements have boundedly finite rank*, Glasg. Math. J. **51** (2009), no. 2, 341–345.

A. Lubotzky and A. Mann

[179] *Powerful p-groups. I. Finite groups*, J. Algebra **105** (1987), no. 2, 484–505.
[180] *Residually finite groups of finite rank*, Math. Proc. Camb. Phil. Soc. **106** (1989), 385–388.
[181] *On groups of polynomial subgroup growth*, Invent. Math. **104** (1991), no. 3, 521–533.

A. Lubotzky and D. Segal

[182] *Subgroup growth*, Progress in Mathematics, vol. 212, Birkhäuser Verlag, Basel, 2003.

A. Lucchini

[183] *A bound on the number of generators of a finite group*, Arch. Math. (Basel) **53** (1989), no. 4, 313–317.

I. D. Macdonald

[184] *Some explicit bounds in groups with finite derived groups*, Proc. London Math. Soc. (3) **11** (1961), 23–56.

N. Yu. Makarenko

[185] *Rank analogues of Hall and Baer theorems*, Sibirsk. Mat. Zh. **41** (2000), no. 6, 1376–1380, iii.

A. I. Maltsev

[186] *On groups of finite rank*, Mat. Sbornik **22** (1948), 351–352.
[187] *Nilpotent torsion-free groups*, Izvestiya Akad. Nauk. SSSR. Ser. Mat. **13** (1949), 201–212.
[188] *On certain classes of infinite soluble groups*, Mat. Sbornik **28** (1951), 367–388 (Russian), English transl. Amer. Math. Soc. Translations **2** (1956), 1-21.

A. Mann and D. Segal

[189] *Uniform finiteness conditions in residually finite groups*, Proc. London Math. Soc. **(3) 61** (1990), 529–545.

D. J. McCaughan

[190] *Subnormality in soluble minimax groups*, J. Austral. Math. Soc. **17** (1974), 113–128, Collection of articles dedicated to the memory of Hanna Neumann, V.

D. H. McLain

[191] *A characteristically-simple group*, Proc. Cambridge Philos. Soc. **50** (1954), 641–642.
[192] *On locally nilpotent groups*, Proc. Cambridge Philos. Soc. **52** (1956), 5–11.

[193] *Finiteness conditions in locally soluble groups*, J. London Math. Soc. **34** (1959), 101–107.

Yu I. Merzlyakov

[194] *Locally solvable groups of finite rank*, Algebra i Logika Sem. **3** (1964), no. 2, 5–16.

[195] *Locally solvable groups of finite rank. II*, Algebra i Logika **8** (1969), 686–690.

[196] *On the theory of locally polycyclic groups*, J. London Math. Soc. (2) **30** (1984), no. 1, 67–72.

G. A. Miller and H. Moreno

[197] *Non-abelian groups in which every subgroup is abelian*, Trans. Amer. Math. Soc. **4** (1903), 398–404.

W. Möhres

[198] *Auflösbarkeit von Gruppen, deren Untergruppen alle subnormal sind*, Arch. Math. (Basel) **54** (1990), no. 3, 232–235.

I. M. Musson

[199] *On the structure of certain injective modules over group algebras of soluble groups of finite rank*, J. Algebra **85** (1983), no. 1, 51–75.

N. N. Myagkova

[200] *On groups of finite rank*, Izvestiya Akad. Nauk SSSR. Ser. Mat. **13** (1949), 495–512.

F. Napolitani and E. Pegoraro

[201] *On groups with nilpotent by Černikov proper subgroups*, Arch. Math. **69** (1997), 89–94.

B. H. Neumann

[202] *Groups with finite classes of conjugate elements*, Proc. London Math. Soc. (3) **1** (1951), 178–187.

[203] *Groups covered by permutable subsets*, J. London Math. Soc. **29** (1954), 236–248.

[204] *On amalgams of periodic groups*, Proc. Roy. Soc. London Ser. A **255** (1960), 477–489.

P. M. Neumann and M. R. Vaughan-Lee

[205] *An essay on BFC groups*, Proc. London Math. Soc. (3) **35** (1977), no. 2, 213–237.

M. F. Newman

[206] *The soluble length of soluble linear groups*, Math. Z. **126** (1972), 59–70.

V. N. Obraztsov

[207] *An embedding theorem for groups and its corollaries*, Mat. Sb. **180** (1989), no. 4, 529–541, 560 (Russian), English transl. in Math. USSR-Sb. **66** (1990), 541-553.

A. Yu Ol'shanskii

[208] *Groups of bounded exponent with subgroups of prime order*, Algebra i Logika **21** (1982), 553–618 (Russian), English transl. in Algebra and Logic, **21** (1982), 369-418.

[209] *Geometry of Defining Relations in Groups*, Mathematics and its Applications, vol. 70, Kluwer Academic Publishers, Dordrecht, 1991.

O. Ore

[210] *On the application of structure theory to groups*, Bull. Amer. Math. Soc. **44** (1938), no. 12, 801–806.

J. Otal and J. M. Peña

[211] *Groups in which every proper subgroup is Černikov-by-nilpotent or nilpotent-by-Černikov*, Arch. Math. **51** (1988), 193–197.

[212] *Locally graded groups with certain minimal conditions for subgroups II*, Publicacions Mat. **32** (1988), 151–157.

[213] *Nilpotent-by-Černikov CC-groups*, J. Austral. Math. Soc. Ser. A **53** (1992), no. 1, 120–130.

J. Otal and N. N. Semko

[214] *Groups with small cocentralizers*, Algebra Discrete Math. (2009), no. 4, 135–157.

B. I. Plotkin

[215] *On some criteria of locally nilpotent groups*, Uspehi Mat. Nauk (N.S.) **9** (1954), no. 3(61), 181–186 (Russian), English transl. Amer. Math. Soc. Translations **17** (1961), 1-7.

[216] *Radical groups*, Mat. Sb. N.S. **37(79)** (1955), 507–526 (Russian), English transl. Amer. Math. Soc. Translations **17** (1961), 9-28.

[217] *Generalized solvable and generalized nilpotent groups*, Uspehi Mat. Nauk **13** (1958), no. 4, 89–172.

[218] *Generalized soluble and generalized nilpotent groups*, Amer. Math. Soc. Transl. (2) **17** (1961), 29–115.

K. Podoski. and B. Szegedy

[219] *On finite groups whose derived subgroup has bounded rank*, Israel J. Math. **178** (2010), 51–60.

H. Prüfer

[220] *Theorie der Abelschen Gruppen*, Math. Z. **20** (1924), 165–187.

A. Rhemtulla and H. Smith

[221] *On solvable R* groups of finite rank,* Comm. Algebra **31** (2003), no. 7, 3287–3293.

D. J. S. Robinson

[222] *On soluble minimax groups,* Math. Zeit. **101** (1967), 13–40.

[223] *Infinite Soluble and Nilpotent Groups,* Queen Mary College Mathematics Notes, Queen Mary College, London, 1968.

[224] *A note on groups of finite rank,* Compositio Math. **21** (1969), 240–246.

[225] *Finiteness Conditions and Generalized Soluble Groups vols. 1 and 2,* Ergebnisse der Mathematik und ihrer Grenzgebiete, Springer-Verlag, Berlin, Heidelberg, New York, 1972, Band 62 and 63.

[226] *On the cohomology of soluble groups of finite rank,* J. Pure and Applied Algebra **6** (1975), 155–164.

[227] *Splitting theorems for infinite groups,* Symposia Mathematica, Vol. XVII (Convegno sui Gruppi Infiniti, INDAM, Rome, 1973), Academic Press, London, 1976, pp. 441–470.

[228] *Soluble products of nilpotent groups,* J. Algebra **98** (1986), no. 1, 183–196.

[229] *A Course in the Theory of Groups,* Graduate Texts in Mathematics, vol. 80, Springer Verlag, Berlin, Heidelberg, New York, 1996.

J. E. Roseblade

[230] *On groups in which every subgroup is subnormal,* J. Algebra **2** (1965), 402–412.

A. Schlette

[231] *Artinian, almost abelian groups and their groups of automorphisms,* Pacific J. Math. **29** (1969), 403–425.

O. J. Schmidt

[232] *Groups all of whose subgroups are nilpotent,* Mat. Sb. **31** (1924), 366–372 (Russian).

R. Schmidt

[233] *Subgroup Lattices of Groups,* de Gruyter Expositions in Mathematics, vol. 14, Walter de Gruyter & Co., Berlin, 1994.

I. Schur

[234] *Über die Darstellungen der endlichen Gruppen durch gebrochene lineare Substitutionen,* J. Reine Angew. Math. **127** (1904), 20–50.

D. Segal

[235] *A footnote on residually finite groups,* Israel J. Math. **94** (1996), 1–5.

[236] *On modules of finite upper rank,* Trans. Amer. Math. Soc. **353** (2001), no. 1, 391–410.

D. Segal and A. Shalev

[237] *On groups with bounded conjugacy classes,* Quart. J. Math. Oxford Ser. (2) **50** (1999), no. 200, 505–516.

N. F. Sesekin

[238] *On locally nilpotent groups without torsion,* Mat. Sbornik N.S. **32(74)** (1953), 407–442.

L. A. Shemetkov

[239] *On formation properties of finite groups,* Doklady, AN USSR **37** (1972), 1324–1327.

J. A. H. Shepperd and J. Wiegold

[240] *Transitive permutation groups and groups with finite derived groups,* Math. Z. **81** (1963), 279–285.

V. P. Shunkov

[241] *On a locally finite group with extremal Sylow p-subgroups with respect to a certain prime number p,* Sibirsk. Mat. Ž. **8** (1967), 213–229 (Russian), English transl. in Siberian Math. J. **8** (1967), 161-171.

[242] *On the conjugacy of the Sylow p-subgroups in SF-groups,* Algebra i Logika **10** (1971), 587–598 (Russian), English transl. in Algebra and Logic, **10** (1971), 363-368.

H. Smith

[243] *Hypercentral groups with all subgroups subnormal,* Bull. London Math. Soc. **15** (1983), no. 3, 229–234.

[244] *A note on Baer groups of finite rank,* Math. Z. **184** (1983), no. 1, 139–140.

[245] *Some remarks on locally nilpotent groups of finite rank,* Proc. Amer. Math. Soc. **92** (1984), no. 3, 339–341.

[246] *Hypercentral groups with all subgroups subnormal. II,* Bull. London Math. Soc. **18** (1986), no. 4, 343–348.

[247] *Group theoretic properties inherited by lower central factors,* Glasgow Math. J. **29** (1987), no. 1, 89–91.

[248] *Subnormal and ascendant subgroups with rank restrictions,* Illinois J. Math. **34** (1990), no. 1, 49–51.

[249] *Groups with few non-nilpotent subgroups,* Glasgow Math. J. **39** (1997), 141–151.

[250] *A finiteness condition on normal closures of cyclic subgroups,* Math. Proc. R. Ir. Acad. **99A** (1999), no. 2, 179–183.

S. E. Stonehewer

[251] *Permutable subgroups of infinite groups,* Math. Z. **125** (1972), 1–16.

J. Tits

[252] *Free subgroups in linear groups*, J. Algebra **20** (1972), 250–270.

M. J. Tomkinson

[253] *FC-Groups*, Pitman Publishing Limited, Boston, London, Melbourne, 1984.
[254] *FC-groups: recent progress*, Infinite groups 1994 (Ravello), de Gruyter, Berlin, 1996, pp. 271–285.

S. M. Vovsi

[255] *A remark on ascendant subgroups of finite rank*, Sibirsk. Mat. Ž. **17** (1976), no. 4, 936–939.

B. A. F. Wehrfritz

[256] *Sylow subgroups of locally finite groups with min-p*, J. London Math. Soc. (2) **1** (1969), 421–427.
[257] *On locally finite groups with min − p*, J. London Math. Soc. (2) **3** (1971), 121–128.
[258] *Infinite Linear Groups*, Ergebnisse der Mathematik und ihrer Grenzgebiete, Springer-Verlag, New York, Heidelberg, Berlin, 1973, Band 76.
[259] *The rank of a linear p-group; an apology*, J. London Math. Soc. (2) **21** (1980), no. 2, 237–243.

J. Wiegold

[260] *Groups with boundedly finite classes of conjugate elements*, Proc. Roy. Soc. London. Ser. A. **238** (1957), 389–401.
[261] *Multiplicators and groups with finite central factor-groups*, Math. Z. **89** (1965), 345–347.
[262] *The Schur multiplier: an elementary approach*, Groups—St. Andrews 1981 (St. Andrews, 1981), London Math. Soc. Lecture Note Ser., vol. 71, Cambridge Univ. Press, Cambridge, New York, 1982, pp. 137–154.

D. I. Zaitsev

[263] *Groups which satisfy a weak minimality condition*, Ukrain. Mat. Zh. **20** (1968), 472–482 (Russian), English transl. in Ukrainian Math. J. **20** (1968), 408-416.
[264] *Groups which satisfy a weak minimality condition*, Dokl. Akad. Nauk SSSR **178** (1968), 780–782 (Russian), English transl. in Soviet Math. Dokl. **9** (1968), 194-197.
[265] *The groups which satisfy a weak minimality condition*, Mat. Sb. (N.S.) **78 (120)** (1969), 323–331.
[266] *Stably solvable groups*, Izv. Akad. Nauk SSSR Ser. Mat. **33** (1969), 765–780 (Russian), English transl. in Math. USSR-Izv. **3** (1969), 723-736.
[267] *Groups satisfying the weak minimal condition for non-abelian subgroups*, Ukrain. Mat. Zh. **23** (1971), 661–665 (Russian), English transl. in Ukrainian Math. J., **23** (1971), 543-546.

[268] *Solvable groups of finite rank*, Groups with Restricted Subgroups (Russian), "Naukova Dumka", Kiev, 1971, pp. 115–130.

[269] *Theory of minimax groups*, Ukrain. Mat. Zh. **23** (1971), 652–660 (Russian), English transl. in Ukrainian Math. J.,**23** (1971), 536-542.

[270] *Groups with complemented normal subgroups*, Some questions in group theory (Russian), Izdanie Inst. Mat. Akad. Nauk Ukrain. SSR, Kiev, 1975, pp. 30–74, 219–220.

[271] *Solvable groups of finite rank*, Algebra i Logika **16** (1977), no. 3, 300–312, 377.

[272] *Hypercyclic extensions of abelian groups*, Groups defined by properties of a system of subgroups (Russian), Akad. Nauk Ukrain. SSR, Inst. Mat., Kiev, 1979, pp. 16–37, 152.

[273] *Groups of operators of finite rank and their application*, Sixth All-Union Symposium on Group Theory (Čerkassy, 1978) (Russian), "Naukova Dumka", Kiev, 1980, pp. 22–37, 219.

[274] *Products of abelian groups*, Algebra i Logika **19** (1980), 150–172,250 (Russian), English transl. in Algebra and Logic, **19** (1980), 94-106.

D. I. Zaitsev, M. I. Kargapolov, and V. S. Charin

[275] *Infinite groups with prescribed properties of subgroups*, Ukrain. Mat. Zh. **24** (1972), 618–633, 716.

D. I. Zaitsev and L. A. Kurdachenko

[276] *The weak minimal and maximal conditions for subgroups in groups*, Tech. report, Math. Inst. Kiev, 1975, Preprint, 52 pp.

D. I. Zaitsev, L. A. Kurdachenko, and A. V. Tushev

[277] *Modules over nilpotent groups of finite rank*, Algebra i Logika **24** (1985), 631–666 (Russian), English transl. in Algebra and Logic, **23** (1985), 412-436.

H. Zassenhaus

[278] *Beweis eines satzes über diskrete gruppen*, Abh. Math. Sem. Univ. Hamburg **12** (1938), 289–312.

Author Index

Symbol Index

$C_G(A)$: centralizer of A in G, 15

E, E_n: $n \times n$ identity matrix, 38

F_+: additive group of the field F, 38

G': derived subgroup of G, 42

$GL(F, V)$: group of all non-singular linear transformations of the F-vector space V, 36

$GL_n(F)$: group of non-singular $n \times n$ matrices with coefficients in the field F, 36

G^n: the subgroup $\langle g^n | g \in G \rangle$, 28

$G^{\mathfrak{X}}$: $\mathfrak{X}$-residual of the group G, 27

$H^1(G, V)$: first cohomology group, 162

$H^2(G, H)$: second cohomology group, 46

$M(G)$: Schur multiplier of G, 46

$N_G(H)$: normalizer of H in G, 17

$PSL_2(F)$: projective special linear group, 111

$T_n(F)$: group of upper triangular $n \times n$ matrices with coefficients in field F, 39

$UT_n(F)$: group of unitriangular $n \times n$ matrices with coefficients in field F, 38

$UT_n^k(F)$: subgroup of $UT_n(F)$ consisting of matrices whose first $k - 1$ superdiagonals are all 0, 38

$[X, Y]$: subgroup generated by $[x, y]$, for $x \in X, y \in Y$, 13

$[X_1, \ldots, X_n]$: $[[X_1, \ldots, X_{n-1}], X_n]$, 13

$[Y,_n X]$: $[Y, \underbrace{X, X, \ldots, X}_{n}]$, 13

$[x, y]$: $x^{-1}y^{-1}xy$, 13

$\Pi(G)$: set of prime divisors of orders of elements of G, 21

$\acute{\mathbf{p}}_n \mathfrak{X}$: class of hyper $\mathfrak{X}$-groups, 12

$\mho_n(G)$: $\langle g^{p^n} | g \in G \rangle$, 189

$\mathbf{FC}(G)$: $\{x \in G | x^G \text{ is finite}\}$, 222

$\mathbf{l}_p(G)$: p-length of p-soluble group, 194

$\mathbf{s}_p(G)$: p-size of Chernikov group, 102

$\Omega_n(G) : \{x \in G | x^{p^n} = 1\}$, 34

$\mathbf{cl}_G(A)$: length of G-chief series, 53

$\mathbf{cl}_{RG}(L)$: RG-composition length of L, 187

$\mathbf{L}\mathfrak{X}$: class of locally $\mathfrak{X}$-groups, 19

Subject Index

Ranks of Groups: The Tools, Characteristics, and Restrictions
By Martyn R. Dixon, Leonid A. Kurdachenko and Igor Ya Subbotin
Copyright © 2017 John Wiley & Sons, Inc.